To my friend

I hope you enjoy

military history of my unit

in Vietnam 1966-67

Bill Comeau

5/29/2023

Christine C Comeau

Editor, Researcher

05/29/2023

DUEL WITH THE DRAGON AT

THE BATTLE OF SUOI TRE

DUEL WITH THE DRAGON AT
THE BATTLE OF SUOI TRE

BILL COMEAU

Deeds Publishing | Athens

Published by Deeds Publishing in Athens, GA
www.deedspublishing.com

Cover design Jim Nelson, used with permission.

ISBN 978-1-950794-85-0

Books are available in quantity for promotional or premium use. For information, email info@deedspublishing.com.

First Edition, 2022

10 9 8 7 6 5 4 3 2 1

Dedicated to the men who showed me the true meaning of courage and dedication; my fellow A/2/12 veterans who served with me in Vietnam. This includes the men who were at my side and those who arrived to take our place after our year had ended. We were pretty tired by then and appreciated their arrival as they continued writing new company history after we left.

"We should not look back unless it is to derive useful lessons from past errors, and for the purpose of profiting by dearly bought experience."

—George Washington

CONTENTS

PREFACE

On a balmy day in March, 2000, my wife Christine (Chris) and I were making a journey that would shape our lives for the next twenty years. We had no idea what awaited us as we traveled from our home in New Bedford, Massachusetts to the home of a veteran of the army unit that I served in during 1966-1967. The occasion was a reunion of some of my company soldiers who were gathering at the home of Henry Osowiecki, a well-respected squad leader, who lived in Thomaston, Connecticut. I had learned of the reunion online in January through a fellow I had served with in the third platoon, Porter Harvey.

It had been thirty-three years since I last saw any of the men who served with me in Vietnam under the A/2/12 (Alpha Company, 2nd Battalion, 12th Infantry Regiment, 4th Infantry Division) guidon. I was apprehensive as I made my way down I-95 to a gathering in the hills south of Hartford. Chris was more adventurous than I was and was open minded about the meeting. I wasn't so sure, and told her that if this turned out to be a weekend of old war veterans sitting in a circle with bandanas on their heads, smoking joints, and listening to old eight track tapes of The Doors, we were gone.

I was pleasantly surprised to learn that in fact, the men I met that day were very similar to the men I left behind in 1967. Certainly, we were not in the shape we were in back then, quite a bit older and wiser, but we were still the regular guys who were brought together in 1966 to serve in the jungles of Southeast Asia. In retrospect, we all had the same apprehensions about going back to that time of our lives. We were all victims of the media's portrayal of the former Vietnam infantrymen. None of that proved to have been true. Most of these men went back home to continue very average existences, in spite of the burden of the memories of that cruel war. Most buried the good with the bad, so much so that the men never realized much of our accomplishments and honors.

After a very pleasant weekend with the twelve men that attended that reunion, Henry and I came to the same realization. He offered this suggestion, "Let's go find the guys." So began a series of steps that not only located the men from our unit, but also answered many of the questions posed from our service in Vietnam, including the one that was pivotal. "What did our sacrifices and losses in Vietnam accomplish?"

Without a detailed search of our history in Vietnam, the answer seemed to be, not much. We were wrong. Politics be damned, our Company served valiantly and wrote Regimental history that compared very well with that written in earlier, more popular conflicts. For their heroics at the Battle of the Bulge in World War II, our battalion (the 2/12th Infantry Regiment) earned the Presidential Unit Citation (PUC). This rarely awarded honor is equal to every man in the Unit receiving the Distinguished Service Cross. Our battalion, twenty-three years later would also be honored in earning the PUC for our action during the Battle of Suoi Tre. In addition to this tribute, we were also honored to have within our ranks Donald Evans, the first recipient of the Medal of Honor from the 4th ID and was also the first Medic to earn this badge of courage. Surprisingly, for many of our men, neither of these achievements was realized until we told them about it 35 years after the fact. Oh, they knew that they were in a very large battle, and knew that Donald had performed heroically in a jungle encounter on January 1967. Unfortunately, the medals were not issued for many months after the fact. By that time, most had left the unit and never looked back. I accidentally learned of the Medal of Honor awarded to Donald.

Many Vietnam veterans of the infantry units that trudged through the jungles of Vietnam back in the 60s shared similar stories of glory and accomplishments that went unnoticed by the public and the press, who were eager to put that period of turmoil behind them. To all the men who served in the fields of Vietnam, especially my brothers of A/2/12, I dedicate this book. I encourage others who are now in the process of coming to terms with their troubling history in Vietnam to share their stories with the public that years ago just wanted the problem to go away. If we don't tell them the truth, we allow the non-informed to paint us with the broad stroke that prevailed for decades; that we were all losers who could not adjust back into the American mainstream. Nothing could be further from the truth.

The Year Of The Dragon 1964

The Wood Dragon Characteristics:
Innovative • Enterprising • Flexible • Self-assured
Brave • Passionate • Conceited • Tactless
Scrutinizing • Unanticipated • Quick-tempered

In July of the year 1964, the first Viet Cong (PLF) Division, was formed; the notorious 9th Viet Cong Division. It was formed in Tay Ninh province, northwest of Saigon. This division was originally organized from the 271st and 272nd Viet Cong Regiments and additional supporting units. The 273rd Regiment was added soon after.

By December, the Viet Cong 9th Division seized the Catholic village of Binh Gia east of Saigon. During the battle, the division ambushed and destroyed a South Vietnamese Ranger and a South Vietnamese Marine battalion.

It was the first time the National Liberation Front (NLF) launched a large-scale operation, holding its ground and fighting for four days against government troops equipped with armor, artillery, and helicopters, and aided by the U.S. air support and military advisers. The Viet Cong demonstrated, when well-supplied with military supplies from North Vietnam, they had the ability to fight and inflict damage even on the best ARVN units. For the first time in their history, the NLF was able to control a government stronghold for several days and inflict heavy casualties on the regular units of the South Vietnamese military in a large set-piece battle.

The Viet Cong suffered light casualties with 32 soldiers confirmed killed, and they did not leave a single casualty on the battlefield. The South Vietnamese and their American allies lost the total of 201 personnel killed in action, 192 wounded, and 68 missing.

This unit would become our rival when we arrived in Vietnam, nine months after we began training at Fort Lewis, Washington.

We arrived at Fort Lewis in December, 1965, just as the Vietnam buildup was growing in leaps and bounds. The draft totals had tripled leading up to our arrival for training. My basic training company, A Company of the 2nd Battalion, 12th Infantry Regiment, 4th Infantry Division, had an established number of one hundred and eighty men in its ranks. One hundred and seventy-eight were draftees.

This is the story of our brigade, the 3rd Brigade of the 4th Infantry Division and our pursuit of the soldiers of the 9th Vietcong Division. That challenge reached a climax on March 21st, 1967 at the Battle of Suoi Tre. That engagement involved the bulk of the soldiers from our brigade pitted against twenty-six hundred soldiers of the 9th VC Division. It is reputed to have cost the largest number of deaths to any enemy unit in a one-day event of the war. All the units of our brigade who were engaged that morning were awarded the Presidential Unit Citation. This was equal to every participant receiving the Distinguished Service Cross, the United States Army's second highest military decoration.

The last time my battalion received this honor was for the action at the Battle of the Bulge in World War II. That's not bad company to be associated with.

ALPHA ASSOCIATION

As is the case with most wars, the combatants return with a deep sense of sadness, loss and the guilt of survival. They want...no need...to go on with their lives with a new sense of purpose and direction. It was no different for the men who served with me in a line infantry company in Vietnam. What was different for us who came back in 1967 was the poisoned atmosphere that was present when we returned. For those of us who grew up in the atmosphere of the American pride of the 1950s, it was a bitter disappointment to receive the treatment that awaited our return.

Who cared what these nut cases thought about our service in Vietnam? Who cared that the silent majority out there remained quiet while the peace at any cost folks set the agenda in the news media? Who cared when some of our guys went to join their local Veterans of Foreign Wars organizations and were told that they did not qualify because they hadn't served in a 'real' war? We did, and we cared very much.

In dealing with the sense of loss when one of our buddies became a casualty in Vietnam, the line infantrymen had a saying that kept us going; "It don't mean nothing." We understood it to mean, "It wasn't important, don't dwell on it." The mission was important, our personal survival was important, getting home in one piece was important. Nonetheless, it was this 'it don't mean nothing' attitude that empowered us to try to get our lives together and allowed us to go on with our lives productively.

Unless, you have been in a desperate struggle for survival, it is very hard to explain the contradictions that play into the brotherhood of combat veterans. When it all hits the fan, you realize very quickly how much our common defense was vital to our survival. If you cannot depend on the men on your right and left, you are alone in a struggle that threatens to end your life. Yet, when the smoke clears and one of those defenders is hurt or killed, you are in a state of shock and disbelief. You very quickly realize that you could have been the one on that evacuation chopper. That man leaving the field of battle could have been a very important player in trying to maintain your sanity while these incredibly chaotic situations played out around you. You mourn the loss, followed quickly by a prayer of thanks. Repeatedly in the course of the year, this begins to play tricks with your mind. We would question ourselves asking, "What made me so special to have survived while so many were killed or maimed?" By the time you leave, you have deeply implanted in your psyche a sense of guilt to have made it when so many others were less fortunate.

Late in our year in Vietnam, Skip Barnhart, the Company clerk, prepared a listing of all the men with their hometown addresses. The decision was made then that we would all reunite at the resort town of Lake George in upstate New York on the 4th of July weekend following our return to the States. It would be decades before we would learn the fate of that meeting. Invitations were never sent out and only one man answered the call. Paul LaRock, a third platoon veteran, was the only one to make the trip to the resort. He spent three days riding around the resort and bitterly left with a feeling of sadness. How could the guys not come? Here is the long and short of it. We could not. It was just too soon.

Even after 32 years had passed, it was still difficult to take that step in reuniting with the men of our Company. Each one of us had very similar feelings as they considered meeting the men that they served with in Vietnam. The first concern was that no one would remember us. We need not have worried. It all came rushing back as if a day had passed instead of decades. At that first meeting of veterans in Connecticut, there were Ron Bergeron, Fred Van Amburgh, Jim Deluco, Walter Butkus, Jim Heys, Henry Osowiecki, Jim Shulsky, Gary Barney, Tom Bremner, Steve Sanborn, and me, Bill Comeau.

At the conclusion of the weekend, the men decided it was time to establish a veteran's group with the goal of finding the men who served with us and setting up annual reunions. I volunteered to organize a quarterly newsletter that would allow us to share stories and update the expansion of the club. We chose the name Alpha Association as our organization. We agreed to allow anyone who had served in our company during the years 1966 through 1967 to join us. In addition, we undertook the mission to find the families of the men of our Company who were killed in Vietnam. Ten are presently in our association. We set annual dues to help support our group. Honor Roll families were exempted from paying anything, though many would chip in anyway. We also resolved that no one could be denied membership because of financial problems. They were there for us then, and we would be there for them now.

I published an internet site (www.alphaassociation.org) to coordinate the effort and to share history and photos from our adventure in Vietnam. Eventually, over three hundred and thirty photos were published on the site of men who served in my one hundred-and-eighty-five-men Company during our period in Vietnam. I don't believe that is duplicated at any other unit website anywhere on the net.

The newsletters became a powerful force in bringing the men together. We were astonished by the richness of the memories of the men, even after all these years. These men may not have recalled much of recent events, but the recollection of their time in service to our Country was as vivid today as it was when most of them hung up their uniforms in 1967-1968.

Our story together began in 1965. Most were just out of high school and we understood that we had a military obligation. With that in mind, we set out to emulate the heroes of our lifetimes, our fathers and uncles who won the big one in the 1940s. We came home disheartened by the futilities of the war and the general lack of progress that seemed to be the perception. The big questions seemed to be, "What was that all about?" After twenty years of collecting history and personal stories, we realized that our story was one of bravery, accomplishment, and a very unique epic that in some cases were even more glorious than the men who served in our company during World War II. For the first time, our men had a sense of pride in what we accomplished as a Company back in our youth. Finally, it mattered and it all came about through a veteran's newsletter entitled appropriately, ALPHA'S PRIDE.

In order to get a clearer picture of our history back then, I took on the challenge of assembling whatever I could research from the National Archives in College Park, Maryland. After eight trips there paid by the association, I had a clear picture of not only the company's history, but also, our regiment's story and the operations conducted by the brigade during our time in Vietnam. Years later, I was honored to serve as the 12th Infantry Regiment's Historian for the regimental monument that was established on the Walk of Honor at the National Infantry Museum at Fort Benning, Georgia. That was due to what I learned during those expeditions to Maryland.

As an avid fan of history throughout my school years, I understood the significance of my military service in historical terms. We may not accomplish much, but I knew from my first day in uniform that my tour would be historical. As such, I became a prolific writer of what I experienced while serving in my unit during the Vietnam War. The book that I am writing was made possible with the help of the 330 letters that I wrote home during this time. I need to acknowledge my family who kept every letter I wrote, but also my first wife, Linda, who was gracious enough to send me all the letters that I wrote to her in real time about what I was experiencing. She gave them to me years after our divorce and my remarriage and had no idea that I would eventually write a book about this period. All she knew was these letters may be important, and she was right. Some of what I wrote, I never shared with family to prevent them from becoming alarmed by what we were undergoing in Vietnam.

I thank my wife of over 35 years, Chris, for helping me sort through all the letters, including those written by Linda, which couldn't have been easy. She catalogued them and set them all in chronological form for use in the newsletters and now in the book.

Much of our story was published over the last twenty plus years in our newsletters. Hundreds of hours of interviews and personally written memoirs from our veterans were used to tell their personal side of our story. This book would have not been possible without those valuable recollections, and from others from the brigade who had specific recollections of historical events. I will be forever grateful for their participation in this effort. To assure accuracy, most of, if not all, of the narrative was confirmed through the collective memories of the soldiers who shared my history. Official documents provided details like Brigade Day Reports, After Action Reports, and Lessons Learned documents. The newsletters and this book would not be as accurate without knowing the specific details depicted in this book. Without the official documents that I assembled from the National Archives in College Park, Maryland, much of the book would only characterize certain locations and times as we may have remembered them; "We were in the jungle somewhere south of basecamp, sometime in early January." We only knew about what was important to accomplish our mission in Vietnam. I later learned that was not only the experience of the enlisted men, but many of our leaders as well.

THE FORTIES AND FIFTIES

My mother's family belonged to the poor working class of New Bedford, Massachusetts. Her mother and father came down from Canada in the early twentieth century to find work which was plentiful in the mill city of New Bedford.

She was the oldest daughter of a family of eleven and very early in her childhood, she had to attend school and come home to help with the rearing of the younger siblings. Those were coming along at a once each nine-month clip. Needless to say, she couldn't properly do both and at the end of her 4th grade with poor grades, she was forced to quit school and go to work in the sweatshops of the city. This was in the days before child labor laws and the children of the poor were brutally exploited by the cruel factory owners from that period.

My father had a similar background, but his family arrived in the 1920s, again looking for work. Like most immigrants he ended up in the sweatshops.

Trying to make a few extra dollars, my father went to work at a local bowling alley which he eventually managed. It was during this time that he met my mother and four years later they were married. After all, this was during the Great Depression. Who had money to get married?

They had three daughters over the course of ten years. They were named Theresa, Annette, and Cecile (Cookie). I was the last of their offspring, being born almost nine months to the day after the dropping of the bomb on Hiroshima on August 6, 1945. I was born on May 4, 1946. Think there's a connection there?

My dad looked forward to attending little league baseball games with me. He finally had a son. My dad never did feel the prosperity that came to America after the war. He never owned a car, or even a license. He walked everywhere, including to work at a factory a mile from our home.

In April of 1949, a month before I turned three, my dad suffered a tragic accident at work. While transporting a heavy roller with another worker, he dropped it on his toe. It swelled terribly, but dad couldn't have it looked at by a doctor. In those days, there was no workmen's comp and he needed his job. When told by friends that he should take some time off to let his toe heal, he just replied, "I need to work, I have four children to raise." He cut the leather off around the toe and still continued walking to the old factory and working for that check.

Two weeks after the accident, at home, he complained about feeling funny and soon after he fell to the floor. My mother realizing the emergency told my oldest sister, Theresa, to stay with dad while she went running to the corner of the street where a doctor lived. There were no emergency rooms or ambulances in those days. My sister, Annette, then eleven years old, was told to take me and my sister Cookie, to our aunt's house three blocks away to stay with them until things settled down. I was only two and my sister Cookie was only six years old, so we didn't realize what was happening. Annette cried all the way to our aunt's house. When we got there, our aunt was not at home and we had to return to our house. By that time, my dad was gone. He had suffered a massive heart attack due to a blood clot that had travelled from his foot to his heart.

My mother, then 37 years old, was left with no insurance. Who could afford that working in the mills? She had four children and she raised them by herself after that tragedy. She never did remarry.

As I was too young to go to school, my mother drew a widow's pension that barely paid for necessities. When I finally got into kindergarten, she returned to the factories at night while my older sisters took care of me. We probably should have felt deprived, but our community contained many struggling families.

In September, 1951, I began attending St. Anthony's Parochial School in the kindergarten class. Mom found employment in an electrical component factory named John I Paulding's. She worked as an assembler and though the work was tedious and demanding, she was pleased at having the opportunity to earn her own way after two years on public assistance. My sisters also attended St. Anthony Parochial School. The classes were free to parishioners and superior to the public schools in the area.

Unfortunately, the attitude in my community hadn't changed much since the thirties on the role of women in society. Before the decade was out, my two older sisters left school at sixteen. They applied and worked at Paulding's, waiting for their men to come by so they too could assume their place in the home. Unfortunately, it was a different world evolving, which required the work of both partners to get ahead. Although their lifestyles were dramatically enhanced by the new found prosperity of the jobs, my sisters, Teresa and Annette always suffered by the lack of education that followed them when they went to work to help out mom in those days. My younger sister and I were the beneficiaries of the extra cash coming into the household but it was many years before I came to appreciate their sacrifices.

Life was no picnic for us, but it was certainly an improvement compared to the period right after dad's death. My family was particularly close, not only because of the dire straits that fate gave us, but also because we became self-reliant in ways that we only now can appreciate. When my father died, it seemed that all our relatives sort of disappeared for the most part. My mom's family, who forced her to surrender her childhood to tend and support their upbringing, kept their distance for the most part, with few exceptions. My father's family similarly was rarely seen after my dad's death. In retrospect, I better understand how such things happen and the bitterness that I felt in my early adulthood years has dissipated into a feeling of resignation. It probably was human nature playing out. They were in struggles of their own and never wanted to look back, especially if they didn't think that they were in a position to help out. They were all struggling in those days. My aunt Doris (my mom's sister) and her husband were the exception. They had children our age and frequently took us to beaches and picnics. They were special folks and we appreciated them for being there for us.

The education that I received at St. Anthony was outstanding. I will forever be indebted to the good Sisters of the Holy Cross who suffered through our foolishness and still managed to provide us with an outstanding primary education. Whatever writing that I later turned to in archiving the story of my company was learned during those first nine years in parochial school. Our classes were broken into two distinct sessions, morning and afternoon. In the mornings, the class work was presented completely in French in reverence to our French-Canadian ancestry and the parishioners who came from Canada and paid for the schooling though generous donations. The classes were in religion (always of paramount importance in continuing the propagation of the faith), French Grammar, Reading, Composition, Spelling, and Art. English classes followed in the afternoon and consisted of English Grammar, Composition, Spelling, History, Geography, Arithmetic, and Nature Studies. The nuns were dedicated and hardworking and had the strength of character to maintain a much-disciplined atmosphere where it was understood that it was their way or out the door. To be banished from parochial school to the city's public school was a dishonor that was not to be tolerated by student or parent. The nuns taught, we learned.

The fifties were a unique decade for our country and as the victors of World War II, we enjoyed a tremendous advantage over the rest of the world. Most of their industrial base had been destroyed during the campaigns in Europe and Japan. These were heady days for Americans who benefited immensely in the immediate years following the war. I can still recall the early years of the fifties where we would joke about the 'MADE IN JAPAN' logo on the bottom of low value goods that preceded the high value product development that led them out of their demise. We stopped laughing in the 1980s.

Although Americans felt justly proud of their victory in the World War, they felt threatened by a new adversary; the spread of Communism. The U.S.S.R. was also benefiting by their success against the Axis Forces of World War II and was expanding their reach in the world. They achieved the skill to detonate an atomic weapon by 1949 and the fear that resulted in America was terrifying. That same year, China joined the Communist community when Mao Tse-tung wrestled control of the country and inserted its own form of communism on the poor country. The following year, 1950, the Communist North Koreans crossed the 38th parallel and invaded the Democratic South Korea. It must have seemed to the democracies that their world was spinning out of control. Fear permeated the country.

It wasn't long before this manifested itself into the political fabric of America. It was presumed that Russia was heavily involved in espionage in America and must have had very high political and military figures in their fold to be as successful in their political and nuclear advancements since the end of the war. The witch hunt began in earnest at the behest of Congress, led by the ambitious congressman Eugene McCarthy. It would be a while before the cooler and more level headed leaders prevailed and the dust settled down, but the threat of atomic warfare was real. In a country like America that had never seen their cities bombed, the atmosphere was bordering on panic.

As a young student in school, this fear manifested itself by the many air raid drills that became commonplace in school. I recall the overpowering shrilling sounds of the air raid sirens emanating from the local fire stations throughout the city. We were drilled into jumping under our desks at the first recognition of the alert. We would face down with our arms covering our heads to protect us from flying glass from the windows. Good old 'duck and cover' was the order of the day. If you were caught in the open when the alarm was sounded, you were told to enter into a nearby doorway or store entrance and 'duck and cover'. It would be years before the folly of this exercise was revealed with the question, "and then what?" There really wasn't anywhere to run, now, was there? Some creative entrepreneurs came up with an idea; the fallout shelters. These were the rage for a while, which led to the moral dilemma for the lucky owners of these survival bastions. The question was, "What would you do if your neighbor arrived at your shelter, pounding on the door asking for entry? Would the extra demand on your supplies compromise the survivability of your family?" The question was never satisfactorily answered but it didn't matter much as the real question was the same as the 'duck and cover' question, "And then what?"

Needless to say, Communism and its corresponding threat was the source of much anxiety during those days in the fifties. It was not so much for us growing up during the period. What threat did this pose for us as young people? There wasn't much that we could do about it anyhow. It was best to let the grownups face that challenge. We had complete trust that America would end up winning in the end. We always did, didn't we? America needed heroes and we had ours, which included our fathers, uncles, and in some cases, our aunts who won World War II.

Heroes in America were not hard to find as the new medium for the masses were coming into play, namely television. I was told that the first public places to use the televisions were the bar rooms. Made sense actually, as the TV would serve as the bait to lure you into the establishment and keep you there, buying those beers and shots. They couldn't lose. Once they made the investment, the rest was free. It would be a few years before the masses could afford the televisions, but by 1953, the programming was expanding considerably and the market exploded with the addition of the popular *Today Show* with Dave Garroway, newsreader Jim Fleming, and who could forget announcer Jack Lescoulie or J. Fred Muggs, the mascot monkey. *I Love Lucy* with Lucille Ball and Desi Arnaz came on TV that year and the country fell in love with the Ricardos. Slowly but surely the masses were drawn into purchasing this ticket to all that free entertainment. It would be a few years before enough money could be saved to purchase our second-hand television, but when it came, it was amazing. I can remember my sisters and I spending an hour watching a test pattern that came on an hour before the shows began. The test patterns were there so that you could adjust the screen size. They would place a circle with the station identification logo on it and you could adjust the lateral and longitudinal controls to make the circle round. We'd come back the next night and sure enough it was off again. We used to say that someone at the station was alternating elliptical with round logos to play tricks on us.

By the time I was ready for high school, there was no question that I would get a high school diploma. We all knew that I couldn't go to college, but I had the grades which punched my ticket to the local vocational school. At thirteen years old, I had to choose a profession that I planned to use for the rest of my life. The choices were, Carpentry, Welding, Electrical, Electronics, Steam Engineering, Machine Shop, Automobile Repair, Drafting, Industrial Design, and of course, Household Arts for the girls; what else could girls excel at? Remember, it's 1960.

I chose machine shop, as I heard that machinist and especially tool and die makers, were in high demand and commanded the highest salaries. What did I know?

I graduated with average grades and went to work as a Maintenance Technician at a local bakery that produced Sunbeam bread in 1964. I left there after a year to work at a factory that produced spare automobile transmission parts. While there, I received my draft notice. After so many of my extended family served, it was now my time to help defend our country. My mother was very broken hearted. Her only son was set to be placed in danger in Vietnam.

DIAMOND IN THE ROUGH

The young boys of Kailua-Kona, Hawaii lived near the US Army base. They looked up to the soldiers who they saw each day marching about with such pomp and grace. One boy in particular, Sidney Raphael Kalamakuikalani Springer, had already made up his mind as to how he wanted to spend his life. He aspired to be just like those men and it seemed that he was destined to become one of the most highly decorated soldiers of the 20th Century.

In 2001, he wrote in an article in his hometown newspaper of that decision. "I used to hang out near where Hale, Hawaii stands now and watched the soldiers camped out in tents play cards and wait for the post exchange trucks to snag candy bars. This is the life for me," I thought.

On one particular Sunday morning, when he was ten years old, he and some friends noticed a gray shadow overhead looming in the distance towards his home island. When it came into focus, they made out an array of foreign aircraft racing low overhead. As a young boy, he recalled clearly making out the pilots of the planes and waving to them as they passed overhead towards the next island in the string. He was young and had no idea the seriousness what was about to take place. It was December 7, 1941, and the Japanese forces were attacking the US Pacific Fleet stationed at Pearl Harbor.

World War II began and Sidney waited patiently for his chance to join the fracas. It was not to be, as the war ended by the time that he was old enough to participate. He would have his chance for glory when he was 18 years old and the North Koreans flooded across the borders into South Korea, commencing the Korean War. It wasn't long before he was in the uniform of a US Army Infantryman. He was a nineteen years old PFC when he was part of the risky landing in the Inchon Peninsula during the Korean War, miles behind enemy lines. Later he took part in the house to house fighting that led to the liberation of Seoul, Korea, where Sidney earned his first Purple Heart and his first sergeant stripes. Later, he was part of a second amphibious landing behind enemy lines, where he earned his second Purple Heart.

This was Sid's first war and he looked back at it sadly in later years. The memories were still fresh even decades after the period. He told me that he still had nightmares of the first American he saw killed in combat. "We were being hit pretty hard by North Korean artillery early on in my time in Korea. I jumped into a nearby foxhole for cover. There was already a young soldier leaning face first on the side of the hole. At first, I thought he had been injured and I tried to help him. When I called to him and he didn't react, I grabbed his shoulder and leaned him back. What I saw horrified me. His face was completely blown away with nothing but a skeleton starting back at me. I still see that image today." War is hell, and every man who experiences the trauma has their own story to try to forget. It isn't easy.

Sidney was very early on recognized for his bravery and leadership qualities battle tested during the Korean Conflict.

He was sent to Fort Ord, California for a number of years to train recruits for the US Army. During this period he was part of the Army Human Research Unit which developed many of the procedures and plans used in Basic Training Centers all over the country for the US Army. In 1955, Sid somehow wangled a temporary assignment at Fort Lewis, Washington where the movie, 'TO HELL AND BACK', a movie about Medal of Honor recipient, Audie Murphy's life, was being filmed. He managed to be filmed in a scene of soldiers jumping into a foxhole. Sid was very proud of the fact that he 'made' it in the movies.

In 1962, Sergeant First Class Springer was chosen to participate in the US Army's Tactical Mobility Requirements Board that was created at the request of Secretary of Defense Robert McNamara. This group had the assignment to review and test new concepts integrating helicopters into the United States Army. It gave birth to the idea of air mobility. It was led by General Hamilton Hawkins Howze who valued Sid's recommendations. This era proved to be a catalyst for his upward mobility, as he was promoted twice to the rank of First Sergeant.

With the approaching storm taking place in Southeast Asia, 1SG Springer was assigned to Fort Lewis, Washington where he would be instrumental in the training of newly inducted young men who were destined to hold the line against Communist expansion in South Vietnam. It was there that we caught up with the "Pee Bringer" ("My name is First Sergeant Springer and you need to know a couple of things about me. First of all, I am the Pee Bringer. If you screw up in my Company, I will bring pee upon you. Second, I want you to note my stripes and in particular the diamond in the middle. Around here, diamonds are trumps!"

Just about then, a young recruit named Walter Kelley from

Rhode Island smiled. Springer spotted him in the back of the room. He went up to him, stood face to face with him and told him, "What's your name? You're on my shit list." Walter would remain on that list until one fateful day in the jungle in Vietnam a year later when Springer turned to Kelley and said, "Okay, Kelley, you're off my shit list now." Walter was amazed that he still remembered the incident from the first day that they met. He was not much relieved as Springer never treated him any different than anyone else in the company. He just wanted to 'trump' Kelley for the benefit of the rest of the new men coming into his company. The message took.

THE PRISONER OF ZENDA

Allyn, 'Jon' Palmer was born in Elkhorn, Wisconsin on March 22, 1942, where the county hospital was located, about 20 miles from his father's dairy farm. The town nearest the farm is Zenda, named for the classic novel "The Prisoner of Zenda". The population of Zenda was about 100 when he was growing up and it still about that today. His folks wanted to name him Jon Allyn Palmer, but decided against it when they realized what his initials would spell. It was just too soon after the Pearl Harbor attack.

His dad owned 35 milking cows when Jon was born. By the time he reached high school, his father grew weary of the milking two times a day routine and sold off the dairy herd. He decided to raise calves instead. He would buy the heifers (young female cows) from the farmers in the area who were still in the dairy business and raise them until they were springing (about due to deliver a calf themselves) and sell them back to the dairy farmers. In the later years when he was on the farm, his chores consisted of feeding the calves out of a bucket and cleaning out the pens on Saturday.

Jon attended Zenda State Grade School, which was situated

in a two-story building that housed grades one through four on the first floor and fifth through eighth on the top floor. He had the same teacher for the first four grades and another for the last four levels. Each of those teachers taught six or seven subjects in each class. The greatest number of kids in any of his classes was either seven or eight. The enrollment each year was based on how many migrant workers were needed for a particular year and how many children they had to educate. Therefore, of the children that enrolled in the beginning class with him, only four of them graduated together from grade school.

One distinctive feature of his town is the railroad that runs through it. In fact, the tracks ran right through their farm, splitting it in half. When he was a kid, he would play on box cars that were sometimes parked for three days near a feed company waiting to unload fertilizer and feed grain. He would have a great time climbing all over them. It was their playground. On weekends, when the feed company was shut down, they would use the boxcars to reach the roof of the feed mill. Once there they could climb through the unlocked upper window and climb inside and gain access to this giant operation where they would play hide and seek for hours at a time.

He attended school at a consolidated regional high school in Walworth, Wisconsin, about ten miles southwest of Elkhorn. The school served a district probably twenty by ten miles. The high school had approximately 400 total students attending the four grades. His class graduated around ninety in the year he completed high school.

In early 1960, during his senior year, he was already signed up to attend North Central College in Naperville, Illinois, twenty miles west of Chicago. He had plans on majoring in engineering there because of the reputation the school had developed in that field. Notable amongst its alumni was Frederick

Louis Maytag. You guessed it, the founder of Maytag Washing Machine Company. One night his dad was reading the local newspaper when he came upon a notice that his Congressional Representative, Gerald T. Flynn, was giving Civil Service exams in anticipation of selecting nominees for the US Military Academy at West Point. Flynn could nominate two candidates for the Academy that year. It was an offer of a free education, so he thought, why not?

He took the exam and a month or two later, he received a letter that indicated that he was selected as the third alternate to go to West Point. He assumed that ended that chance to go to the academy. Two weeks before he graduated from high school, he received another letter notifying him that he was selected to attend West Point and needed to report there on the fifth of July. He immediately notified North Central of this development and a short time after graduating from high school he took a train from Chicago to West Point.

On that same train, he met Tom Erdman who was from Racine, Wisconsin. It turned out that he was the second alternate from my congressional district. Later he learned that the principals that the Congressman chose had chickened out and refused to go. The first alternate flunked the physical and that opened it up for Tom and him to use the two selections that his Congressman was allotted that year. Tom and he rarely saw each other while attending the academy, but sometimes met up for trips back to Wisconsin for Christmas or summer leave. They both graduated from West Point in 1964.

He was opened minded about going into the military. There wasn't a tradition of military service in his family. He had an uncle that was drafted into the army during World War II and that was about it. His dad was a farmer, who was essential for the war effort and remained home during the war. He decided to give it a go and if the military turned out to not be what he wanted, he would serve his time and leave. The original commitment was four years after graduation before it got restructured to five years once the Vietnam Conflict began to escalate.

Here is the story of how Sandy, his wife, and he got together.

Sandy's mother and his mother were in the same class in high school. He lived in the country and Sandy lived in the city. Her folks owned a grocery store in Walworth, Wisconsin where Jon's family shopped. He knew Sandy from the time when he was very young. She attended his high school and was one year behind him. In the first two years in high school, she dated his best friend, who was also a farm boy. They didn't have any interest in each other in those days. He could only think of one time he saw her outside of school during high school and that was when he went to a party in her house and he was with another girl then.

Things changed when he came home for Christmas leave during his senior year at West Point. Her folks had sold the grocery store by then and were running a gift shop named The Honey Bee. A couple of days before Christmas, he went to shop in search of Christmas presents for his parents and three brothers. Sandy was the clerk that day and helped him chose some gifts. She asked him if he wanted them wrapped and he said that he would be happy to have her do the wrapping. She brought the gifts to the back room, through some swinging doors, and he followed close behind. She was startled by his boldness. They talked while she wrapped the gifts and it occurred to him that this was a pretty nice-looking girl.

A couple of days after Christmas, he was in Walworth and decided to call her up to ask her if she would like to join him and see a movie. She said that she would enjoy that, and asked what time he would be over. He told her he would be by to pick her up in ten minutes. He wasn't taking any chances that she would change her mind. Unknown to him, Sandy was seeing another man a couple of years older than him and she felt that he was taking her for granted. He didn't care what her motivation was back then. Jon picked her up as planned and they went to the movies to see Elvis Presley in "BLUE HAWAII". It would be years before he learned that Sandy had gone to see the same movie with the man who was taking advantage of her on the night before.

They had another date over that holiday period and then he returned to West Point to continue his studies. He came home for Spring Break, dated her again, and finally gave her his West Point pin before he returned to the Academy.

He invited her to attend the graduation ceremonies at West Point. She came out with his folks and his brothers. After graduation, he had some leave on the books so he returned to Wisconsin. Sandy and they dated during that break. One date in particular remains in his memory. That was when she accompanied him to a wedding in Ohio for a classmate of his from West Point.

After graduating, he left home to attend Airborne School from early August until November 10th. He returned home and on November 14th, 1964, Sandy and he were married. Two days later, they set out on their honeymoon on the road heading to Fort Lewis, Washington by car. The first branch selection for him was for the Infantry and he chose Fort Lewis, Washington as his initial destination. He succeeded in getting both preferences. Sandy and he arrived at Fort Lewis in late November, 1964, just after Thanksgiving. Along the route they got stranded in a blizzard in Big Timber, Montana and actually had their Thanksgiving dinner in a bowling alley. They ate bowls of chili for the holiday.

Upon his arrival at Fort Lewis, he learned that he was assigned to B Company of the 2/12th to serve as the Platoon Leader for the First Platoon. His Company Commander was Captain Bob Strickland. The Battalion Commander at that time was Colonel Atkins. He served as Platoon Leader for about five months. One Saturday night, the battalion had a game night and was attended by the Officers from the 2/12th Infantry and their wives. They played a number of different games and he won the Scrabble tournament. Colonel Atkins took note of that victory and on the next day he called him over to tell him that he noticed his skill at scrabble and he decided to choose him as his Adjutant. He left B Company to become the new S-1 Officer.

Here is an explanation of the four staff positions that were used when we served. We were told that we could recall what each staff position was responsible for by remembering the key word 'PITS'.

S-1 – P – Personnel

S-2 – I – Intelligence

S-3 – T – Training (or later tactics once we were in combat)

S-4 – S – Supply

He served as Battalion S-1 until the first week of December, 1965. It was then that he was given command of A Company. He just barely had time to get rid of the deadwood in preparation of the arrival of the trainees who were due to arrive in late December. According to the first morning report, he had one Officer and maybe 45 ranked below Officer. Upon signing that report, he went straight to work to eliminate problem soldiers who may have had a harmful influence on the large contingent of draftees that would shortly be arriving. The next morning, he had one Officer (him) and 19 Enlisted Men or NCOs remaining on the rolls of the Company. Included in that listing was an armorer, a clerk, a supply Sergeant, and a couple of administrative people. There were two NCOs for each of the first three. Sgt. Ziebart was the lone NCO for the Fourth Platoon. Lt. Bob McFarland joined them shortly before the trainees arrived and he was made the Company XO.

First Sgt. Sidney Springer (Diamonds are trumps) was in the 8th Infantry at this time. He arrived in our Company shortly before our arrival. His first impression of Springer was very positive right from the very first time that he met him. He was very pleased to have him as the First Sgt. of the Company and he recognized he was just who we needed at that time.

These were the two major figures waiting for us as we arrived at Fort Lewis in late December, 1965.

CIVILIANS TO SOLDIERS

Greetings

November, 1964, New Bedford, Massachusetts

"Greetings,

You've been selected by your friends and neighbors to represent them in the Armed Forces of the United States of America . . . "

It was no great surprise when I received my induction notice. I had gone for two physicals at the Boston Army Base previously and I knew it was just a matter of time. I was 19 years old when duty called and I knew that I would be stepping into the shoes of most of the men of my family who served in World War II or Korea. Hey, it was my time.

Things were going pretty well for me in the fall of 1965. I had a car, a good paying job, and a high school sweetheart that had been with me since my sophomore year. Life was about to change radically for me.

At 6AM on the chilly morning of Monday, December 13, 1965, I arrived at the local draft board where the mayor of

New Bedford met us. There were nineteen draftees, counting me, there. A representative from the local veterans' organization bid us a fond farewell and hailed us to "Give those commies hell."

After receiving our gift of shaving gear from the veterans' rep, we were registered and sent outside to wait for the bus to take us away to Boston Army base.

As we were waiting for our bus to arrive, the mayor approached my mother who was very upset about what was happening. He told her, "Don't worry, lady. The army is going to take good care of your son." Hmmm.

After a while the bus arrived and the nineteen chosen by our friends and neighbors to represent them in Southeast Asia boarded it. Each of those draftees ended up serving in Vietnam.

As we proceeded to our destination, the Boston Army base, the bus was fairly quiet. I sat with a high school friend named Roger Davignon and others softly conversed with similar acquantainces or new friends. None of us knew what awaited us, but, heck, this was not something new for young men growing up in our era.

We arrived at Boston Army Base around 8AM and went though processing before being transported to a reception center elsewhere. The processing incuded answering questions to determine whether Uncle Sam had chosen wisely. For the most part, it was similar to what was asked previously during our predraft physicals. One question that stood out to me, even today, was the question of whether or not we are or were a member of the John Birch Society. I would wager a weeks salary that few of the nineteen in my group even knew what the John Birch Society was or what they promoted. Politics was foreign to us young men.

After we completed that segment of the process, we waited around a couple of hours before we went through a final physical. There was a Spec 4 medic who was there moving us along. He caught our attention when he looked at us, shook his head and pronounced that, "If I had to go through what you guys will be going through, I would put a gun to my head and pull the trigger." We thought, "What does he know that we didn't?"

By 2PM, our group and others were assembled in a meeting room. There were approximatley fifty young men there waiting to take their oaths. Before long an officer of the U.S Army entered the room with two NCOs from the Army and two more from the Marines.

The officer explained that we were about to take an oath to serve in the military. He explained that it was a choice. You could step forward and take the oath, or step back and you were chosing a future that led to incarceration at Levenworth Army prison. Everyone stepped forward. He then ordered us all to stand and raise our right hand.

He then began, "Repeat after me."

"I, *'name'*, do solemnly swear that I will support and defend the Constitution of the United States against all enemies, foreign and domestic; that I will bear true faith and allegiance to the same; and that I will obey the orders of the President of the United States and the orders of the officers appointed over me, according to regulations and the Uniform Code of Military Justice. So help me God."

There was one last ritual that remained from that ceremony. The Marines stepped up to the front of the room and asked if anyone wanted to be drafted into the Marines. No one stepped forward. They said, "That's okay, we'll choose who goes into the Marines." With that they began walking through the ranks and choosing around six or seven of the tallest, strongest of the lot.

I need not have worried. I weighed 130 pounds and was 5 foot 8 inches tall. They didn't want me.

We were then about to learn one of the maxims of life in the the miltary; 'hurry up and wait'. We spent the rest of the afternoon waiting for busses to transport us to our first duty station, Fort Dix, New Jersey. The buses arrived around 6PM and it took about an hour before all of them were filled and we were off down Route 95 and heading to the Fort Dix Reception Center.

Fort Dix Reception Center

We pulled into the bus parking lot around 1AM. Looking out our window we could see the imposing 14-foot Ultimate Warrior Monument. We didn't know it then, but it had been erected in 1957 as a symbol of the American fighting man. It was purposely erected in front of the Reception Welcome Center to inspire new arrivals coming into the Army.

The bus driver entered the building to alert the greeters that the new recruits had arrived. A couple of ornery Non-Commissioned Officers followed them out with clip boards in their hands. As soon as they reached the bus, they shouted for us to get off the bus and make our way over to the building. As we alighted the bus, we called out our names and they ordered us to run, not walk, to the nearby building. It was the first time we were referred to as 'shitheads'. I thought, "Hey, I'm not a shithead...whatever that means." It would be a while before we 'got it'.

In the welcome center we were broken into rosters. The draftees from New Bedford were all sent to Roster 13. I thought, "Great, drafted on the 13th and sent to Roster 13."

This was not starting off well for anyone inclined to agree with superstitions.

From there we were marched to what looked like a warehouse building. There we formed a single line leading into a supply area where army clothes were handed out to us by supply soldiers who looked at us and said things like, "Hmm, size, I don't know, probably medium, here." In that way we were handed an assortment of clothes. None of it fit very well, but it would keep us warm in the cold New Jersey climate of December. At the end of the line, we were given wrapping paper and told to take off our civilian clothes, wrap them up and address the package to our people back home. We wouldn't be needing them for a while.

By the time our group completed the process, it was then 3AM. We were then sent to an old wooden army barracks that was heated by an old fashion pot belly stove that had to be fed periodically. We drew lots and men rotated to stand fire watch and feed the stove in half hour increments. Exhausted by then, we chose beds and laid down for a long winter's nap, with the emphasis on nap.

Less than two hours later, at 5AM, Reveille was being blasted through giant megaphones. I awoke and mentioned to the nearest man, "They can't mean us. We just got in two hours ago." Like clockwork, the doors were thrown open and a pair of new NCOs (Non-Commissioned Officers) rushed into our barracks and grabbed garbage barrel tops and banged them off the barrels and ordered us up and to prepare for a formation in a half an hour. (Whatever became of "Don't worry, lady. The Army is going to take good care of your son?").

We rushed to get cleaned up and shaven and headed out our building in the cold winter morning wearing those long woolen overcoats that reached to the floor. As I rushed into formation, lit up by bright floodlights, and us all dressed up in those poorly fitting coats, my mind raced to old World War II movies. The image compared to my recollections from the POW movie 'Stalag 17'. I had nothing else to compare it to.

After a calling of the roll call, we were released and told that the mess hall would be opening up at 6AM. The trap was set for all new arrivals. Being all young men who were famished after going without eating since 5PM the day before, we all rushed to the mess hall. I was about fifth in line outside and thought myself very lucky on that cold morning.

At 6PM the doors opened and a couple of mess sergeants came out to escort us into the hall. The first twenty men or so were escorted to the back of the hall where the walk-in freezer was and told to hang our coats on the hooks in the locker. I thought, "That's pretty nice that we get this kind of personal service." We walked out of the freezer and with a smile on his face one of the mess hall cooks slapped a heavy lock on the freezer. This can't be good.

He then began passing out aprons to us trapped recruits and told us we would be eating after we served everyone else breakfast. We were learning. Every day after that, we waved on new guys at the mess hall ahead of us until we were placed just outside the net.

Our very first responsibility was to get our hair cut. One of the guys (who would later become one of my best friends in the army) had much longer hair than the rest of us. His name is Jim Heys and he was one of those drafted with me on December 13th. (You can see him in the earlier group shot). He was tall and at six-foot four he had just slipped under the six-foot five height limit that would have kept him out of the draft.

Heys had that long hair and we teased him a bit as we went to the barber shop to get our army crewcut. As we were waiting our turn, some enterprising soldier decided he could make some money and set a lottery to see if Jim would cry when they chopped off his hair. Only nineteen-year-old teenagers can come up with these bazaar schemes. The money was collected and the odds leaned towards that he was going to break down. When Jim made his turn to the chair, everyone inside had their eyes fixed on him. Outside the building, guys crowded around the windows to see the event. Jim was a trooper and took it like everyone else. There was a lot of grumbling as the winnings were paid out.

I got my haircut which took about 30 seconds and as I got off the chair the barber asked for two dollars. It was okay if I could not pay for it in cash. They would just take it out of my first army paycheck. I looked at him in disbelief! 'I' have to pay for this haircut? Couldn't they draft some barbers and order them to cut our hair? After all, as Pvt E-1 soldiers we were only making $89 a month in 1965. It was not my last bewilderment that I faced in the military. There were a lot of those, 'You've got to be kidding me' moments.

Here are a few other recollections from our time spent at the reception center. We were issued military ID cards with our new haircuts that stayed with us throughout our time in the military. Each had a number attached to it. My military number was US51-538-499. Jim's number was US51-538-500. He was right behind me in line. The US prefix was given to draftees. An RA designation was assigned to those who joined the army for at least a three-year commitment—it stood for Regular Army. The first two digits (51) were assigned to soldiers from Connecticut, Delaware, Maine, Massachusetts, New Hampshire, New Jersey, New York, Rhode Island or Vermont. The next three digits (538) represented the series designation and the last numbers (499) were our personal numbers in that series.

The next day, Wednesday the 15th, the tests began. There were some specialty skills that were useful to the 'New Action Army' as it was called in those days. Mostly though, they needed men, lots of men, to comb the jungle of Vietnam searching out Vietcong. One skill needed that perplexed me was Morse Code talent. I was pretty good at that in Boy Scouts, but I guess they met their quota before they read my test. I never tapped a key in the army.

On Thursday morning, the 16th, there were more physicals followed by the running of the gauntlet of Army Medics with needles. Four on each side, each injecting us with different vaccines to fend off historic afflictions that devastated past armies. Uncle Sam had made an investment in us and he was going to protect that asset.

That afternoon we were sent to an assembly hall to view some Army produced films. While we were waiting our turn into the room, we were assembled in a lobby lined with posters promoting volunteerism into elite fighting units like Airborne, Rangers, and Special Forces.

It's funny the things one remembers even 56 years later. Music was being piped into the lobby for our enjoyment that day. I heard the song, "Michelle" by the Beatles for the first time that afternoon. I thought, "How unusual, the Beatles are singing part of that song in French. "Sont les mots qui vont tres bien ensemble." (These are words that go together well). It just wasn't your usual Beatle song and stuck in my head. There were a quite a few songs I associate with particular events during my military time today. This was one of them.

Finally, it was our time and we were led in to view the movies. The films were slick productions that encouraged the adventurous to join elite forces. After the films ended, they passed around paperwork to fill out for the convinced. I didn't see many takers, certainly not from my New Bedford draftee friends.

We spent the 18th-19th weekend mainly in our barracks. The bolder of us slipped out of the base Saturday night and visited nearby Wrightstown, a mile from the fort. The small town was, and still is, dependent on the nearby military establishments of Fort Dix and McGuire Air Force Base for survival. Most remained at the base. Sunday morning, I attended mass at the base chapel, while the daring few were sleeping it off or still working their way back to the reception center.

By Monday, the 20th all the written tests and physical scores were completed and we were once more assembled into the largest hall yet for a final speech. A company grade officer spoke to us about our future in the army based on our test scores. At one point he said, "Gentlemen, there is a mistaken belief that the Army doesn't care about how its men would prefer to serve their country. There are a number of occupational positions that requires filling and we would be very happy to oblige you if at all possible. You will be interviewed by a recruiting officer and he will see if we can satisfy both the Army's and your needs". Wow! Who knew?

Lou Berard, whose photo can be seen in the earlier draft day image, said to us before he went in. "My dad was in the army in World War II and served as a cook. He told me if I had a chance to choose how to serve, I should elect to be a cook." That's just what he chose when asked and in fact he was given that role. What his dad did not know was the Vietnam War was not your typical war. There was no front to serve behind. The enemy was everywhere. Lou at times faced the same dangers in camp that we in the jungle faced, and sometimes more.

George Gobeil, who is also seen in that draft day group photo,had tested high in automobile skills. Why wouldn't he as he went to the Vocational school, as I did, and studied automobile repair? He was perfectly suited to be a driver and repairman of army vehicles. They accommodated him and he did serve in that capacity.

Eventually I went into the room and after a quick review, the officer told me there was not much use for machinists in the 'New Action Army'. He never told me what that would mean, but I kind of figured it out.

Jim Heys followed me into the room and sat down across from the Officer. "Tell me, son, what would you like to do in the Army?" Without batting an eye (nor much thought) he answered, "I would like the army to send me back home." The officer smiled for a few seconds and grabbed a rubber stamp, dropped it on an ink pad and stamped Jim's paperwork. Jim looked down and saw the word 'INFANTRY'. He looked up at the officer, who still smiling said, "That's all soldier." Jim shrugged and left.

On Tuesday, the 21st and Wednesday, the 22nd they kept us busy in close-order drilling and other details on the base which were mostly 'keep busy' diversions while waiting for our assignments.

When we formed up on the company street on Thursday, December 23rd we were alerted about a new directive ordered by President Lyndon Baynes Johnson. Apparently, he was coming under a lot of heat for drafting troops so close to the holidays. To cool the temperature, he ordered the army to release all those waiting for basic training to begin an extended weekend pass. We were given Thursday, the 23rd, Friday, the 24th, Saturday the 25th and Sunday, the 26th off. What a deal!

Meanwhile out at Fort Knox, the draftees who would form the second half of our company at Fort Lewis were going through the exact same process there.

Christmas And Back

There was a race to get to the fort's bus station. The clock was ticking and no one wanted to waste any more time than necessary. Surprisingly, not all of the New Bedford draftees traveled the same way that I did. I traveled with two men from Rhode Island named Ed Smeed and Walter Kelley. We were both from Roster 13 but arrived differently. Both would end up in my training company and sent to Vietnam with me. Walter and I became very close once we got to Vietnam for a reason that I will explain later.

As we had mailed our civilian clothes home, we had to travel in our dress greens. After only ten days in the army, we didn't have anything to show on them, but that was okay. It allowed us discount pricing for the trip.

It was the first time we travelled this far and we had to put our heads together to figure out how to make it to where we were going. We figured we could make it to New York via a bus that traveled to the New York Port Authority Bus Station. Once there we could take the ten-minute walk over to Penn Station where we could take the train from there to Providence, RI, where our folks could pick us up by car. Simple enough for even us to figure out.

Everything went as planned and by late afternoon we boarded the late afternoon train from Penn Station to Providence,

Rhode Island. As might be expected, the train was packed with people trying to make it home for the holidays. I never forgot that ride as it was the first train trip of my life. We couldn't find seats at first but as the train traveled east, seats opened up as people alighted the train. As we crossed into Rhode Island, day turned into night. I spent all my time looking out the window in awe. This was all pretty cool for a young kid who had never traveled this far before.

The train pulled into Providence Train Station somewhere around 6PM and before the three of us split up we took each other's telephone numbers so that we could reconnect for the trip back to Fort Dix.

My family and girlfriend were there to greet me and it was a chatty trip to New Bedford. So many unfamiliar experiences had taken place and I was eager to share them.

As you can tell by this photo taken of me at my sister's home during this Christmas furlough, my uniform had nothing but a name plate on it. That would change soon enough. I weighed 128 lbs. at that time. By the time that I was sent to Vietnam the army got my weight up to 156 lbs. That was not unusual as the theory goes, "Once Uncle Sam takes you in, he will add or trim weight to arrive at whatever shape gets you into fighting condition." That was true for all the men I served with in the army.

I could say that it was an eventful, happy Christmas at my house, but I would be lying. The dangers laid ahead dampened the normally festive family mood of the holiday.

Note: in the year 2005, I threw a Christmas party for the surviving members of the original 19 draftees from New Bedford. Most were there with not only their wives, but with their families and siblings. I told them I organized that party to make up for that 1965 Christmas that we were deprived of back in the old days. We had a great time!

We needed to be back at Fort Dix for Reveille Monday morning so that meant traveling on Sunday, the day after Christmas. Walter, Ed, and I called each other on Saturday night and planned on meeting at the Providence Train Station on Sunday morning.

We met on the platform and everything went like clockwork as we retraced our steps back to the base and arrived just in time for dinner.

Monday was used to make sure everyone made it back in time before they would send us on our way to our next duty station. Stragglers did indeed arrive throughout the day.

On Tuesday morning, rosters were posted on our barrack's bulletin board indicating our next duty station for basic training. Most from our roster were assigned to the 4th Infantry Division, stationed at Fort Lewis, Washington. Some of the guys figured that wasn't too bad, only a couple of hundred miles south. They were told, "No silly, that's Washington *State* on the west coast, three thousand miles away". Well, at least it wasn't Fort Hood, where we were told was inhabited by the dreaded SNOW SNAKES. Were we naïve? You bet.

My high school buddy, Roger Davignon, was the only one of the original New Bedford draftees sent elsewhere. He was sent to the 9th Infantry Division at Fort Riley, Kansas. Both the 4th and 9th Infantry Divisions were being built up for Vietnam deployment and were later sent to South Vietnam.

Roger and I didn't meet again until years after we got home from Vietnam. It would be decades before we made the connection of two major battles that we fought in against the 9th

VC Division less than a day from each other and only fifteen miles away.

We boarded busses that evening for a trip to Philadelphia International where we would catch a redeye flight to Seattle-Tacoma Airport. We were on our way to basic combat training (BCT) and advanced individual training (AIT) in preparation for deployment to Vietnam.

We arrived at the Philadelphia airport at approximately 10PM and waited in the terminal for our chartered DC8 jet to arrive. At Fort Knox, that second group that would make up the second half of A/2/12 recruits was also traveling by air on a DC7 a couple of days after our arrival at Fort Lewis. They would travel from Cincinnati by that propellor driven aircraft that would require more than twice the time to arrive at McChord Air Force Base outside of Fort Lewis.

We would have a two hour wait for our plane to arrive and during that time we milled about in small groups in the boarding area. I recall there was a lot of horseplay and the usual bravado associated with young men on an adventure that they couldn't comprehend. After an hour or so, an incident took place that I never forgot. It was as if it was scripted for a movie. A plane arrived at the adjacent gate and flyers were unloading. In the front of this group was a soldier being pushed in a wheel chair. It was a Spec 4 soldier who had a leg missing and he was being pushed by a non-commissioned officer. The crippled man stared straight ahead as he made his way through our midst. A chilling silence engulfed us as we began to realize the dangers that we would be facing in the coming year. It was relatively quiet for the remaining period until we boarded the aircraft for our flight to SEATAC.

When our DC8 arrived at our gate, I could not believe how huge that plane seemed to me. In my hometown regional airport in New Bedford, they couldn't accept jets and so this was the first time that I actually saw a plane this large. "How do they get those things to fly?" I thought.

Having never flown in any type of aircraft, I boarded the plane apprehensively. I took my seat and watched as the plane left the gate and headed towards the end of the runway. Once we got clearance to take off, the giant engines came to life and we raced down the runway. Not knowing how long it would take before the plane had enough lift to leave the ground, I thought it was taking forever. Once it reached sufficient ground speed, the plane lifted off the tarmac and we were airborne. I watched as the plane climbed skyward and the ground took on a surreal appearance as the structures grew smaller and smaller. By the time we reached cruising height, the ground appeared black with only faint 'spotlights' indicating major cities. I was in complete awe. This flight, more than any other, encouraged me to train as an Aviation Technician using my GI Bill credits I earned while in the army.

The DC 8 that we flew on had problems that resulted in the cabin being well below comfortable temperature. The stewardesses passed out blankets, which helped but we were still pretty chilly. As we passed over the Rocky Mountains, the plane hit some turbulence that awoke those of us who were able to sleep. That kept most of us awake until when we approached Tacoma.

In the distance to our right, we could just barely make out in the darkness a silhouette of a huge mountain covered with snow. That was Mt. Rainier, which we would later name our very first basecamp after. As the plane descended low enough to approach the landing runway, I looked out the window. It was not yet dawn and I recall thinking that the main street, which ran parallel to SEATAC airport, took on the aura of a carnival midway with booths, instead of the shops whose lights uniformly adorned the avenue.

We landed a little after 4 AM. When we left the plane, we were met at the terminal by NCOs from the 2/12th Battalion Headquarters. Later when we arrived at our barracks, A/2/12 cadre were there to greet us. They were none too pleased with having to GI (clean and polish) the barracks and make our bunks before our arrival. They immediately began the traditional ritual of removing any semblance of normality we had in our young lives. "Welcome home, boys. You are ours now!"

With cold air encompassing the jet cabin for much of our trip to Seattle, I developed a cold and a bit of a fever. One of my buddies introduced me to the term 'sick call'. As there would be some preliminary introductions and classes for the first day, I thought that I wouldn't be missing much. Why not? After all, didn't my mayor tell mom that the army would take good care of me?

There were about eight of us reporting for sick call that morning. We were greeted by a burly looking man who called himself First Sgt. Springer (and "don't you forget it"). He allowed us to go to the infirmary and upon our return we were all placed on KP. Fact of the matter is, I would have gotten more rest had I ignored sick call and just went with the Company. This lesson wasn't lost on me for the rest of my Army tour. I shunned sick call from that time on.

Basic training was scheduled to begin the following Monday, January 3rd. The time before that was occupied with informal meetings and cleaning the barracks.

During our initial time at the barracks, we were alerted to the fact that Fort Ord, California had just experienced a troubling number of cases of spinal meningitis that winter. This resulted in deaths for a few soldiers. It was ordered that the barracks needed to have the windows kept open partially at night for ventilation. When the Day Room Orderly would make his rounds, he would open the windows. When he left, we'd close them. At that age, we felt ten foot tall and bullet (*and disease*) proof.

On Friday, December 31st, we were sent to bed at 10PM. Most of us remained awake until midnight to usher in the new year at midnight. At exactly midnight, the barracks came alive when a number of us began rattling our dog tags to serve as noise makers. Little did we know that exactly a year from then we would be hitting the sack early to awake at ten PM to commence a night operation to encircle a village south of Dau Tieng.

Unit Esprit de Corp

The weekend of January 1 to January 2 was one of those 'acclimate the troops to being soldiers' periods. We were allowed to go to church on Sunday, but we were marched there and back. In the vicinity was an Army PX, an Enlisted Men's Club, a movie theater (The Evergreen Theater), and a donut shop near a bank of outdoor pay telephones. All were off limits to us during basic. In fact, for all eight weeks during this initial training, we were pretty much confined to barracks, except for weekly Sunday visits for church services.

Even those that had no particular interest in religion became devout church attendees, just to break out of our barracks prison.

Just before leaving for Vietnam, I read an article in the *Seattle Times* which said that before the 4th Division began its buildup in the Fall of 1965, the division was at 20% strength.

By the time we deployed to Vietnam, we were at 110% strength. Of course, that was as a result of those massive draft quotas in the Fall of 1965 and later filling in key leadership positions with experienced cadre and officers.

The Commander of the 4th Infantry Division at that time was Major General Arthur S. Collins Jr. He was born in 1915 in Boston Massachusetts. He graduated from the US Military Academy in 1938.

During World War II Arthur Collins served in the Pacific Theater as the battalion commander of the 1st Battalion, 130th Infantry Regiment. In May, 1944, the 130th Infantry deployed overseas to Finschhafen in New Guinea. In August 1944, Collins assumed command of the 130th Infantry, commanding the regiment during the Battle of Morotai and Battle of Luzon, where the 130th Infantry participated in the capture of Baguio. His unit later served during the occupation of Japan in 1945.

General Collins assumed command of the 4th ID as it was building up for Vietnam deployment. He led the division until January, 1967. At that time, he was appointed to command I Field Force, Vietnam.

However, on this day, January of 1966, he was commanding a division made up of many draftees and leaders with little combat experience. To welcome his men in to the Division he sent out letters to all the parents of the recruits who were new to his command.

Our Battalion Commander was name Lt. Colonel Atkins. My family and I saved everything from my Army days, as you will realize reading this book, and I have no copy of any letter from the battalion commander.

On the walls of our barracks were a number of illustrations

which depicted our unit's history since it was first organized. There were a few that stood out to me because I knew so little about these battles. One of them portrayed a battle that took place in Cuba during 1898 at a place named El Carne during the Spanish American War. Another illustration depicted action that took place during the Philippine Insurrection, whatever that was. It would be decades before I researched that American engagement. Still another showed a landing of troops on D-Day on Utah Beach in Normandy, France.

The poster that most mystified me was an 18-inch sign that only had the words "E COMPANY", nothing else. When we asked about it, we were told that that was our company's designation in World War II. A, B, C, and D Companies of the 12th Infantry Regiment were designated later as 1/12th Infantry. E, F, G, and H Company were now known as the 2/12th. That meant our company (A/2/12) was actually E Company during WWII.

We were told that our Company stood out in the war, but nobody seemed to know why. It took me decades to discover that our Company of the 12th was the last to hold back a German battalion attacking Luxembourg during the Battle of the Bulge. Luxembourg was saved, but our Company had to surrender because they ran out of ammunition after days of resistance. Luxembourg was saved and Patton used that strategic hub to relieve Bastogne and save the day.

On Monday, these sheets were passed out to recount our unit history. Somewhat.

In the photo section of this book, you can see General Collins letter that was sent to our families and the unit histories that were passed out to us on that day.

Basic Training

"May the south shine with new victories.
With many more Dautieng, Daubang, Pleimi, Danang
May the north fight heroically
The higher the American aggresors escalate."

—Ho Chi Mien's New Year's message to the people of North Vietnam posted January 1, 1966, Associated Press release of January 3, 1966, our first day of basic training at Fort Lewis

We began our basic training on the same day that the Associated Press published Ho Chi Minh's annual address to the people of North Vietnam. He understood that the Americans were preparing to send major forces into South Vietnam and he was inspiring the North Vietnamese to stiffen up for the coming tempest. Dau Tieng would be where our brigade would set up our basecamp in Vietnam. That Dau Tieng victory that he was boasting about in his address, referred to a battle that took place a month before.

On the evening of 27 November 1965, the ARVN 7th Regiment, 5th Division was overrun in the Michelin Plantation by the VC 272nd Regiment. The 9th VC Division unit ended up killing most of the Regiment and five US advisers. Once the ARVN unit ran out of ammunition and surrendered, the Vietcong went up to their foxholes and murdered them. Very few escaped to tell the tale. This is where we were training to go and where we would eventually face this same notorious regiment in one of the biggest battles of the war. Oh, yes, one more thing, no one in my company had a clue what was happening in Vietnam at that time and who we would be facing in the upcoming year. As we began basic training for combat, many of our men couldn't believe that we would end up in Vietnam.

In the year 2006, our second Company Commander in Vietnam, Ed Smith helped me to research what basic training in the 1960s was like as we prepared for combat. We came up with a typical training schedule for each basic training day:

0430 Wake Up
0530 Physical Training
0630 Personal Hygiene/Breakfast
0730 Move to Training
0800 Training
1200 Lunch
1245 Training
1630 Return to Barracks
1700 Dinner
1800 Preparation for Next Day/Counseling/Extra Training
2000 Soldier Preparation Time
2100 Lights Out

It was more difficult to acquire the exact curriculum for the 1966-1967 'NEW ACTION ARMY' Basic Training recruit. I scratched my head. Ed Smith did the same down in Tennessee and we think we organized a complete listing of classes for our period. Here it is:

- Individual Proficiency Testing
- Close Order Drill & School of the Soldier
- Manual of Arms
- Military Code of Justice

- POW responsibilities
- Physical Training
- Basic Rifle Marksmanship Course (BRMC)
- Bayonet Assault Course
- Map Reading Course
- First Aid Course
- Infiltration Course
- Chemical Training including the Gas Chamber
- Communications Course
- Confidence Course (Obstacle Course)
- Hand to hand Combat Course
- Hand Grenade Course
- Bivouac (35 days)

All of this within a very demanding eight-week schedule.

Larry Walter was a drill instructor awaiting our arrival at Fort Lewis. This is his memory of the period period the draftees arrived:

> "Before I cover our basic training period, I need to share with you an important bit of information. I was in Alpha Company a few months before the trainees arrived. I was a Spec 4 the month before their arrived. Lieutenant Palmer brought us into the battalion meeting hall three weeks before basic began. He told us that we should kiss our wives goodbye for a while. We would be spending most of our time on base in preparation of the arrival of the recruit and we would need to learn how to prepare them for their deployment to Vietnam. In addition, most of you will be joining them in Vietnam, so train them well. Your life may depend on it.
>
> With that, I was promoted to E-5 sergeant and began taking drill instructor classes. During our training period, there were times

that I only learned the subject matter on the night before I was expected to teach the class."

None of us had a hint that was going on while we went through our training. It was a tremendous credit to them that our instructors were able to perform so well while learning on the fly. It was even more impressive that they were able to keep that a secret for so long.

Larry Walter was a former Marine who had put in his four years in the Corps and decided to try his hand in the US Army. That was not the case with Sgt. James Harris and Eugene Barton,. Both were longtime veterans of the US Army.

Each served as Platoon Leaders in the company during basic training in anticipation for officers, who had yet to arrive.

One officer who was in our company was 2nd Lieutenant John Concannon who did, in fact, lead a Platoon and later served as Company Executive Officer.

John was a Bostonian who graduated in Jon Palmer's West Point Class of 1964. He was prolific in a few languages, which propelled him later into a staff position as Intelligence Officer for the 2/12th Infantry Battalion just as we left for Vietnam.

As we had few leaders during this period, it was not uncommon for John to serve as an instructor during our basic period.

During our very first week of training, John had the mission of training us for bayonet combat. Preparing us for that exercise, John explained the importance of moving quickly. In bayonet fights there are only two contestants; the quick and the dead.

It went along well until the end when John directed us to remove our bayonets off our M14 rifles, our weapon at that time.

John explained that the bayonets were sometimes hard to release from the mounts of M14s. He explained that the best way to release it was to hit the bottom of the bayonet with the base of your hand and eventually, it would pop out. One of our men, Charlie Neyman, forgot to place the scabbard over the bayonet. He had a hard time to get it off, but he was determined. He kept increasing the force until the bayonet finally released from the mount abruptly and unfortunately hit him near the eye. He was lucky, and only ended up with a few stitches. John felt terrible about the accident. First week on the job and already a casualty.

John Concannon also taught the hand-to-hand course for us in a small clearing in the forest that was designated for that exercise. Much of the activity stressed the importance of protecting and attacking the vulnerable groin area.

Johnny Martel took this hand-to-hand training very seriously. At the end of basic training, we were given a two week leave to return home. When Johnny returned to base, he was displaying a prominent shiner on one of his eyes.

When John Concannon saw this during an early training exercise, he went up to Johnny and asked him what had happened. Johnny told him that he had gotten into a fight with another Maine guy when he went home and used what he had been taught by John earlier. It didn't work out well with him and he was beaten badly. John asked him where the incident took place. Johnny responded, "In a bar" John, tongue in cheek, responded, "In a bar! Not in a bar. In a bar you crack a beer bottle on the bar and stick it in the guy's ear."

There were a number of memorable moments during our basic training. These last two examples were iconic for our group. There was a young conscript from the Midwest named Freddy Chamberlain in our company. As soon as we began training at the firing ranges with our M14 rifles, Freddy refused to even carry his weapon. I can only assume that he was a conscientious objector and this was beyond him. 1SG Springer, sensing this could be a problem with others in the company made Chamberlain carry only his rifle stock with him, which he assumed would encourage Chamberlain to come around and 'get with the program'. That was not to be and so just before the end of basic training arranged for Chamberlain to be transferred to one of the base fire departments.

1SG Springer and SSG Homer Davis cooked up a scenario for the exit of Chamberlain from our company. On the day that he was due to leave, the entire company was assembled on the company street by platoons. Freddy was in front of the group holding his duffle bag with all his gear. To his side was Springer who addressed the assembled group. I can't recall exactly what he said, but it was essentially that Chamberlain was leaving our company as he no longer held up to our standards.

At the end of the speech, Springer and Chamberlain marched over to the First Platoon. Sergeant Homer Davis, sitting on the barrack's steps, began playing a slow drum cadence played on an inverted waste paper basket with a drum stick.

Company Commander Palmer hearing the commotion stepped to his office window to see what the commotion was. He saw that one of his men was being 'drummed out of the company', a practice outlawed many years before. He thought to himself, "If Colonel Atkins ever hears about this, I will surely lose my command." The battalion commander never did speak to Jon about the incident, but it's hard to believe he didn't find out.

As Chamberlain stopped at each platoon, the command of 'about face' was given. This ritual was repeated in front of each platoon and by the end, the entire company had turned their back on their former associate. With that exercise completed, we remained with our backs to Freddie as he made the long march across that large assembly field leading away from the barracks towards the main street.

The epilog to this story was funny, in a way. Freddie did manage to serve at one of the base's fire stations for his Army tour. On the day that we were brought to the Tacoma docks to board a troopship taking us to Vietnam, we assembled all of our gear on that large assembly area in front of our barracks. When it was time to leave, we boarded deuce and half trucks and traveled off base, passing by the fire station where Freddie worked. We didn't know which station he was sent to but Freddie made sure we found out. He was standing out in front of it with a big grin on his face as we passed by and gave us the one finger salute. We laughed hysterically.

At the very end of basic, we were brought out to the hand grenade range, again located in the forests of the base. When we arrived, we gathered on the ground in front of some foxholes that were used after we threw our first and only live grenade.

Sergeant First Class James Harris was teaching us the proper way to handle and throw hand grenades on this day. A number of other NCOs were there to assist and man the two-man foxholes where we threw the grenades from. Perhaps Jim felt the need for us to pay close attention to the instructions when he made these startling statements.

There was a big blackboard there and Jim wrote the figures 184,300 and 2,268 on it.

He turned to us and paused as he gauged our attention. He then asked if anyone knew what those numbers represented. No one answered.

He then proceeded to tell us, "Those numbers represent the number of American troops in Vietnam and how many had been killed up to that time." He looked about the group to assure that he had our undivided attention. After a few seconds, he went on, "I sense that many of you are not taking your training seriously. I'm here to assure you that as I stand here, you can be assured that next Christmas you will be celebrating the holiday in a foxhole in Vietnam." This is when it all hit home. We were not preparing to serve in Germany, Korea or some other relatively safe location. We were going where some of us would not survive. The biggest non-secret at Fort Lewis during this period was that the 4th ID was being prepared for combat in Vietnam. The NCOs and the Officers understood that. Now the skeptics among us enlisted men were assured of that ultimate assignment. We all finally 'got it'.

A week later we graduated from basic training and were sent home for two-week leave.

ADVANCED INDIVIDUAL TRAINING WITH THE MEN OF THE 2/12TH

Donald Evans, Moh Recipient

As we returned to Fort Lewis in March of 1966, there were many more soldiers from around the country who were preparing to join our company. New officers and non-commissioned officers would be arriving in the latter part of the Spring of that year. In addition to these leaders, a host of new soldiers were chosen to serve as medics and they were assembling at Fort Sam Houston, Texas. Sam Houston is known as "The Home of the Combat Medic." Since World War II Fort Sam, as it is known, was the only army base where their medics trained and it was there that Donald Evans learned his role in an infantry company. It was here that he amusingly learned that he would not be known as Donald Evans, but become 'Doc' Evans at his new duty station.

Prior to his induction into the army, Donald was an outstanding track star in California. While attending his hometown high school, Covina High School in 1961, Donald set state track records that weren't broken for 25 years. He ran the 660-meter, 860-meter, 1500-meter races, and the mile, and mile relay races and also ran cross country races.

In 1961 he was awarded the Athlete of the Year Award at Covina High School.

Illustrating how shy and humble Donald was in high school, Richard, his brother, shared this story with me:

"Donald was terribly shy in high school. He was very popular and talented and should have been flush with confidence. His bashfulness was just part of his personality.

His coach informed him that he was chosen to receive that Athlete of the Year Award in the school auditorium in front of the school assembly. Don told his coach that he couldn't do it. The award was presented to him in the coach' office, with only his brother present. Only then did he appear in front of the assembly.

All that athletic notoriety was noted in the state's colleges and he was offered several scholarship opportunities. UCLA, Berkeley, and San Jose State had outstanding track traditions, but Donald elected to stay near home and attended Mt SAC (Mt San Antonio College). Not a bad choice actually as the college has an excellent reputation for producing outstanding graduates that have made their mark nationally in government, sports, business, education, and entertainment.

Donald had an excellent athletic record in college and with the help of John Norton, another outstanding runner, San Antonio College won the state championship.

While at Mount San Antonio, Donald became part of the outstanding relay team made up of the following men: 1st–Tom

Iredale; 2nd–Ron Copenhaver; 3rd–Don Evans; 4th–John Norton.

This team won the L.A. Times Indoor Mile Relay, Mt. San Antonio College Invitational Mile Relay, Southern Cal. Mile Relay, Long Beach Relay and came in second in the California State Championship and the Coliseum Relays. In addition, they garnered dual undefeated meet seasons.

Bonnie Linn, his future wife, came into Donald's life while he was attending college. Bonnie was working at The May Company in California in West Covina. The store was a five-story establishment on the east side of the then recently opened mall. A two-level W.T. Grant anchored the west side. She began in the restaurant of the store and after attaining a certificate for interior design, she was promoted to the Art Department. She was also doing some modeling work on the side. She was working at the department store when Donald was introduced to her by a friend who also worked there. He was twenty years old and she was twenty-one. In preparing for this book, I interviewed Bonnie in a Zoom Meeting that took place in April of 2021. Her daughter, Kristy, Jim Olafson (who worked so hard to get Donald properly recognized for his gallantry in Vietnam) and Jim Deluco, a Company member of Donald Evans unit joined in for an hour and a half Zoom meeting.

Bonnie shared this during the meeting: "We both fell head over heels for each other from the first meeting. What attracted him to me was his kindness and gentleness. He was kind and loving. What wasn't to love?

Donnie loved to race. He raced cars at the former Riverside International Raceway in Moreno Valley, CA."

(*Riverside track was one of the very few tracks where the cars raced clockwise. This necessitated an additional fuel filler pipe to be located on the left-hand side of the car to speed up the refueling of the vehicle during a race. The racetrack was built in 1957 and was originally on the U.S. Grand Prix circuit. It was not long before it served as a NASCAR site. The track was known as a treacherous course for drivers with a long, downhill back straightaway and brake challenging turns at the end. In 1965, IndyCar great A. J. Foyt suffered a brake failure at the end of the straight, shot off the road and went end-over-end through the infield at high speed. He barely survived the calamity. The track closed in 1989 and the site was developed into Moreno Valley Mall which is still in operation today).*

"I recall one event that took place while Donnie and I were still dating. Donald always loved to race and I can recall one incident where he and his best friend had driven out to Glendora to race each other. Glendora had many hills and cliffs and driving at high speed there could be challenging. While navigating the hills, Donnie lost control of his car and off he went down into the valley. He wasn't hurt and was able to call me telling me that he had just run off a cliff in his car and would be late to see her. It didn't seem to be a big deal to him, but I was pretty shaken by the episode.

It wasn't terribly long after we started dating before he proposed to me. He took me far away in a secluded spot up in the mountains. I thought to myself, 'OK, what's going on?' He shut off the car and abruptly asked if I would marry him. Stunned, I answered, 'What? Yes, yes, yes.'

With that we drove away and he brought me right to my mom's house. We walked in the house and he told my mom that he wanted to marry me and asked my mom if that was okay. Of course, my family loved him so they were very pleased to hear of our intentions."

Here is where a twist took place. Bonnie details it here:

"On August 19th of 1964, we ran off to Las Vegas and were secretly married. No one knew. They pretended that they weren't married and went along as if the marriage never took place. He brought her home each evening to her house and he went to his home. This charade went on even after December 5th of 1964, when they had their church wedding in California.

After graduating from Mt. SAC in June of 1964, Donnie worked as an architectural draftsman in an engineering firm. In May of 1965, he received a raise of $50, from $400 a month to $450 a month. Things were looking up for Donnie, but not for long. In December of 1965, he received his draft notice.

He was sent to Fort Bliss, Texas, for basic training on Friday, January 7, 1966 and began training on Monday, the 10th. I did not join him for that period. On March 6, he received orders to report to Fort Sam Houston where he would be trained as a Medic. In the directive it was advised that wives not follow them or private cars transported to the new assignment. We both drove to Fort Sam Houston in our car.

When we got to his new assignment, we found a house in San Antonio where other veterans and their wives were staying. It was a huge house and it had seven bedrooms, one for each of the military couples that shared the house.

At night, Donnie would come home from training and we would spend the evenings together. All except one night when he didn't come home. He said that he couldn't come home that night because he messed up and was ordered to stand guard duty all night.

To occupy my time while Donnie was training, I babysat for all the neighbor's children. When Donnie had time off, we would take some of the kids to the park where they could play. Donnie loved children and they loved him back.

After Donnie finished his medical training, he was sent to Fort Lewis in early June where I once more followed him. We once more found a house that we rented with others where I would stay while he was training.

On September 22, 1966, I was on the dock with others to wave him off as his unit sailed for Vietnam."

Jim Olafson, Leader, Third Platoon

Jim Olafson arrived in A/2/12 during our company's Advanced Unit Training (AUT). Our company was in need of leaders and others to fill out essential positions. Up until May of 1966, many of the officer's roles were taken up by senior Non-Commissioned Officers. It worked for training, but would not for combat deployment. Jim came to us in a roundabout way.

Jim was born in Jamestown, North Dakota, on November 27th, 1940. He grew up hunting and fishing and never participated in sports while in school. He was more interested in cars and girls. When he was 14 years old, he lied about his age and joined the local National Guard unit. This outfit was an artillery battery and he thought it was really cool to be able to fire those big cannons. A year or so later, his unit became a 155mm self-propelled unit, which was even better, because he could drive where he was needed and didn't have any set up in order to fire. While at summer camp one year, they were having an evaluation of his unit. On that day a 2nd Lt Observer had concealed his jeep behind a very large bush. Jim's unit was on the move and he was driving a self-propelled Howitzer. The Gun Commander, an NCO, told him to take a direct route. A self-propelled howitzer is just like a tank, they need not go around bushes and small trees. When he hit the bush, it was like hitting a rock, as the self -propelled started to climb up and over the jeep. The 2nd Lt hiding in the bush came running and screaming out but it was too late. The outcome was they filed charges on everyone in the vehicle and they had to split the cost of the jeep. There were six of them and the cost came to $300 apiece.

After graduating from high school in 1958, Jim had completed his three years in the Guard unit and he enlisted in the US Navy. He was sent to boot camp at San Diego, California for 11 weeks. After boot camp, he went to Bayonne, New Jersey to become a Fuels Handler. Finishing that school, he was assigned to the USS Oriskany CVA 34, an aircraft carrier which was stationed at San Diego. He made one Far East cruise with the carrier. Upon returning, the ship went straight into dry dock for repairs and overhaul. The rest of his time onboard was making short cruises out of San Diego, mostly for the purpose of qualifying pilots for carrier duty.

In 1961, he married his wife, Janie. In November of 1961, he was discharged and Janie and he went back to North Dakota. Being as it was then winter, and in a farming area, jobs were not very plentiful.

As a result, two months later he enlisted in the Army and was sent to Fort Carson, Colorado for basic training. Following that he was sent to Fort Bliss to become a Missile Repairman.

Completing that, he was sent up the road to White Sands Missile Range where he went to a Nike Hercules site and later to a Hawk Missile site. White Sands Missile Range was where the Trinity Site was situated. That was where the first atomic bomb was tested on July 16, 1945, before it was used on Japan.

One day while at the Nike site, his unit was tasked with putting on a demonstration for a number of high-ranking officers and congressmen. On White Sands there is a mountain, Mount Stallion, to the north housing radar units. On the day of the demonstration, they made ready to launch a Nike missile. There was to be an unmanned drone as a target. When the drone was locked on the radar, we launched the missile. Mt. Stallion notified us they had radar contact. Seconds later, they notified us that radar contact was lost and the missile had not intercepted the drone. Shortly thereafter, there was a phone call from officials in Juarez, Mexico, one hundred miles down range, about a rocket landing in their town square. Along the way, the missile flew directly over El Paso, and Fort Bliss, Texas. Had it landed just nine miles closer, that missile could have landed in on an American city or base. Needless to say, all our high-ranking officials were not impressed.

After leaving White Sands, he was assigned to Germany to an Ordinance unit responsible for maintenance of Hawk Missile batteries. He got there about the same time that Kennedy was assassinated on November 22, 1963 and everyone was on high alert. Everyone was in full battle gear and issued live ammunition. He thought it somewhat strange as they told us that if the balloon went up, meaning that the Russians were coming, their life expectancy was 11 minutes. He thought, "So much for a loaded M-14 rifle."

In 1964, his wife gave birth to their first child, a little girl named Vicki Lynn. They lived off post at the time in a neighboring small German village in a one-bedroom apartment. At that time, he was still a PFC and on payday after deductions he took home $111.00. With that they bought groceries, gas for the car, fuel oil for the house heater, and still managed to go see a movie or go bowling now and then.

On April 22, 1964, Jim reenlisted in the army for an additional six years while serving in the 183rd Ordinance Detachment, stationed in Schweinfurt, Germany.

In 1965, he went before a promotion board for Sergeant. During that board, the Battalion Executive Officer asked him to apply for OCS (Officer Candidate School). Later the word got back to his CO and the commander started pestering him to apply for OCS. He finally got tired of the nagging and put his paperwork in. Nothing happened for a while.

When he went before the board again, prior to his being sent to OCS, they wanted to know why he chose the Infantry instead of the Artillery branch. His background would have seemed a likely choice. He told them he wasn't real keen on all the math required for the artillery school and would rather be an infantry officer. They weren't impressed, thinking that it was a waste of his experience, but they did recommend him as he requested.

He finally received his orders for OCS and left Germany in the middle of October, 1965, en route to Fort Benning, Georgia. Another note from that period: It was said at the time if you went to Germany you would return with either a baby, a cuckoo clock, or a Volkswagen. In Jim's case, he came back with all three as Janie was expecting their second child, Robert Alan.

He started OCS on November 1, 1965. With all the harassment and bull that transpired, it did not take him long deciding that he didn't need to put up with that discipline. One day, after about five weeks, he went to the local grocery mart and bought a 12 pack of beer and took it up to his room where he commenced to drink it. His roommate got all excited saying they would both get kicked out of OCS. He told him to have a beer or leave. He left and Jim finished the beer himself. He never did get caught. The next week they held a formation and told us all that anyone wanting to quit could do so after the 8th week. He thought that if he could make it through eight weeks, that's one third of the way; he might as well stay.

Here is another story he told me from that OCS training period. One night they all got together and placed a food order to be delivered. Of course, that was strictly prohibited and called "pogey bait". Jim ordered the food and had collected all the money. It was late when the delivery van pulled into the parking lot and flashed its headlights. Jim and another cadet ran out, paid for it, got the food, and went back to the dorm. They tried waking up people to tell them their food was here but most just rolled over and went back to sleep. The other cadet and he said they weren't going to waste the food so they took it into the latrine to eat. About the time they got started on some cheeseburgers, in walked a Tactical Officer. He had each of them take a pizza and get in the shower. They had to adjust the showers and then stand there under the water and eat the soggy pizzas. After that, the Tac Officer took the remaining food and left. After the 24 weeks, they finally graduated on April 28, 1966. After graduation, he walked up to his TAC officer and said, "Hey Brown, you couldn't run me off, could you?" He commenced to inform Jim that he was a 1st Lt and out ranked him and to address him as sir. "Yes, Sir!"

Following commissioning, he was assigned to the 4th ID at Fort Lewis. After a 10-day leave, he, Janie, and the kids jumped into their Beetle Bug car and began the 2,700-mile journey cross country. Today that trip would take 42 hours. I can't imagine how long it took them, but I'll bet Janie and Jim never forgot that adventure in that small German car.

At the time of his arrival at Fort Lewis Washington, in May, 1966, the 4th ID was beefing up and training in preparation for its deployment to Vietnam.

Jim had orders to report to the 2/12th Infantry Headquarters. After examining his paperwork, he was assigned to A Company to be the 3rd Platoon Leader. Things were moving fast, which is what usually happens when a major military buildup takes place.

Jim would later serve as the Company Executive Officer in Vietnam, return to Vietnam for a second tour as a chopper pilot, and eventually serve as a Texas Ranger for twenty years.

He arrived in the platoon just about the same time that Donald Evans was sent to the Third Platoon. Awaiting to greet them both was me. In early Spring, I was chosen as the RTO for the Third Platoon Leader.

Both Jim and Donald didn't know it at the time, but they would forever be linked together as a result of the actions of January 27, 1967. That event led to Donald becoming the first Medic and first 4th ID soldier to receive the Medal of Honor.

Advanced Infantry Unit Training

We returned from basic training leave and awaiting us were lists which showed us where we would be sent for advanced training. The majority of the basic graduates remained in the company for our next phase of training; Advanced Infantry Training. Another group was formed and they would learn to handle heavy weapons, such as mortars and recoilless rifles. After a few weeks of training, they would spend a couple of weeks at the Yakima Firing Range, 92 miles east/southeast of Fort Lewis. There they would apply what they learned in live fire drills using heavy weapons. Upon return from that exercise, they would join the rest of the company and would become 4th Platoon of A/2/12, better known as the Heavy Weapons Platoon.

The rest of us formed the three rifle platoons, 1st, 2nd and 3rd Platoons. We specialized in light weapon tactics. I was one of those who only had to move my duffle bag from the 1st floor north bay to the 2nd floor south bay, where I became part of the 3rd Platoon.

Still others were sent to the 2/12th Infantry's Headquarters Company to form the then non-existent Battalion Reconnaissance Platoon (led by Lt. Bill Bradbury) and the 4.2 Mortar Platoon (led by Lt. Ron More)

Ken Eising, who was one of those sent to Headquarters Company, explained his arrival in the Recon Platoon and what their mission was during his time in the battalion. This was his experience returning to Fort Lewis from basic training leave.

"When Brian Neal and I returned after basic training leave, we couldn't find our duffel bags. Someone told us they had been sent to Headquarters Company. We both thought that sounded good but then they told us we had been assigned to a Recon Platoon and then we were not so sure. I doubt if either of us knew exactly what we would be doing except the word recon was derived from reconnaissance. Lt. Bill Bradbury was our Platoon Leader and Sgt John Raymond was our Platoon Sergeant at that time."

More on Sgt. Raymond shortly.

"The Recon Platoon was composed of three sections of twelve members each. Scout Section, Weapons' Section, and Rifle Section, led by the Recon Platoon Leader and the Platoon Sergeant. Medics and Radio Operators made up the rest of the platoon.

We had four scout jeeps with M60 machine guns mounted on them, four jeeps with 106 recoilless rifles (reduced to two when we shipped to Vietnam). In addition, there was a rifle squad with a 3/4-ton pickup. As I recall, we had five M60 machine guns, six radios, and we carried 90 MM recoilless rifles in the field.

Our duties consisted of some of the following: scouting patrols, a couple of LRRPs (Long Range Reconnaisance Patrols), mounted security details for Medcaps in the villages, and platoon size patrols. At other times we were attached to other battalion companies to support them.

Yes, we were your 5th Platoon. Early on we would lead the battalion. Later we would serve as security for the Battalion Command Group when the whole battalion was out on operation. We considered ourselves as the 'Palace Guard.' We also served as a reactionary force at times when needed."

Now to the story about extraordinary figure, SSG John J. Raymond, who was the Platoon Sgt for Recon Platoon. John was a remarkable soldier who few forgot once they met him at Fort Lewis for training.

I researched his story for one of the Alpha's Pride newsletters that I published a few years back. It was not too difficult as his hometown was 20 miles north of where I live. Here is that story that I published back in October of 2016.

2/12th Recon Platoon Sergeant, John J. Raymond

Sgt. Raymond trained us to fire the M14 rifle on the rifle ranges of Fort Lewis. We knew nothing about him in those days other than he had a strange last name that was usually a first name and he seemed awful old to be in the Army. It was almost 50 years after he trained us at Fort Lewis that I researched his story for an Alpha's Pride newsletter. As I learned, the research was not too difficult, as he was raised in Taunton, MA, fifteen miles north of my hometown and was somewhat of a celebrity in the old Silversmith City.

We were right that he was fairly old comparing him to other soldiers at the fort. John was born in December, 1917. That was the same year America declared war on Germany in World War I. John saw which ways the winds were blowing in Europe in 1941, so he joined the Army in September of 1941, three months before the Pearl Harbor attack. He was then 23 years old, which would have still been considered old as we had few in our enlisted ranks at that age in 1966. Most of us were 19 or 20.

He served with the 5th Army in Africa, Algiers, Tunis, Casablanca, Cassino, Anzio, where he was wounded during the amphibious landing, and finally Rome.

On December 3, 1945, he was discharged from the Army and returned to Taunton.

He returned to his old job he had before the war at Reed and Barton. Reed and Barton was a prominent American silversmith manufacturer in his hometown. The work there was hard and dangerous, especially because of the chemicals that he was exposed to. After the excitement of war, John returned to the Polishing Room of the factory. Working on a manufacturing floor left him uninspired and restless. He put up with that for two years and then rejoined the army.

As time went on, he ended up in the 2/12th in 1966, some 25 years after he initially joined the Army.

As units were being rushed to Vietnam as quickly as possible, the 2nd Brigade of the 4th ID was the first to leave for Vietnam from Fort Lewis, then a month later the 1st Brigade, needed seasoned NCOs quickly in their ranks. That was the reason that Sgt. Raymond was sent to the 2nd Bde. They left for Vietnam two months before us. He ended up in the Central Highlands. Our brigade was sent to War Zone C, 40 miles northwest of Saigon.

The 4th Infantry Division published a weekly newspaper in

Vietnam entitled "The Ivy Leaf". Although the 1st and 2nd Brigades of the 4th ID were up in the highlands and we were 225 miles further south, we could follow their activities pretty well. In an issue published in February 1967, we learned that Sgt. Raymond was killed in action on February 15, 1967.

It's strange in a way, we only knew him from his training us at the rifle ranges, but somehow it had an impact on us. Maybe it was like learning of the death of an old friend. Maybe we thought, 'What was he doing in Vietnam? He did more than his share protecting the country already.' What did we know? How could we understand the dedication of the professional soldier? Today we better understand.

That's what I wrote in 2016.

We did not know the circumstances of SSG Raymond's death until 2017 when I linked up with Ed Northrop. That year, Ed and I were part of the committee that worked on the 12th Infantry Regiment's Monument project. That mission resulted in a monument being built along the Walk of Honor behind the National Infantry Museum at Fort Benning. Ed was the Project Coordinator and I served as the Project Historian. As luck would have it, unbeknown to me, Ed was the Company Commander of C/1/12 when Sgt. Raymond was killed while serving in his company. He was able to fill in the details on how Sgt. Raymond lost his life in Vietnam.

I sent this article to Ed and this was his reply:

> "Wonderful article. The only thing I knew about John was that he was OLD. I had been in command of C/1/12 for five weeks, and remember meeting him, but that's about all. He, the Platoon Sgt and Platoon Leader were all shot at the same time. All died.
>
> "They were planning the early morning squad size patrol. The beginning of two long days. Ed Northrup".

A few days later he wrote a more detailed description of SSG Raymond's loss in Vietnam:

> "I have tried to relate the events of 15 Feb 1967, involving the incident when SSGT Raymond was KIA. To be clear, I did not witness the event, only learned, tragically, that we had lost within seconds, a Platoon Leader, a Platoon Sgt and a Squad Leader, all from the same platoon. In the midst of all this chaos, I had to revise the chain of command and move a couple more senior NCOs into this platoon. Reflecting back, I am sure this caused some initial confusion morale wise, but I couldn't have an E-5 Buck Sgt leading the platoon! Please remember, all hell was breaking loose at that time, and I am trying to figure out how to survive this NVA onslaught. Enough of that.

- On 15 Feb, 1967, after an estimated, four-day, 35 kilometer walk through the jungle and two river crossings, we settled, albeit too late, one afternoon into a patrol base/defensive position, which later was referred to as 501N. In our case "The Battle for 501N".
- LT Ted Glick was the 1st Platoon Leader, SFC (E-7) Richard Carver was the Platoon Sgt. and SSGT (E-6) John J. Raymond was a Squad leader.
- Although I have been unable to determine an exact time of their deaths, I believe it was around 6:45 am.
- Both SFC Carver and SSGT Raymond were KIA instantly. Lt Glick was WIA and died several months later in Japan.
- All three were in an open area, behind a bunker within our perimeter. They were planning their platoon's assigned squad sized patrol (two to twelve men with two machine guns). Per our company

routine, almost daily, but not regularly, each platoon sent out a patrol. The patrols were dispatched in a rotating sequence. Each squad explored out to a distance of about 500-700 meters, depending upon terrain.

- According to my best sources, all three were shot by a single sniper located in a tree about 75-100 meters away. That day we did encounter several enemy snipers in trees. With the help of napalm and artillery fire, the Air Force efforts, and obviously intense fire from our perimeter, most were killed. One of the soldiers, a Sp4 Tom Hedin observed some of the incident, but was pre-occupied with the enemies repeated assaults on our perimeter.

All three were pulled back into a lower area where the medics could evaluate, but only Lt Glick was treatable.

I remember meeting SSGT Raymond and was honestly was struck by how old he looked. However, for all who knew him, he was loved and was a respected soldier.

Soon after returning from basic training leave, a picture was taken of most of the men of the Third Platoon, where I was training. It was shot by one of its members, Paul LaRock. We were preparing to be marched to the chapel for Sunday services, which explains our dress green uniforms.

Henry Osowiecki, Squad Leader

It was not too long during training that certain trainees were recognized for their leadership qualities. There were a number of them, but I will mention two in particular.

The first one was Henry Osowiecki ('Ozzy' to us A/2/12 men).

To understand his story, you have to understand the family military history that preceded him.

Henry's paternal grandfather arrived in America in the early twentieth century when American industry required many more workers than the American population could provide. A young Alexander Felix Osowiecki and his family arrived in America from Poland and were eventually blessed with citizenship.

From his maternal side of the family, his grandfather, Joseph Kubish, arrived with his brother Martin from Czechoslovakia and they also earned American citizenship.

Each family settled in Connecticut. The Osowiecki settled in Thomaston and the Kubish family settled in New Britain, CT.

This was a typical story for many of our ancestors who arrived in America on the east coast. They struggled greatly until they became successful. Alex Osowiecki worked in the Plume and Atwood Brass mill in Thomaston which produced rolled brass for the clock movements for the world-famous Seth Thomas clocks. Seth Thomas is the oldest clock manufacturer in the United States, dating back to 1813. The clock at Grand Central Station in New York is a Seth Thomas clock.

Joseph Kubish worked in a button factory in Bridgeport, CT when he met his wife Agnes (Tibor) Kubish. He was a hard worker and was determined to make his way out of the mills. When he got married, he bought a small farm in Milton, CT. He cut wood on his farm and sold it to the factories in Bantam, CT for their needs. Once all the wood was chopped down, he used the proceeds to buy another much larger farm in Bantam and proceeded to do the same there. Eventually the wood was gone and Joseph began a dairy farm there.

Both families would be profoundly affected when the Pearl Harbor attack brought America into the war.

The Osowieckis had seven children. The oldest son, Chette, was a Seabee who participated in the Battle of Guadalcanal. Soon after the battle, he died of malaria. He left behind a wife and two daughters.

The second oldest was named after his father, Alexander. He was working at the same brass mill as his father in Thomaston when he joined the army in 1942. He served in North Africa and Australia for three years before he was discharged at the end of the war.

Alex died in 1953, eight years after the end of WWII, leaving his widow and three children behind.

Henry's father, Henry Sr. was the next to serve, joining the US Army Air Corps in 1942. He trained as an aircraft mechanic and a turret gunner on a bomber.

While he was training stateside, he married Henry's mother Lillian.

Henry was born while his dad was stationed in Greenville, SC, before he shipped to Africa where he served on a bomber used to bomb German forces.

Henry Sr. would serve in over 50 combat missions, earning ten Air Medals and a Purple Heart for wounds received in action. He returned home on July 15, 1945.

He returned home and worked as a carpenter and soon established his own construction business. *Henry would also do the same after coming home from Vietnam. He first worked as a carpenter in his dad's business and later formed his own construction company.*

The youngest of the siblings, John Osowiecki, enlisted in the Marines and served in the Pacific. He participated in battles at Kwajalein Atoll, Saipan, Tinian, and Iwo Jima. He was one of 19,127 US military wounded during the Iwo Jima invasion.

John came home with deep emotional scars and Henry remembers his uncle as a man who tried to drown his memories in drink. When he drank, he became out of control and people feared to go near him. In 1952, his body no longer able to tolerate the large consumptions of alcohol, gave out, and John left behind a wife and four children.

Henry's maternal grandparents produced three sons. John, the oldest, ran away from home when he turned 16 and was never seen again. Joe stayed on the farm and raised six daughters. Ludwig, the youngest son, served in World War II in the European Theater and was highly decorated.

Luddy, as he was known, joined the army at 20 years old. After training, he joined the 307th Airborne Engineering Battalion of the 82nd Airborne Division and was sent to England.

On D-Day, Luddy and his division was dropped into France in the darkness preceding the landing craft's arrival on June 6, 1944.

Ludwig survived the D-Day Invasion and in September 1944 he participated in Operation Market Garden which was fought in the Netherlands and immortalized in the movie "A BRIDGE TOO FAR'. It was the largest airborne operation up to that point in the war. The goal was to surround the heart of the German industry, the Ruhr, in a pincer movement. It did not work out well for the allies.

The operation began on September 17, 1944 and was predicated on airborne units parachuting into the target area and holding critical bridges for the assault.

Right from the beginning, the stiff resistance from German forces and bad weather delayed the advance, jeopardizing the mission.

Two days into the operation, Ludwig would have his finest moment in uniform.

For his bravery Ludwig Kubish would receive not only the Silver Star for Valor but also the Bronze Lion Medal for Valor. It is the second highest award for valor awarded by the Dutch. This was what was written on his Silver Star Citation:

Silver Star

The President of the United States of America, authorized by Act of Congress July 9, 1918, takes pride in presenting the Silver Star (Posthumously) to Private Ludwig Kubish (ASN: 11045306), United States Army, for gallantry in action while serving with Company D, 307th Airborne Engineer Battalion, 82d Airborne Division, in action on 19 September 1944, near ****, *(location redacted for security reasons)* 'Holland. Private Kubish, Assistant Squad Leader, was a member of a force of two platoons advancing into enemy territory to establish a Main Line of Resistance. During the advance the force was pinned down by intense machine gun and small arms fire. After about a 45-minute fire fight, the platoons were still unable to advance. Private Kubish, with complete disregard for his own personal safety, advanced approximately 50 yards across an open field and attacked the enemy strongpoint. Armed with a Thompson Sub-Machine Gun and a Gammon Grenade, Private Kubish succeeded in killing four of the Germans, causing the rest of the enemy group to surrender, thus enabling the platoon to advance and complete their mission.

General Orders: Headquarters, 82d Airborne Division, General Orders No. 61 (November 30, 1944) Action Date: September 19, 1944

Operation Market Garden ended in failure on September 25, 1944.

You will note in his Silver Star citation that it was awarded posthumously. Here is the story that led up to his tragic loss:

The 82nd remained in the Netherlands after the disastrous operation and that was where he was killed.

On November 21, 1944, Luddy, then a Corporal and a Squad Leader, when he noticed one of his men was pointing his weapon at one of the other men in the unit. Luddy rushed over to grab the weapon to aim it away. The man who foolishly pointed the weapon told him that it wasn't loaded and pulled the trigger. The round went off and killed Ludwig instantly.

Ludwig was buried in the Epinal American Memorial Cemetery in Lorraine, France.

A similar incident would take place in Vietnam when 'Ozzy', as Henry was known to us A/2/12 veterans, saw one of his men fooling around with a pistol and grabbed it from the man. Just as happened in WWII, the man with the pistol told Henry that the gun was not loaded. Ozzy pointed it up in the air and pulled the trigger and a round went off and put a hole in their tent.

As can be imagined, the family was devastated by this senseless loss. It weighed heavily, especially on Luddy's sister, Lillian, Henry Osowiecki's mother. Both families had paid a steep price during WWII.

The Kubish family did more than their share to serve our country. In addition to Luddy, Lillian's sister Anna served 20 years in the WACs, retiring as an SFC. Agnes, another sister, served in the Air Force in the 1950s. Her husband, George Widding also served the Air Force with his wife.

Lillian's cousins also served in the military. Here are the names and service:

John Kubish—WWII, Army Pvt.
Frank Kubish—WWII, Army Pvt.
Martin Kubish—WWII, Army Sgt.
Steven Kubish—Korean War, USMC Cpl.

This is how Henry explained how World War II affected his family:

> "You have to understand how much living through World War II and the consequences of that war played out in our families. My dad suffered from what left a mark on him from his service time and like many returning soldiers would sometimes drink more than he should have, although he was not an abusive drinker. In those days, men returned from war and were expected to leave all the horrors behind and move on with their lives. There were no support groups in those days and they had nowhere to go to let go of their troubling memories.
>
> "After Dad returned from the war, he went back to work as a carpenter, his former occupation before he was drafted. I was the oldest of five children in my family. Following me was Ludwig, named after my Uncle Ludwig, then my brother Greg, followed by my sister Rebecca and finally Jimmy, my youngest brother.

As a result of what happened in World War II, my mother was very protective of her children. That is why she became furious when the local draft board came calling for me right after I turned eighteen years old. She dashed over to the draft board office and demanded that her son be taken off that call up list, invoking the history of the Osowieckis' and Kubish past sacrifices in World War II. She told them they could have him after he turned twenty-one years old. Her admonishing worked and I wasn't drafted until I turned twenty-one. In fact, they must have been waiting because the week I turned that age, I got my 'Greetings' notice."

> *(Note: The Lord works in mysterious ways. Had Henry been drafted when he was 18, he probably never would have served in combat. His*

two years of service would have passed before President Johnson committed American ground troops to Vietnam.)

As the son of a father who was too young for World War I and too old for World War II, I always wondered what the war veteran fathers told their sons as they marched off to war themselves. Ozzy answered this way:

"You know, it's funny that you ask. My father had few words to share with me as far as advice, other than 'don't volunteer.' Apparently, he saw too many men lose their lives, sometimes for foolhardy risks. He wanted me back safe and sound and I guess this is the best advice that he could share without walking down a minefield of memories that he preferred to keep to himself.

"Apparently, the advice didn't stick because it wasn't long before I was being offered leadership positions and I jumped at the chance to prove my mettle. It led to my being awarded three Purple Hearts in Vietnam.

"I came to better understand my dad's mindset by what took place on the day that I was drafted. He drove me to the Draft Board Station in Oakville, CT, just outside of Waterbury, and not much was said that day as I boarded the bus sending me to the Fort Dix Reception Center. However, he didn't return home then but followed the bus until it left Connecticut, more than 65 miles away. He was very concerned for my well-being; he just couldn't bring himself to tell me why he felt that way.

Larry Walter, DI & Squad Leader

Larry's grandparents emigrated from Germany. They were farmers who came to America in the latter part of the 19th century, looking for a better life. Larry's father was born in 1899.

Larry, the youngest of six siblings, was born on March 6, 1943, to a very poor family. His childhood home was located in Eau Claire, Wisconsin, 100 miles east of Minneapolis/St Paul, MN. Eau Claire had a population of 30,745 people, according to the official 1940 census. The latest census available, taken in 2020, shows the population at 68,866. Eau Claire's population has more than doubled since Larry was born.

His family lived in a small two-story house that measured 24 X 24 feet. The only heat they had in the winter was one kerosene heater in the living room on the first floor. The bedrooms were upstairs and had no heat except for what seeped upstairs. On many occasions in the winter, there would be a quarter of an inch of frost on the 'inside' of the bedroom windows when they got up and looked out the window. For breakfast they had coffee with bread and butter. A main meal for his family at night was cooked macaroni and milk or tomato juice.

Larry's father was a carpenter and a millwright and he never made more than $6,000 a year in his life. Many times in Larry's life his father was laid off and living off of welfare. Larry recalls picking beans at 12 years old at a nearby farm to earn money to buy his school clothes.

Larry's older brother, Clint, was drafted in World War II and served in an armored division in Europe. He was wounded twice and shot out of three tanks before he was twenty years old.

His next oldest brother, Glenn, joined the Navy at the end of the war. There were three other sisters in the family.

When Larry was thirteen and fourteen years old, Larry's brother, Clint, would take him to see World War II movies. Larry asked him why he did that and Clint replied, "I did that

that so I could laugh. The depiction of war on the screen was so far out of reality to what he knew about going to a real war."

At fifteen years old, Larry learned about the benefits of making a career of the service. He thought to himself, "free lodging, free meals, free clothing, free health care for himself and his family. In addition, there were no layoffs. He thought, "What a deal! Now that's for me. I'm not going to hang around Eau Claire and struggle like my dad."

His father died when Larry was sixteen years old. He didn't want to actually quit school but he began to fail in every subject. His days were numbered at school from that period on. He spoke to the School Counselor about his eagerness to join the Marines. The counselor told him that he would never make it. That was all he needed to hear. That motivated him to excel in the Marine Corps.

In May, 1960, he joined the Corps. After boot camp he was assigned the MOS of 0311, a rifleman. He was then sent to the Marine camp in California named Twentynine Palms. Twentynine Palms Base is located within the Morongo Basin and the High Desert region of the Mojave Desert, in Southern California. Summer temperatures can peak at 120 °F and bottom at 15 °F in the winter. It was a great place for the Marines to test the endurance of their Marines.

Unfortunately, there were no infantrymen at the base when Larry arrived so he was assigned as a Wireman in the Communications Platoon. He had a new MOS, 2511, Wireman but he also kept his primary MOS as an Infantryman. He was not happy serving stateside, so he put in a transfer request for Okinawa, Japan.

His wish was finally granted and he was sent to Camp Henson. His friend was serving with the 3rd Marine Recon Battalion at Camp Schwab and would tell Larry about all the different places he had been. That sounded like what he was looking for as a Marine, so he put in a request to join the 3rd Marine Recon Battalion at Camp Schwab. He was granted the transfer and spent the rest of his time in the Marine Corps with his chosen unit. He spent over two tours of thirteen months each in that unit in Okinawa.

He would have stayed in the Marine Corps had they had given him his reenlistment option. He had requested to remain in Okinawa for one more year. The Marines denied the request and told him he had to return stateside.

He returned to San Diego to process out of the Marines in October, 1964. During the process he was asked why he did not reenlist. He told them the reason and left the processing center. Shortly after he left the center, a jeep arrived where he was staying and picked up Larry. He was brought to a Major who assured him that if he reenlisted for another tour in the Marines, he could get him on a flight that night headed to Japan. Larry told him that he didn't want to go to Japan. He wanted to be sent back to Okinawa. The Major told him he could not do that. Larry thought that the reason that he couldn't go back was because they thought he was becoming too Okinawan. In fact, Larry was becoming very acquainted with the Okinawan language and practiced with flash cards that he made up to become familiar with the language. He could read and write their language by the end of his tours. With that refusal to return him to Okinawa, his Marine career ended.

Larry returned to Wisconsin and found a job working in a local factory. Before long he was involved in an occupational accident and hurt his thumb badly. That was it. He decided to join the Army right then and there. When asked why he joined the Army after four years in the Marines, his answer was, "They didn't give me my reenlistment option." He never forgave the Marines for that slight.

He reported to the local Army Recruiter and after he took an assortment of tests, he posted high scores. He was asked what in the Army interested him. He told the recruiter that he wanted to go to school for Electronics to learn the internal guidance system for the Nike Hercules missile. That program entailed a 52-week electronic course. The recruiter was reassuring and told him once he got over to Fort Leonard Wood, all he needed to do was to tell them exactly what he wanted. He was qualified for the course and he should have no problems.

When he arrived at his first Army duty station at Fort Leonard Wood, Missouri, a young Spec 4 was doing all Larry's paperwork. He told the clerk what he was earlier told by his recruiter. The clerk left Larry and spoke to his supervisor about the request. The Supervisor returned with him to Larry and told him that Larry had more than four years in the infantry and he would be going back to what he was doing before. He would maintain the Wireman MOS with a secondary MOS of Infantry Rifleman.

Larry was assigned to Headquarters Company of the 2nd battalion of the 12th Infantry Regiment. There he would serve as a Wireman in the Communications Section.

Captain Robert E Kavanaugh commanded the Company at that time. He received his first and only Article 15 (Commanders use this option to resolve allegations of minor misconduct against a soldier when a court-martial is not needed) when Larry drove his car on base without a sticker and got caught.

During the Thanksgiving weekend of 1965, the entire 2/12th was sent to Mount Rainier to locate a missing elk hunter lost on the mountain. Kavanaugh had his Company go over a big rock slide three times because they were not staying in line. When he did that the HQ Commander yelled sarcastically to Larry "How does the Army compare to the Marine Corps, Walter?" Larry responded that he thought the Marine Corps was better organized. That affront punched his ticket into A Company and out of HQ.

When he arrived in A Company, the entire Company was so small that it could assemble itself around the rec hall pool table. He knew most of them by name. They were told what was due to happen. New recruits would be arriving in the Company soon to go through Basic Training. It was their job to make sure everything was set up before they arrived in late December, 1965.

One of the requirements was to make up all the beds for the new men. They were not happy about that and they shortsheeted a number of the beds. (To be honest, most of the guys probably didn't know what a shortsheet was and figured that was how the Army made their beds).

The men who were to train the incoming group spent a month going to classes to teach them what they needed to know. It was a very busy, hectic period. Larry was then a Spec 4 and was made an Acting Sergeant during this month in preparation of our arrival on the days right after Christmas.

There was a severe shortage of cadre initially. As I said before and will repeat, the Army was unprepared for the rapid Vietnam buildup of forces. Most platoons had two or three NCOs to direct and train the new arrivals. Larry was sent to Second Platoon with Sgt. Harris and that was it for leadership of that platoon. In fact, Larry spent almost the entire time he served in A/2/12 in that platoon.

Bob Livingston, New Arrival

As time went by, more men arrived in our Company. By May, additional NCOs and Officers appeared to fill in the leadership positions that would be needed before deploying to Vietnam. A former Marine by the name of Bob Livingston was sent to the Third Platoon to help lead the unit as an NCO. As soon as he reported into the company, 1SG Springer noticed on his paperwork that he had served four years as a Marine. A soon as Bob was sent to his new assignment, Springer called in Larry Walter to alert him that a prior Marine had arrived. Larry raced upstairs to greet him. Bob wasn't hard to miss as he was struggling to figure out how to assemble his web gear, which was nothing like the Marine web gear.

Larry introduced himself and he showed him how the elements of the field gear were attached to the webbing. Larry became a patron for Bob as Bob transitioned from Marine service to Army duty. Thus began a friendship that continues until today.

So, who was this man who was placed in my platoon? I will let Bob speak for himself: "Originally "Pop's' family was Scottish and came from Lexington, Scotland. They were referred to as "Transportation Irish." How did they get that name? That was because they were ordered by the British Crown to migrate to the northeast of Ireland to establish a colony. That area of Ireland included the villages of Roca, Caru, and Newtonberry. There they went, there they settled.

In 1843, the mackerels stopped running and the potatoes were blighted during the Potato Famine and the Irish from the colony were sent around the world. Some went to Australia, some to Canada, and some to America. Members of their family tree were already established in America by then, so that's where my ancestors were sent.

At one time or another, the Livingston's had a family member serve in every American military engagement going back to the Revolutionary War. My great-great-grandfather William E.S. Livingston was in the 5th New York artillery and lost his life in the 3rd Battle of Winchester, Sept 19, 1864. His son, William Allen Livingston, was with the 6th New York Artillery, was badly wounded at Reams Station, Virginia, He survived and became a POW at age 19 but was exchanged in a prisoner swap. He lived to be quite old and very cantankerous.

My eldest brother, Bill, was in the Army Infantry at the invasion of Okinawa. He was only 18 when he went ashore. His unit was involved in the capture of the capital Naha. The Japanese propaganda was instrumental in influencing many simple, honest peasant farmers and their families to commit suicide by jumping from the heights of Shuri Castle. It was a strange time for the Japanese as the Americans got closer and closer to the Japanese mainland. My Uncle Pete was a gunner on the USS Okaloosa, a troop ship which was also at the invasion of Okinawa. There was a book written about that attack on Okinawa. It was called "AWAY ALL BOATS" Jeff Chandler starred in the movie. Uncle Pete said the typhoons and the Japanese kamikaze attacks were the worst of the Pacific campaigns. My Uncle Ed was an Army Engineer and a graduate of Notre Dame who built bridges while fighting the Japanese in New Guinea. He and a black Sgt were the only survivors of a company overrun by the Japanese infantry. The Army had told my Aunt Dot, a terrific Jewish lady, that he was missing in action for four months and to reconcile the fact that he was probably dead. She said "No" and that he would turn up. Sure enough, she was right. He survived, they had many happy years and two kids. He was a great help to me to make my recovery possible when I came home.

My uncle Walter Livingston joined the Merchant Marines when WWII broke out. His ship was torpedoed off the coast of Dakar, Africa. He, four Lascars (militiamen from the south African sub-continent) and two British nationals spent 31 days in an open lifeboat before they were rescued. They were repatriated in Florida. My uncle was taken to Chicago and was arrested for failing to register for the draft. He called my dad to say he was in a Chicago jail even though the shipping line had certified the circumstances. Judge Otto Kerner Jr. (later Illinois Governor) threatened Walter with jail unless he enlisted in the service. He enlisted in the Army Air Corps and served as a waist gunner on a B-17. While a gunner, he was wounded in the legs and lost his hearing. After the war, my uncle worked as a lithographic platemaker for RR Donnelly Press. They were famous for being the printers of the old telephone directories that were given out each year to people who had telephones, starting in 1886. He was a speed reader which helped him in his job. What astounded me was he would say, "Bobby here's a book, pick a page, any page and read the first line." I would and he would repeat the rest of the paragraph verbatim.

The war affected him badly. After the war, he disappeared and my Uncle Bill and my dad set out to find him. They found him living in Kentucky on skid row. They brought him home.

He lived a long life with my Aunt Agnes. My brother Bill and Walter, Pete, and Ed all had a tough time after returning home from war. Today we recognize the symptoms as PTSD.

I was born in Evanston, Illinois in 1942. My mom's name was Rose and my dad's name was Bill.

My family moved to the North side of Chicago soon after I was born. I asked my mom why they had moved to the north side of the city and she told me, "It's none of your business!" Pop was more forthcoming, "You have to understand, we're not southside Irish. We're northside Irish. We're Cubs fans, not White Sox fans. We were strong supporters of Sinn Féin of the Irish Republican Army. The Germans established many beer breweries in the district and we got along well with them."

At one time, during Prohibition, he was a driver for Joe and Louie Monday, big time bookies on the North Side of Chicago. He ran whiskey from Detroit to Chicago. At times he had to make bookie cash deliveries to a hotel where Al Capone occupied three floors. He dreaded that job, but he had to support his family (not an unfamiliar situation in the thirties for the parents of some of the families of our poorer A/2/12 inductees).

Pop later worked as an iron worker on high steel. He spent 20 years in that field. While working at Otis Elevator Company, he got hurt when an elevator car came down on him in "the pit" He survived that episode because the heavy springs at the bottom kept him from being crushed. He took his doctor's advice and chose another field after two near death experiences while working in the iron working field. With his experience in the iron worker field, he was finally able to land a good job with the government. That was his big break.

During WWII, he got a position that sent him to Africa to retrieve destroyed German tanks and vehicles for hauling back to the U.S. to melt down and reuse in the war effort. He retired as the GS-15 Deputy Chief (the same grade as a Colonel in the military) of procurement for 5th Army Ordnance. He did really well for a kid from Brooklyn with an 8th grade education. My brothers and I always admired him.

I went to Catholic schools during my grade and high school years. Immediately after I attended Loyola for one semester, I dropped out and went to work for a few years. I worked in manufacturing for a bit and later went to work in rigging like my dad. When my dad found out I was working on these high risers, he asked me if I was crazy. "Do you realize one mistake and you are dead?" I dropped that job, which was probably a good thing. I joined the Marines in 1961. My MOS was 0331 which meant I was a Machine Gunner. The Marines were the last to get the M60 Machine gun, so at first, I trained on the Model 1917A1, 30 caliber machine gun. Then later on the M60. I was lucky to become an assistant gunner then gunner, then squad leader, and eventually a Corporal in a weapons platoon.

The Marine Corp, at the order of the President, maintains approximately 1,500 Marines in the Pacific Fleet Force and 1,500 Marines in the Mediterranean.

In my time in the Marines, I served in Hawaii, Guam, Okinawa, where I served with Suicide Company, 2nd Battalion, 3rd Marine Division. That Division was the first ground unit to be sent to Vietnam in March of 1965.

At one time we did a special six-month afloat phase where we were assigned to wherever the President said we were needed. It was an interesting assignment.

At one point we went ashore on Mindoro in the Philippine chain to chase the Huks. Huks referred to The Hukbong Bayan Laban sa Hapon, a Philippine insurrection army that formed during WWII by the farmers of Central Luzon. Their name literally means: 'People's Army Against the Japanese' By the 1960s, they had transformed into a socialist/communist guerrilla movement.

We thought that it was regular exercise, but before we went to the landing craft, we exchanged our blank ammo for live ammunition. This operation was called Tulangan which we understood to mean Friendship, in the Tagalog language. The Philippine Marines and Rangers and us found out it was not very friendly. For the most part, the Huks retreated into the jungle at our arrival, but maintained guerilla war against us when it was advantageous to them.

I had served three years in the infantry when my First Sgt David Moore (he served on Guadalcanal as a fraudulent enlistee at age 16), said we have another assignment. You're going to a security assignment on the Eastern seaboard. I reported to Sgt Major Umlauf who assigned me as Corporal of the Guard of the Interior Guard. When my Top-Secret clearance came through, I was transferred to "M Group" where I was guarding Nuke weapons and transporting them to vessels at the longest piers on the Eastern Seaboard.

Peter Filous, Squad Leader

Peter Filous, had a life that screenwriters in Hollywood couldn't dream up. The man is a success story, but what a price he had to pay. His youth was marred by a series of tragic personal losses and a struggle to overcome ignorance and abuse.

Peter's grandfather was a university student when he was arrested after protesting against Hitler's regime. They grabbed him on the street and he was sent to the infamous concentration camp in Dachau. The camp was the first Nazi concentration camp in Germany, established on March 10, 1933. Adolph Hitler had been Chancellor five weeks at that point. It was built at the edge of the town of Dachau, about 12 miles north of Munich. It became the model and training center for all other SS-organized camps.

Not long after he was sent there, a letter was received by his family telling them that their son had died of pneumonia. That was nonsense.

Peter was born on May 8, 1946, in München, Germany, a year after the German surrender ending World War in Europe. Times were hard in Germany and so it was for his family.

There were four of them in the family. Dad was 15 years older than his mother. He had an older brother, Otto, and he was the youngest of the offspring. In explaining the socio-economic condition in Germany after the war, he said that those who were wealthy before the war, made out well after the war. However, his family, which was in the lowest economic class of Germans, struggled terribly after the war.

His dad worked in accounting and was also a Certified Restorer of old documents stored in town halls, recording them into script. It was difficult for his dad to find work in his fields and they travelled extensively from 1946 to 1954. His father worked odd jobs to pay the bills, but it was a humbling life that they lived, traveling from place to place. It was not unusual for his family to move on after living in one place for only three or four weeks.

When he was six years old and ready for school, they moved to Oberammergau in the Bavarian Alps. Oberammergau is known for its once-a-decade performance of the Passion Play at the Passion Play Theater. This passion play is the oldest passion play in the world. The play has been performed every 10 years since 1634 by the inhabitants of the village. Additionally, this town is known for their great craftsmen and wood carvers.

This is where he began his schooling at a Catholic School. There was much resentment between the Catholics and Protestants in Germany. His father was a Catholic and his mother was Protestant. This made him a source of much ridicule by the nuns who taught there.

His brother, Otto, by then was fourteen years old, helped with the bills by going door to door selling odds and ends like shoestrings or razor blades. By his third year of schooling, they moved north to Hamburg where there were much more economic opportunities. That was where they finally had a house that they could call home, rented from a farmer. There was no running water inside the house and they had to get their water from a pump down the road. There was no indoor plumbing and they had to use an outhouse in the back of the building. Still, it was home for the family.

It was about that time that his mother and father separated. While his father was away, Peter found himself alone a lot and he learned to fend for himself. The economic condition were still extremely difficult and basic necessities such as a square meal per day was often not available. He told me that there were weeks when the family needed to gather mushrooms in a nearby forest and collect potatoes from potato fields left behind after a recent harvest to supplement their diet. The ability to be self-sufficient would build up his self confidence that would serve him well later in life. Still, he sorely missed his dad.

His father came back in 1960, wanting to reconcile with the family (especially Peter), and he tried to make a go of it. By that time his mother was improving her economic conditions. She had found a decent job as a Department Head in a major department store in Hamburg and was doing okay.

After returning into the fold, his father found a job in Hamburg and things were moving along splendidly. During this time, they bought their first television and they were beginning to be able to afford more conveniences in life. However, they were still in the lowest economic social conditions then in Germany. Germany and most of Europe, for that matter, has a different class system compared to America. It was difficult to move up the ladder compared to America.

His grandmother on his mother's side was well set. She

owned an apartment house in Heidelberg and his uncle was a State's Attorney there. They wanted nothing to do with Peter's family because of the different station in life that they were in.

Soon after his father's return, he was once more laid off. This created great stress in the family. Eventually he was able to find work based on his strength and knowledge and things settled down in the family, for a while. Acquiring steady work was really difficult for him and he would come and go out of the family's life as the layoffs reoccurred.

Peter built on his personal skills as he learned more about dealing with life's problems during the periods of his father's absence. He did well in school, which he attributes to some very dedicated teachers who understood that he was different from his classmates. He had a heavy southern Germany accent when he spoke and that made him unlike his classmates. They understood how challenging his life was and they went out of their way to see that he was successful in school.

Even with the different dialect, he had no difficulty making friends and had many. Still, he never connected to the fact that his family was poor, mainly because he lived in a dysfunctional family that was in constant turmoil. Still, he had a television and friends. To him, they were doing well.

He was 16 years old when on February 12, 1962 everything changed. He awoke that night around 9PM during a heavy storm and stuck his hand down the side of the bed and felt water. In those days there were no advance warning of catastrophic storms heading towards them. They were unaware until the danger arrived.

He got out of bed and was standing in water up to his knees and it was rising very rapidly. They had just installed a new TV antenna a week before and Peter helped by adjusting it on the roof for the best reception. He was able to get to the roof by climbing out a window, pulling himself up to the ledge and then he climbed on the roof. That saved the day because that was the only way that they could get up to the roof. So, he pulled himself up to the roof, then pulled his mother, his father, and a little dog they owned, up to the relative safety of the roof. They made it to the peak of the roof which was not very strong. The water kept rising and a section of the roof covering a small addition to the building suddenly collapsed. The wall of that addition also began failing and quickly affected the integrity of the main roof. Roof pieces began breaking off. Peter was able to hold on to his mother and he tried to save his father. His father, knowing that the remaining roof could not support all of them for long, jumped into the freezing water and tried to make it to a nearby tree. He didn't make it. All Peter could do was watch his father get swept away in the churning water.

Peter and his mother, with the dog between them, clutched to each other in the swirling snowstorm that had developed. They could see farm animals and debris floating past them as time on the roof turned from minutes to hours. Twenty hours passed by before they were rescued.

Germany took care of the flood victims after that flood and they were given an apartment from February 1962 until August of 1963.

Peter's brother, Otto, was not with them during this catastrophe because he had left for the United States four years earlier. Otto was a craftsman carpenter who specialized in repairing antique furniture. In addition, he did sheet metal work on the side, so he was doing well.

Peter resolved to overcome his loss by being the best at whatever challenges were presented to him. He graduated from High School that same year and was accepted into an apprenticeship in Electronic Engineering with the German Telecommunications Corporation, which was part of the German National Postal Service. This was no small feat for the young man.

After that tragedy, his brother wanted Peter to come to America. However, Peter would not move without his mother. Both would make the trip.

To do so in those days, they needed a sponsor to become permanent residents and his brother found one for him and his mother. He attained what was called permanent resident visas. This meant they were issued identification cards and had to report each year to the consulates. His sponsor was responsible for them during this time.

They arrived in America on September 3, 1963, when they were greeted by his brother. Peter stayed with Otto and his mother first stayed with Otto's fiancé's grandmother, both in New Milford, CT. His mother found a job at a dry-cleaning establishment and Peter found employment as a carpenter's helper.

On his first day on the job, October 10, 1963, he was picked up for work by the owner of the small construction firm where he began working. It was very cold and Peter jumped in the back of an old open pickup truck. He was given a blanket to help him keep warm while he was transported to the construction site. The construction business owner would later become his father-in-law.

The work was challenging as he spoke no English. He primarily carried lumber around and worked as a laborer at first. He eventually learned English and his craft quickly and eventually led his own work team, where he managed three workers doing exterior trim and framing. In the fall of 1965, he got his military draft notice. Everyone told him that they couldn't draft him and send him to Vietnam. However, history told him that, in fact, they could do that. Peter was so proud and happy to be in America. The whole idea of 'America' in his mind seemed Utopian. Although he had the option of returning to Germany, it was out of the question for him. He loved our country and couldn't contemplate deserting it. Had he returned to his former country, he would have probably returned to the apprenticeship program and become an Electronic Engineer, but he would not be American.

On December 14, 1965 he was drafted into the army and joined the rest of us at Fort Lewis.

During his time in the service, he was sometimes hounded by a few, not many, from our unit who teased him about his immigrant status. Ironically one of those who harassed him the most in our company would later save his life in Vietnam. After Peter was wounded, this man crawled up to the forward position and bravely dragged Peter to the rear during intense enemy fire. Peter will always be grateful to him. In spite of the man's personal prejudices, he disregarded his own welfare to save Peter's life. Funny how things work out in combat…sometimes.

Discipline came easily to Peter and he always followed orders strictly. He was also bright and it wasn't long before he was recognized and placed into a leadership role.

By the time we left for Vietnam he had earned Sergeant stripes as an E-5. That was only nine months after we went through basic training. That was quite an accomplishment for the young man.

Norm Smith, Squad Leader

Norman W. Smith was born June 29, 1946, in Nelsonville, Ohio, a small coal mining community in the southeastern part of the state. He was the third of five children born to Hollie &

Mildred York Smith. They lived in Pineville, KY, where Hollie worked in coal mining. He handled heavy equipment that was used to remove coal from the ground. That eventually led to the family moving to Nelsonville, Ohio where his father continued his coal mining job for a while before becoming a crane operator.

Norm had a childhood reminiscent of a Mark Twain character; playing in the woods and on the river, and swimming away summers. Norm loved to hunt rabbit with his father and shoot marbles with his friends. He was a natural athlete and loved to play baseball, basketball, and football. At Nelsonville High School, Norm played guard for the basketball team and center for the football team (at a fighting 5'11", 150 lbs.).

As a high schooler he spent a lot of time with his friends at the local pool hall working on hand-eye coordination and making a little extra cash in the process.

Upon graduation in 1965, Norm went to work to save money for college. He signed up at Ohio University for the spring 1966 session but was drafted into the army in December of 1965.

Norm understood and accepted that it was his time to serve his country. His father had been an artillery training sergeant in WWII. His uncle, Ballard York, was an infantry soldier in WWII serving in France. His mother's cousin was none other than the famous WWI hero, Sergeant Alvin York. The country was now calling for him to serve, and he would heed the call. College could wait.

Norm was ordered to report for duty at Fort Knox, KY in December of 1965. He was assigned to Fort Lewis with a group of 80 other young men from the Midwest.

He and the other men were added to another like sized group from the east coast that was used to make up the soldiers of Alpha Company, 2nd Battalion of the 12th Infantry Regiment.

While at Fort Lewis, Norm, like the rest of the brothers of A/2/12, received basic training & AIT. Norm would soon be recognized for his leadership trait and was sent to the 6th Army NCO Academy for training. He began leadership training in May, 1966 and graduated in July, 1966. Eventually he became a squad leader in the Second Platoon, which was his role in Vietnam.

Norm remained in the A/2/12 company until late in August, 1967. At that point in his Vietnam tour, he and the members of the 2nd platoon were sent to the D/2/12 company, which was newly formed as the Army expanded infantry battalions with an extra company. The thinking then was to merge seasoned combat veterans with the novice soldiers just arriving.

A tour in Vietnam for army personnel meant one year in Vietnam. In order to achieve the imperative of getting all the men of the brigade back to the States by the annual date of deployment, all of the original men of the brigade all had to be back stateside by September 21, 1967. To reach that goal, they needed to start sending soldiers home weeks before their DEROS (*rotation date*). Some left as early as six weeks before they needed to be home. I left four weeks early.

Sergeant Norm Smith, serving as a non-commissioned officer in the infantry, was highly valued by the army in Vietnam. He was one of the 'boat people' who left with the brigade for Vietnam on September 21, 1966. (Boat people refers to the soldiers who left Tacoma with the original deployed troops in September, 1966.) He and most of the other NCOs remained in our brigade in Vietnam until the very last day that the army could keep them there; September 21, 1967.

Thanks to good training, good buddies, a lot of luck and his good hand-eye coordination from the pool hall, he survived the experience, earning two Bronze Stars; one for gallantry and one for meritorious service.

After returning home, Norm was finally able to enroll in Ohio University with the help of the GI bill. He soon married a girl from Nelsonville. His daughter, Stephanie, was born December 17, 1968. He worked in road construction doing everything from operating a chain saw to driving a bulldozer to support his family while going to college. He also worked hard at his studies but suspected that college would be a piece of cake after his experiences in Vietnam. For him it was.

In 1972, Norm earned a bachelor's degree in Geology, Summa Cum Laude. The university offered him a position as an assistant teacher which helped him to earn his master's degree in 1974.

Union Oil Company sent out recruiters all over the country looking for gifted students to be employed by their oil exploration company. When Norm was being recruited, there were only two geology engineers needed by the company. He had to compete for one of those two positions with others who graduated from more prestigious universities in the country, including Ivy League schools.

While Norm was being interviewed for that position, the recruiter asked about his military service. As luck would have it, the recruiter came from a long line of military veterans and was impressed with Norm's leadership experience in Vietnam. As a result, Norm was one of the geologists that was hired for the company. With that opportunity opening up for him, he packed up his family and headed for Midland, Texas to begin a career as a geologist.

Norm's job entailed oil and gas exploration activities. Using available data, he would determine where the company would be most successful in oil drilling operations.

While in Midland Texas, Norm became friends with George W. Bush before he became President, as both would work out playing basketball and running together during lunch breaks.

His years in oil exploration were successful and resulted in him being recruited by an old wildcatter named Frank Caraway who had an oil company in San Angelo. Together they discovered many lucrative oil fields in central West Texas.

All his life he continued to love sports. Norm played basketball and softball at the Y. In 1977, Norm took up jogging after he broke his finger playing softball. This activity led up to running marathons. He ran in five Boston Marathons…his best time run in the Boston Marathon was 2:43. After running in the Las Vegas Marathon he ran a 2:38 time which was good enough for a 6th place finish.

After his daughter left home for the University of Texas, Norm's marriage, unfortunately, ended. He moved to San Angelo and continued his work and golfing.

At one point, Norm had to cut back on his running as back issues developed. It did not interfere in his golf game so his new passion became golf.

While in San Angelo he met and married a golfing friend, Elaine, and they began to travel extensively in the United States and around the world together.

To think, it all began with all-expense paid trip to southeast Asia in 1966.

David Cunningham, Private 'Everyman'

After thirteen years of writing about exceptional men in our newsletter, I had figured out that it was about time to write about the more average enlisted man who served in our company. I wanted to detail the background and thoughts of the majority of us who accepted that all-expense paid trip to Vietnam. After speaking to a number of the men in the association over these years, I had a rough idea on who filled the enlisted man's ranks in A Company. Dave Cunningham's story, left, filled the bill perfectly. It turned out that even our average men were exceptional.

Dave's great grandfather was an Irish Immigrant, who arrived in Clark County, Ohio in the early 1900s. He was poor, (a family trait that was very familiar to all the men from our Company) but he knew that he could provide for his family as a tenant farmer. The work was hard and not very well-paid, but there was a sense of pride in the skills that were required to work the fields.

The parents of our men shared a precarious youth in America in the 1930s and 40s. When they successfully survived the era of the Great Depression and World War II, it tempered their outlook on life. At a very young age, they learned the value of hard work and determination. For most of these men turning of age during the 'Greatest Generation' period, there was almost immediate transition period from the lean years of the 1930s to service during World War II. Some had learned teamwork at WPA camps that sprung up all over the country. Many were sent far from their home to serve as laborers on massive government projects and only periodically saw their families. This would serve them well when they entered the military. They understood that there were no guarantees in life and that was ingrained into the children of that generation.

Men like Dave's father were spared that experience because they were farmers. That occupation was not a lucrative enterprise, but people still had to eat and someone had to produce the crops. Once the war broke out, farming became an essential element of the war effort. Dave's father was part of that army of farmers back home that assured that the soldiers fighting the battles in Europe and the Pacific Theater were properly nourished. In addition, they had to assure an adequate amount of food for those on the home front. It was hard, back-breaking work, but they were proud to contribute to the war effort in their unique way.

Early in the war, his father worked as a tenant farmer, but late in the war, David's father and his two brothers pooled their money and bought some land of their own that they farmed later during the war. It was there in 1946 that Dave was born. Their farm was in Springfield, Ohio, the County Seat of Clark County.

Dave can recall his dad working with team horses in the hay and corn fields of central Ohio. He remained a farmer until his mid-fifties. He then spent the rest of his working days working in the plant in a London, Ohio factory that was the home of Brillo Pads.

The agricultural efforts, as it turned out, were only a down payment for the Cunninghams responsibility to the American defense effort after the war. In December, 1965 with the delivery of the draft notice to his son, the family was once more called to attention.

Dave spoke of poverty and hardship throughout his childhood. There were three boys in the family. Edward was the oldest, followed by Dave, and the youngest was named John. All three worked in the fields to help the family during the farming days.

Eventually they were forced to leave the farm where they were living near Springfield and moved to a small place called Newport, Ohio.

Dave attended Madison South High School. He would have loved to have played sports, but this was impossible as he worked in a dairy after school for all four years that he attended high school.

He had a couple of girl friends in high school, but neither was serious relationships. Between school and dairy work, he just didn't have the time for romance. He was becoming a man and he had his responsibilities and he took them seriously.

Dave graduated from high school in May, 1964. He was seventeen at the time. He continued to work at the farm and at the dairy until September, when he turned eighteen and was able to work in a factory.

That was when he went to work at the McChord Corporation factory which made radiators for Chrysler products. It was hot, hard work.

Later he went to work at a Westinghouse factory. I asked Dave if he had someone he knew to get him those jobs and his answer was typical for the times.

"No, I didn't know anyone. In those days, if you woke up, they'd hire you."

Eventually Dave was able to get his first car, a 1957 Chevy. Things were looking up for this young man!

As the situation in Vietnam was heating up, it became apparent that Dave would probably be called into the service. His mother wanted him to join any branch but the Army, but Dave would have no part of that. He just figured it was his turn and he would go willingly to wherever his country needed him.

In late November, 1965, he got his draft notice and was told to report to Fort Knox, KY for induction into the army. On December 15, he boarded a bus in London, Ohio after saying goodbye to his family. He was taken to Fort Hayes in Columbus where he spent the day taking last minute tests and being sworn in.

He remembers that day as mainly a day of screaming and hollering by low grade NCOs.

He clearly recalls that some of his group was drafted into the Marines and the Navy. The Marine Officer asked for four volunteers to join the Marines and four young men stepped forward. The Navy Officer did the same with the same results. Dave decided to hold his hand and was set to be sent to Fort Knox in nearby Kentucky.

It was during this initial period in the service that he got to be friends with the men from Madison County who were destined for the 2/12th Infantry. Marion Vallery, who went to school with him at Madison South High, John Ogan, Ray Glock, and Jerry McClary were among that group.

After a few days, the new soldiers at the Reception Center were granted a three-day pass to go home for Christmas. Dave caught a ride back to London, Ohio, with Marion Vallery's brother who came up to bring Marion home for the holidays.

After the break and a return to Fort Knox, it was off to Fort Lewis for training in A/2/12. Dave was in First Platoon during Basic Training.

After basic training and AIT, Dave was transferred to Weapons Platoon and trained with them at Yakima, Washington.

Dave spent his entire time in A Company in the Weapons Platoon. It was rare that the platoon, numbering somewhere around 20 men, used the mortars. Most of the time, they were just used as another maneuver rifle platoon that patrolled the area looking for enemy forces in the jungle.

He was fortunate in not being one of the men who transferred to other units during the detested Infusion Program (more about that later) that took place during the Spring/ Summer period of 1967. Of the original A/2/12 men who landed in October, 1966, fewer than half left Vietnam from the original Company. The rest had been sent home as casualties or were sent to other Companies in Vietnam which sent them home on their DEROS (Date Estimated Return from Overseas) date.

MAKING THE MOST OF TIME

In May 1966, the 2/12th battalion trained for two weeks in the Hoh Rain Forest on the western section of the Olympic National Park. The Fourth Division was earmarked to be sent up to the Central Highland region of Vietnam so it made sense for us to train for that type of terrain.

Ultimately it didn't work out that way for the Third Brigade, although we didn't know that then. Still, it was a fun period for us and was a relief from the fort's restricted sites on base.

We trained in platoon size assaults on the mountain. The men of the Recon Platoon of HQ Company were our aggressors. Their job was to harass us as we made our way to our mission objectives on the top of mountains.

We remained out in the field after our first week and they provided us some leisure time on the weekend.

Sunday morning services turned out to be special for me. The Catholic Chaplain couldn't make it so the Christian worshipers all went to a Protestant service. I didn't know this minister, but he made a lasting impression on me. It was my very first Protestant service so I was curious to see what it was like. I was Catholic and, in those days, it was a sin for us to attend Protestant services, so this would be all new for me.

The services were not as formal as the usual rituals of Sunday mass that I was accustomed to at home. We sat in a 3/4 circle on the ground around the Protestant Chaplin, Major Gene Adler. At one point the cleric, understanding the apprehension felt by some of the men, spoke of the war. The chaplain framed it in such a way that we could understand the importance of what we were about to accomplish in Vietnam. Fundamentally he said, "We understand the apprehension that you feel going to war. It's important that you understand that there are worse things than war." He didn't elaborate, but I understood his comment as a reference to what occurred during World War II in Europe and in the Far East.

You will read more about this chaplain in a later chapter when he played a significant role during the Battle of Suoi Tre.

After religious services, the company enjoyed some well-earned time off. Kegs of beer were brought out to the site on Saturday and placed in a nearby pond to keep them cool. They were pulled out of the water early Sunday morning and placed in ice, and waited for the conclusion of the religious services. Each man was allotted two beers and for many of us young men, that was plenty.

After lunch it was time for friendly competition between the different platoons. There were group contests like the tug of war and individual matches of arm wrestling. It was all in good fun, but after chugging down the two beers, it was a lively affair, much enjoyed by the men who were enjoying the weekend schedule.

On July 21, 1966, the first of the 4th Infantry Division brigades left Tacoma bound for Vietnam. That brigade, the 2nd Brigade, was being transported to Vietnam on the troopship USNS General Nelson M Walker. It was commissioned in 1944 and used to transport troops both during World War II and the Korean War. After each conflict it was placed out of commission as wars ended and peace ended its usefulness. Following many years of storage after the Korean War armistice, it was acquired by the US Navy to transport troops to Vietnam. We of the Third Brigade would be taking that same ship to Vietnam, exactly two months after the departure of the 2nd Brigade from Tacoma.

In early July, 1966, our brigade began preparing their troops for the last leave that they were due before going into combat overseas. We all had this two week 'Vietnam' leave, which needed to be accomplished in six weeks. To achieve this objective and get everyone back by September, they broke the company units down into groups of thirty. Then they sent the detachments home on leave in succession.

By this time, there was no pretense amongst the ranks. There was no more talk of being sent to Germany or Korea (a favorite belief among the young draftees). We were going to Vietnam, where according to everything we read, was winding up, not down. American KIAs in Vietnam were nearing 2,500 and four-times that amount were being wounded in action. To drive home the point even more, we were all required to make out a will in the event we were killed in Vietnam. I was twenty years old at that time and it felt odd. Nothing matches the naivety of youth.

Just before sending the first men home on leave, A/2/12's First Sergeant, Sidney R.K Springer, gathered the company in the large assembly area outside our barracks. Sid, as he later preferred to be called at our reunions, had some important advice to share based on his history in combat in the Army. As a young enlisted man, he served in the Korean War and had a pretty good idea about what we would be facing in Vietnam.

Ernie Pyle, the infantrymen's champion reporter once said, "All wars are the same. Only the locales change."

Sid expected tough times for the men of A/2/12, who he came to love. Because of this, he began giving a short speech based on what he knew to be true of combat and the men who engage in the fighting.

My memory of that speech may not be verbatim, but I never forgot the finer points that he was striving to make to us novices.

"Gentlemen, I'm bringing you together to inform you that I will be sending all of you home in groups beginning next week for two-week 'Vietnam' leave. Make the most of this opportunity. For some of you it may be the last time that you will return home alive. I am taking you with me to Vietnam and I hope to bring you all home safely. Unfortunately, that will not be the case. Some of you will lose your lives there, others will become injured and take a different route home to the United States.

Make the best of this visit home.

I've been in combat and I learned quickly what has probably always been the case for all war veterans. It is never hard to send a soldier into combat the first time. It is much more difficult to get him to go a second time. (*I'm guessing that was what Sid was feeling about the return to danger*).

With that said, you make sure that you go home when you're allowed and return when you are supposed to. (*He then gave one of those long glares which bushed up his eyebrows for emphasis*). Don't make me come looking for you."

That may have been more of a problem for us than he or we knew then. The airline industry unions struck five major airlines on July 8th that year and the strike would last for 43 days, ending on August 19th. This was the longest strike in airlines history and created huge problems for those needing to make it back to the east coast for that Vietnam leave.

My first day of leave was on Saturday, August 13th. Steve Kawczak, one of my platoon buddies, traveled with me. We followed the lead of earlier platoon friends who had already returned from leave.

We got released at 11 am after inspection and jumped on a base bus that brought us to McChord Air Force Base. We stood in line with approximately 30 other people to sign up for a 'hop' (a seat on a military plane heading east) around noontime. Once we registered and told them where we needed to go, we sat down and waited to be called. We spent all day and night at the terminal awaiting our flight. We finally got called at 11 am on Sunday morning. We got seats on a C-124 Globemaster plane that would fly us to McGuire Air Force base, just outside of Fort Dix, New Jersey. The plane was a propellor driven cargo plane which seated probably 40 people along the walls of the cargo area. Its cruising speed was 240 miles an hour, so this was going to be a long flight, interrupted by one stop in Minneapolis for refueling.

The flight to Minneapolis was uneventful and we landed sometime after dark.

While the plane was being refueled, we noticed a mechanic climbing a step ladder and opening the cowling (covering for the engine compartment) on our side of the plane. We could clearly hear the conversation from the mechanic to the pilot who was shouting through the side window of his seat. We couldn't make out what the pilot was saying, but clearly heard what the mechanic said, "Yeah, it's shot. We got nothing like that here. You're going to have to fly without it."

Steve looked at me and I stared back. We simultaneously said, "Great! What are we flying without?"

We put that in the back of our minds until 30 minutes later while we were flying our last leg to McGuire AB. Whatever we were flying without on the plane we hoped it wouldn't affect the stability of the plane as we ran into a turbulent storm. The plane swung violently and altered plunging and gaining altitude as the air masses took hold of the plane.

The Air Force, in their infinite wisdom, had issued us burp bags that were made of transparent plastic. As the plane fought to make it through the thrashing air, more and more of the passengers, equally made up of military and civilian personnel, began heaving into the bags. Of course, seeing the bags filled up made others sick. Steve and I made it through OK. We were too busy laughing about the clear plastic bags that were used to notice the airsick people.

Eventually we landed in New Jersey on Monday morning and Steve and I took a bus to New York Port Authority and then caught trains in New York headed to close to our homes. I made it to Providence, RI by late afternoon, where my family picked me up and brought me to New Bedford. In essence, that transportation adventure took three days out of my fourteen-day leave, but I figured I could make it up on the trip home. They couldn't blame us for reporting late with this major strike going on. Yes, they could, as the strike was settled on that Friday. "Don't make me come looking for you," rang in my head.

I arrived back at Fort Lewis in plenty of time, but there were a few who didn't. The army never made a big deal out of soldiers returning late.

The army was anxious to get us all back by September 1 because we needed to qualify with the recent addition to our armory, the new M16 rifle. We had all trained with the M14 but this weapon was unlike anything any of us had seen before.

The M16 rifles were fairly new to the army and because of their limited numbers, only infantrymen received this weapon. We received our firearms still in the cardboard boxes they were shipped in, with the printing 'Colt Firearms, Hartford, CT' on the top. We eagerly opened the boxes and were shocked by what we saw. It looked nothing like the M14 rifles that we trained with. The butt and the handgrip of the weapon were made of plastic and it was pitch black in color. It wasn't long before the jokes were flying around about how the weapon was probably made by Mattel, a very familiar toy gun manufacturer to us young men. "It's swell…it's made by Mattel!" was a common comment.

Still, it was much lighter and the firepower was much increased by the ability to switch a lever and change the weapon from semi-automatic mode to fully automatic. We were told that the rounds leaving the weapon had a tendency to tumble when it hit its target, increasing the damage it produced as it tore into its targets.

Some guys were looking down at their M16s and wondering if the rifles were really as lethal as they told us. They were, but still too new to be considered dependable. That unreliability condition would remain with us until it was remedied by the replacement of the buffer assembly, which we didn't receive until nine months into our tour in Vietnam.

Until then, we had a number of inopportune jams that kept us in suspense whenever we fired our weapon. We would complain about the issue and were continuously told that we were not cleaning the weapon properly. They even sent us new cleaning kits to see if that would help. Eventually the army learned that the type of powder in the rounds was burning much hotter than expected and the expended rounds would swell and would not release from the chamber. The soldiers who took our place when we got home had the revised M16s and never had problems like we had. Good for them.

Saddle Up

On the morning of September 21, 1966, my company was ordered to assemble on the west side lawn of our barracks. There we would await transportation to the troopship moored at the Port of Tacoma dock. We took all our personal equipment in our duffel bags and our recently issued M16s and sat on the lawn until it was time to go. By then all the radio operators were also issued the new PRC 25 transceiver radio which replaced the larger and heavier PRC 10s that we trained with. The PRC 25 weighed almost ten pounds less than the PRC 10, was much shorter, top to bottom, and the electronics were solid state except for one tube. This was a much-improved model compared to the PRC 10. It was waterproof and you could drop it 50 feet from a helicopter and it would suffer no damage. The range of the radio was also improved significantly.

Directly in front us was a large assembly area (*which today is used as a parking lot for the modern all-volunteer army soldiers*). There was much activity going on in that open field as more and more transportation vehicles arrived and were loaded with what we needed to face the enemy in Vietnam. Considering that both the 2/12th battalion and the 3/22nd battalion were assembling in this area, it became increasingly congested as we viewed the movement of vehicles into the clearing from our westerly lawn.

The oldest adage in the army is the observation of "hurry up and wait." It never seemed more accurate than on that morning as we waited for our transportation to the docks in Tacoma.

Finally, the busses pulled up and we jumped into the old military busses that would take us past the fire station where our drummed-out man was able to wish us a 'Bon Voyage'. We arrived at the dock and were greeted by a view that we couldn't have imagined.

As far as the eye could see there was tons of equipment and vehicles waiting to be loaded on to the troop ship.

To be certain that we would leave Tacoma on the next day, all the material and men were loaded onto the ship one day early. Still, it seemed like it would take a herculean effort to get that mission accomplished.

They were able to make a lane for the buses and we were allowed entry to the gangplank. As I made my way up the walkway, I noticed something unusual. I wasn't very far from 1SG Springer and I could see a private being coaxed along up ahead of him. We all carried our duffel bags onto the ship and it was no different for this soldier. What was different was I saw that the Private had a handcuff attached to one wrist with the other end firmly attached to his duffel bag. I later learned that this soldier, a recent addition to A Company, was earlier a member of a Second Brigade company. When this man had reached this point in boarding the 'Walker' in July, he jumped into the waterway and swam away. It took them a week to locate him and he was brought back to Fort Lewis where he became part of our company. Springer was determined that there would be no repeat of that escape attempt and if the man tried to jump with us, it would be a suicide undertaking. No such effort took place under Sid's watchful eye. Sid was a wise old owl, especially in those days, and he understood that sometimes men lost some of their nerve under pressure and just needed some encouragement. He was right. This soldier became a very dependable leader in Vietnam. He just had to be properly led in the right direction.

Even before I could join my third platoon on deck, I was approached by someone and told to report to Lt Palmer, our Company Commander. He informed me that he chose me to be his Radio Operator on the Company Net. He told me to get my gear and report to the Company HQ section where I would meet others in that unit. After serving as the RTO (Radio Telephone Operator) for Third Platoon Leader, Lt. Jim Olafson since May, I would be handling the company communications for Jon Palmer. I don't know how that happened as he had someone else before, but I didn't question the move. I was honored. Before I was sent home in late August, 1967, I would serve in that role for three different A/2/12 company commanders.

Our ship contained close to 4,000 troops which made up the bulk of the troops of the 4th Division's 3rd Brigade. The contingent was made up of three infantry maneuver battalions, the 2/12th, the 3/22nd, both infantry battalions, the 2/22nd Mechanized Infantry, and a battalion of artillerymen of the 2/77th Field Artillery. By nightfall, all these troops were on board and getting set into their overcrowded bays.

None of us could have guessed that on a day exactly six months later, our brigade would be caught up in a desperate struggle with major elements of the 9th Vietcong Division. On March 21, 1967, a major early morning attack targeted the 2/77th Field Artillery Battalion and two companies of the 3/22nd. The site was an artillery fire base in War Zone C called FSB Gold. Before it was over, all four 2nd Brigade units would slug it out with over 2,500 very determined Vietcong soldiers. At the conclusion, three and a half hours after it began, 647 Vietcong soldiers were left dead on the field of battle. That battle produced the greatest enemy loss in a one-day battle of the entire Vietnam War.

Each brigade combat unit and one unit of the 2/34th Armor Battalion, which also participated, received the Presidential Unit Citation for their action on that day. This is the highest honor that can be bestowed on an Army organization. The Army, Navy, and Air Force Presidential Unit Citation is the personal equivalent of that service's Distinguished Service Cross. The Distinguished Service Cross is the United States Army's second highest military decoration for soldiers who display extraordinary heroism in combat with an armed enemy force.

When we trained at Fort Lewis as 12th Infantry soldiers, we wore the PUC above our right shirt pocket. We did so to honor the only other time that the 12th Infantry Regiment earned that prestigious award. That was awarded for their defense of Luxembourg during the Battle of the Bulge in WWII. The 2/12th Infantry Battalion earned the award as a battalion for that March 21, 1967 battle, the Battle of Suoi Tre (Fire Support Base Gold). To my knowledge, those were the only two times that the 12th Infantry Regiment received such an honor. Having our unit be associated with the heroes of Luxembourg is very gratifying. To see modern day 12th Infantry soldiers wear that PUC on their uniforms to honor both units, is even more so.

However, for us mostly draftees who boarded the USNS Nelson M Walker on September 21, 1966, that was the furthest thing from our minds. We had two overwhelming emotions at that time. We were apprehensive but resolute about our responsibility to defend our Country from Communist expansion. We also took strength through the example of our kinfolks when they served in WWII. They looked up to us and we looked back at them.

Ever the optimists, we also wondered if the war would be over by the time we got to Vietnam or soon after we arrived. We needn't have given that much thought. Our overseas tour began on that day we boarded the 'Walker'. That would set our individual DEROS (date expected return from overseas) as September 21, 1967. The 2/12th starting, with us boat people, on this date, would remain in Vietnam until April 16, 1971, four and a half years after we climbed the gangplank in Tacoma. The 3/22nd left the same month. The 2/22nd Mechanized and the 2/77th Artillery left in December, 1970, five months before we did. Who knew then? We certainly didn't as we wondered what would happen if the war ended on the way over. Silly boys!

From a letter that I wrote my family on the night of September 22, 1967

"Today is Thursday, September 22nd and at 2pm, we left the dock on schedule. It took until 6 pm to clear Puget Sound and arrive on the high seas. By 9:45 pm, when I'm writing this letter, it raining like crazy and the seas are getting rough.

In July, 2020, fifty years after the fact, just out of curiosity, I decided to look up the weather on that date online. With the help of a website called wunderground.com I was able to pull up the weather for Seattle Washington on September 22, 1966. I nailed it.

We pulled out of the dock at 2 pm. According to the historical weather site, it was 74 degrees Fahrenheit at that time. At 5 pm it reached the high temperature of 76 degrees. The temperature dropped rapidly between 5 and 6 pm as we approached a cold front. The rain began and continued in intensity, hitting a crescendo just as I was writing that letter home.

On the next day, periodic showers continued all day. The temperature never reached more than 67 degrees. It was just another September forecast for Seattle. We lucked out leaving when we did.

The weather would be much different where we were headed.

All Inclusive Cruise Experience

"Hoist up the John B's sail
See how the main sail sets
Call for the captain ashore
Let me go home, let me go home
I want to go home, yeah yeah
Well, I feel so broke up
I want to go home"

Speak to any of the boat people of our cruise to Vietnam and more than likely you will hear words like, "The boat trip from hell" or similar thoughts. It may be stretching it a bit, but there were uncomfortable moments as we traveled through some storms. For those who had a weak stomach from the rolling of the ship (and there were many), it was a nightmare.

With not much else to do, I wrote letters home detailing what it was like for us during this period on the high seas. On my very first letter home written on the 22nd of September, I complained having to stand in line for almost an hour and a half to get fed. A couple of days later, the situation was much improved. The lines were still very long into the mess hall, but they moved pretty quickly. It seemed that many of the men in line would get close to the food and would become overwhelmed with the smell of the food and would run up to the deck so they could heave over the sides.

However, there were days when the weather was fair and the seas were calm. On those days we would gather on deck and pass the days the best that we could. Late each afternoon there was an alert on the PA system "Bingo, Bingo, Bingo!" notifying the men that Bingo would be held in the mess hall that night. Some nights there was entertainment from the troops who had musical talent. Neither diversion gained a large following.

There were numerous fire drills, of course, that began almost as soon as we hit the high seas.

Highlights of that trip were written in letters home:

Friday, Sept. 23

"We are about 350 miles off the coast. Quite a few guys are getting sick. The ship has a nifty newspaper that it publishes each day.

They are feeding us by compartments now and it's a lot quicker getting our meals."

Saturday, Sept. 24

"We're approximately 5500 miles from Okinawa, where they told us we would be refueling. Monday we're due to take our next malaria pill, which we take once a week."

Sunday, Sept 25

"I went to mass this morning and we all received rosaries."

"... Weather changes frequently. Raining one minute and the sun shining the next.

"... Not much to do except take naps. Very little work done today."

Monday, Sept 26

"The temperature is warm enough where we no longer need a field jacket. Today we're having the roughest seas yet. I'll guess that maybe 10 % of the troops are seasick now. The rest of us are just queasy. No movies tonight because it's too windy to set up a screen. I spent the evening on deck singing songs with some of the men."

Tuesday, Sept. 27

"The temperature is 81 degrees. We are traveling further south and it is affecting the temperature.

... We are now just north of the Hawaiian Islands and will pass through the northern islands either tonight or early tomorrow morning.

They're having boiler problems. Actually, we lost the use of one of our screws, (props) and it is slowing us down considerably. We are now approximately 2,500 miles from Tacoma with about 3,500 miles to go to arrive at Okinawa, our refueling stop."

I must have been getting those figures off the ship's newspaper, which was quite informative and published daily.

"Water is becoming a problem now. We distill seawater to get fresh water and we are using it up too fast. They ask us to conserve." Sudsless salt-water showers are the results of this shortage. It would remain a problem for the remainder of our cruise. Strange to rub soap and not produce suds!

International Date Line

Wednesday, Sept. 28

"There is now seven hours difference between where you are and us in the Pacific. By the time I arrive in Vietnam, there will be 13-hour difference and I will actually be a full day ahead of you once we pass the International Date Line."

"... We passed by Hawaii last night while I was sleeping but one of my buddies picked up a Hawaiian radio station on his portable radio. He said they played the same popular music as the stateside stations do.

"We are crossing the International Date Line tonight which means we lose Thursday, Sept 29th. Tomorrow when we wake up it will be Friday the 30th. We'll get that lost day back when we cross it again going home."

Thursday, Sept. 29

"We received our Domain of the Golden Dragon Certificates this morning. This is what it says on the manuscript:

This is to certify that (name) has on this 28th Day of September, 1966, being aboard the USNS NELSON M. WALKER, entered into the Domain of the Golden Dragon at the latitude of 30 degrees and 03 minutes and the longitude of 180 degrees.

Therefore, let it be known that upon doing so, he has satisfactorily undergone all requirements and has been duly initiated into the Brotherhood of the Domain of the Golden Dragon.

Signed: *Captain G.B. Swortfiguer, Ship's Master*
Witnessed by: *King Neptune*

The map at the start of this chapter shows a star laid out on the dateline. That figure represents accurately where we crossed the International Dateline.

**Here is an interesting oddity. If you were really efficient getting connections to flights, and didn't live far from the west coast, you could actually get home before you left Vietnam. I just missed arriving in Boston before I left Vietnam. I arrived at the airport three hours after I left Vietnam (by the clock and date) with the help of that regained day I lost in September, the year before.*

Friday, September 30

"We won't be able to receive mail until we reach Vietnam, but we'll be able to get our mail off to you in Okinawa, which they predict will probably be Tuesday."

My next letter is dated Tuesday, October 4th. In the letter I write about a failure in the ventilation system and the lack of air down in our compartment. Our Company was situated at the lowest troop level, four decks below the top deck of the ship.

Tuesday, October 4

"Our compartment has 300 men cramped in it. There are no port-

holes for air. We have air forced into the bay by a ventilation system and they were not working last night. I was sleeping when my friend Johnny Martel awakened me in a panic. "There's no air down here. I'm getting out of here." It was stifling but I didn't believe it was threatening. As I climbed down from my bunk to join Johnny on deck, I awoke 'Skip' Barnhart, our Company Clerk, to warn him. He was already awake and was hanging half in and half out of his bunk. His response to me was "Go, save yourself. I don't have the energy to leave this bunk."

"There were men moving about all over the compartment by then. Johnny and I made our way to the top deck and slept on deck on a bundle of heavy rope for about an hour. Then we returned. The ventilation system was working fine by then."

Wednesday, October 5

"We're due in Okinawa tomorrow, a day later than what they predicted. The loss of that screw is really slowing us down. We're around 500 miles from there now.

It's raining on and off this morning and it's keeping the temperature nice and cool."

Thursday, October 6

"It is morning now and we are nearing Okinawa. I'm listening to an Okinawan radio station and it's broadcasting the World Series. It comes in loud and clear. (Back home it was October 5th and the first game of the World Series was being played by the Orioles and Dodgers. The Orioles won 5–2, and the Dodgers would not score another run in the remaining three games of the sweep by the Orioles).

We're told we will be arriving in port at 10 am and we'll be there for about 12 hours to refuel.

By Sunday or Monday, we should be in Vietnam."

Friday October 7th

"We just left Okinawa an hour ago at 7 am (it took longer than predicted to complete the refueling). I took a picture of two Japanese Navy ships as we pulled out of port…

…They only let us off the ship to do PT, and boy, did they work us out. The workers on the dock seemed friendly. One of the workers seemed to be taking a tally. 1SG Springer made us pass by him twice to screw up his count."

Here is something that we only learned later. In the past they were letting troops off the ship to tour Naha while the ship was being refueled. They told us the last time that they allowed that port visit it took them hours to round up the tardy soldiers, delaying their departure. We wouldn't be allowed that opportunity to get 'lost'.

"…As we pulled out of Okinawa, the mood seemed to be more somber than before. It's as if we all realize that our next stop will be Vietnam."

Saturday, October 8

"Last night we picked up an escort, which we can see off in the distance. The further south we travel, the warmer it should be, but it isn't much warmer than it's been for the last two weeks. The difference is the intensity of the sun. If you stay to the shady side of the ship, it's rather pleasant."

Monday, October 10

"It's 9:30 am now and we have our first view of South Vietnam. We've been following the coastline of Vietnam for around two hours. Before me I see a huge mountain range. It's clear out here but you can see clouds hugging the tops of these mountains. Now we are seeing a long narrow strip of sandy beaches for as far as we can see.

...As of now we're not sure how long we'll stay on board in the Vung Tau Bay, where we will anchor. They say it may be 4 or 5 days before we get off the ship."

Tuesday, October 11: Arrival At Vung Tau, Vietnam

We spent the next three nights off the coast, as one by one, the units of the brigade were brought ashore on landing craft. We were the final battalion to land. It gave us a lot of time to ponder what laid ahead. The evenings were eerie as we witnessed flashes in the distance hills. These were followed by the 'chunk' sounds that we would later identify with artillery shells hitting the ground. The flashes and far off booms reminded us what lay ahead for us. For these evenings, we were mere witnesses to a struggle. We would be participating soon enough.

During this three-day period, we underwent training to learn how to safely disembark our troopship using nets to climb to a barge. From the barge, we would board landing craft, which would bring us to the shore.

We were told that we would initially be sent to a base camp built by the First Division when they arrived in July of 1965. The name of the base camp was called Bearcat (also known as Camp Martin Cox) and we would be manning it to free up the 2nd Brigade of the 1st Division who were serving as a relief force for the 196th Light Infantry Brigade. The light infantry brigade was building their base camp, near Tay Ninh. The Brigade had arrived in Vietnam in August and by September 2nd were conducting Phase One of Operation ATTLEBORO. They were collecting intelligence indicating a major enemy build up in the area just north of a rubber plantation owned by the French tire makers Michelin, near a sleepy village called Dau Tieng.

Phase one of that operation kept the 196th LIB mainly in the area near the Black Virgin Mountain. On October 17th, in response to new intelligence, the 196th was given the mission to explore the area north of the Michelin Plantation. This region was suspected to encompass one of the enemy's largest storage and shipping sites.

With us serving as garrison defenders, the First Division Brigade at Bear Cat would be sent it to reinforce the 196th if need be. By early November, the situation for the 196th LIB would become extremely dangerous and other units of the First Division would be required to be sent up north into that enemy hornets' nest. The operation in the Tay Ninh/Dau Tieng area was developing quicker than MACV had anticipated. I believe that is why as we neared our destination drop off point in the Central Highlands, our brigade was diverted to the south where we would participate in the operations in War Zone C.

As we stood on the deck on these evenings viewing the hills surrounding Vung Tau, the stage was already being set for our mission in Vietnam. Operation ATTLEBORO would be later known as the Battle of Dau Tieng. The Third Brigade of the 4th Division would be sent to this remote area 90 miles north of Saigon to engage and disrupt the activities of the 9th VC Division, the major Vietcong force threatening the country's capital, Saigon.

We were ignorant of all of this activity as we watched the intermittent firefights taking place in the distance.

All we knew for sure was that our brigade had a change of plans. Along the way to Vietnam, we were told that our destination had changed. We were not going to the Central Highlands as planned. Our destination was further south where we would be placed under the operational control of the 25th Infantry Division, which was headquartered in Cu Chi.

FIRST MONTH IN VIETNAM

"Welcome To The Party"

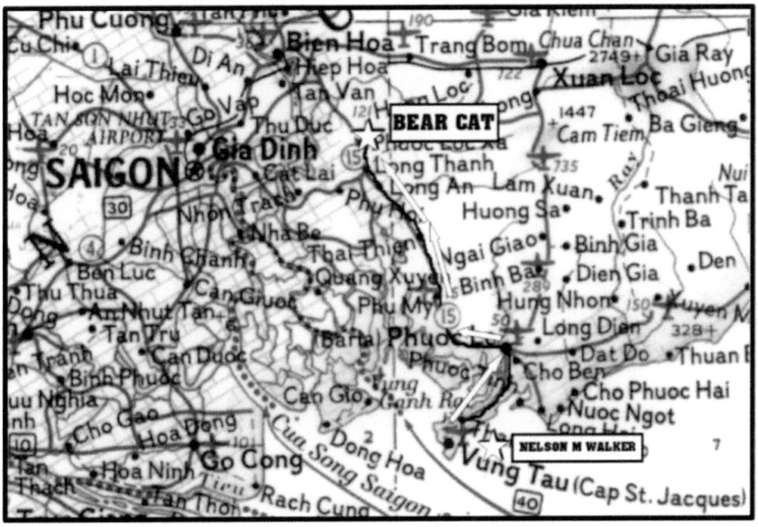

Each day starting on October 12, another battalion unit of the brigade would be sent ashore. On October 14, the 2/12th would be the final battalion to disembark.

We had been training to climb down ropes to barges correctly, holding only the vertical ropes. This prevented the man above you from stepping on your fingers as he maneuvered down the netting. Small point, but important.

Once we were in the craft, we couldn't see much looking forward as our view was blocked by the ramp. Basically, we didn't know what to expect once the ramp was dropped, after watching so many John Wayne movies when we were kids. We were pleasantly surprised when as soon the ramp hit the beach a military band began playing John Philips Souza 'The Stars and Stripes Forever'. We also saw two lines of welcomers cheering us on. We passed through the greeters to much applause.

There were young ladies all dressed in 'ao dai' attire. These dresses were a tight fitting, split tunics worn over trousers. Most wore white with matching trousers. We marched past the cheering greeters and assembled near our transportation vehicle parked nearby. We had a few minutes of last-minute instructions from our leaders.

We were told that there would be a number of infantrymen along our route to Bearcat to protect our convoy. (*In fact, we were protected by a number of men from the 1st and 25th Divisions and the 173rd Airborne Brigade, who we saw at intervals along our route.*) We were each given one bullet. We looked down at the single round in disbelief. They finally explained to us that should we be attacked in an ambush; we were to dismount the deuce and a half and to jump into the ditches on the sides of the road. The men riding shotgun and nearby infantrymen would deal with any attack. We were told that if there was an attack there may be a panicked reaction from us new arrivals and they didn't want to deal with a lot of friendly casualties. That seemed like a prudent move after we thought about it a bit.

We traveled on mostly dirt roads heading north, passing by a combination of densely populated districts and isolated villages along the way. Every time we passed some Americans soldiers on the side of the road, we heard numerous, "Welcome to the party" proclamations.

Somewhere near the base, we followed the road around a bend to our right and spotted a small group of huts on our left that made up a small village. There was an old lady standing in front of her hut chewing on something that I would later learn were betel nuts. Those nuts were, and still are, a popular drug that alongside nicotine, alcohol, and caffeine, are believed to be one of the most popular mind-altering substances in the world. I can still picture her today, chewing these substances vigorously and shouting at us and shaking her fist at us. It was obvious to us that this elderly woman was not too pleased with our arrival. How could that be? We were told that the South Vietnamese were very happy that we were arriving in Vietnam to prevent a Communist takeover in the south. The answer would be a lot more complex than what we could reason with so little background into where the South Vietnamese attitudes in the small villages were when we arrived.

When we arrived inside Bearcat, the first thing I noticed was a huge American flag flying high near our tents. That flag planted on land in a different country really hit home. Here I was as part of an American expeditionary force to help establish American foreign policy. The Communists of North Vietnam had dreams of expanding Communism in South Vietnam and we were here to halt the spread. To me, that was how I looked at it. It was one of the prouder moments that I felt in Vietnam.

On the first evening at camp, I was sent to relieve a First Division radioman in the Communications bunker. There I would maintain contact with company ambush patrols which were sent out at night to alert us to enemy movement near the base.

We got into some small talk and I asked him what he thought about his time in Vietnam and the Vietnamese people.

He looked at me for a moment and said, "I've been here in Vietnam for six months, but I finally figured out how to win the war. You just collect all the Vietnamese who are supposed to be on our side and put them on ships and place them offshore. Once they are safely in place you send over B52s and pulverize the country with bombs. Once that is completed, you go offshore and sink all the boats."

He waited to see what my reaction was and I just had a blank look on my face. He finally said, "You'll get it after you're here a while." Obviously here is a man who learned to be suspicious of the Vietnamese citizenry. I didn't pursue it anymore and shrugged it off. This man must be serving a very long tour in Vietnam with that attitude.

From a letter I wrote home on October 22,1966:

> *"We received jungle boots today." I made no mention of jungle fatigues.*
>
> *...At 9:45 PM, the men who manned the berm in the C Company sector were engaged by enemy who penetrated the barbed wire outside the berm. A red star cluster was shot in the air indicating an attack was taking place, the whole camp was put on alert, and everyone was sent out to man the perimeter. I went out with the CO carrying my radio on my back to maintain communications. We remained online until almost 1AM in a driving rain.*
>
> *While this was taking place at camp, something was going on in front of our sector. An ambush patrol from our company's second platoon made contact with an enemy force and maintained action with an enemy force of 20 or so Vietcong. They said the ambush was triggered early and they were hit from the rear. Had they not set up rear security they would have been in deep trouble. No casualties*

on our part except for a member of that patrol getting the wind knocked out of him after not throwing a concussion hand grenade out far enough at the enemy. He's fine now. Later it was determined that there probably was a platoon of enemy sent to harass us that evening. They were split in half, one sent to the C Company sector, and the other two squads sent to A Company's area. "Welcome to the party."

Thursday, October 25th

"... Guess what. My company is going to march in a parade in Saigon on November 1st celebrating National Day. That's like our 4th of July back home."

Sunday, October 30th

"It's now 10 AM and at 12: 30 PM my Company is moving out for Saigon. We're going to be in the National Day Parade and they say we may have free time to tour the capitol. We're still the Honor Company of the battalion and perhaps that was why we were chosen for this mission. We may not be back until Tuesday or Wednesday."

Tuesday, November, Written Upon Our Return

"... I don't know where to begin to tell you about the amazing experience I had in Saigon over the last three days.

We arrived in Saigon after stopping at Long Binh to pick up other Companies chosen to march in the parade. We were allowed to go to the PX there. We bunked up at an Air Force base that had Quonset huts that were used to send men on

R&R and house men who had completed their tour and were going back home. (*Camp Alpha at Tan Son Nhut air base*).

On Sunday evening, we practiced right on the parade route where we would march. We assembled in front of the Majestic Hotel, near the Brinks Hotel, which has been a frequent target for saboteurs over the years and had often been bombed.

I noticed that there were many patrol boats cruising the Saigon River that night. No wonder, as the Capitol was on high alert after Vietnamese intelligence agents in Saigon uncovered 1,400 pounds of TNT secretly stored in the city on this evening by terrorists.

On Monday, our Company Commander made the decision to allow the men of the Company to spend a few hours in Saigon. The one stipulation was that we were required to be back at the base by 7:30 PM. Johnny Martel, and I went into Saigon together.

Needless to say, the men in the Company went a bit out of control and many had difficulty in meeting the deadline. Johnny and I were 15 minutes late because of cab problems and were greeted by the CO who took our names and told us to turn in. Of course, we had no intention of hitting the sack and headed out to the BX to have more fun. Once there, we teamed up with some Australian troops we had met.

You should have seen the Company when we were awakened at 2 AM on November 1 to get ready for the parade. Although all the men did in fact return by this time, some were not in very good condition. In fact, one of our men passed out when we began marching. He was left behind.

During the parade a mortar and recoilless rifle attack took place aimed at the reviewing stand where President Ky was standing.

Actually, the rounds landed two blocks past the stand. *(Actually, this was not entirely true, but was the scuttlebutt then.)*

I didn't even know until it was over because we were a few blocks away when that happened. We were told that eight were killed in that attack. They were packed in like sardines. We heard the rounds going off but from where we were, we couldn't tell what was going on. I was told that was the first time that Saigon had been mortared.

After things settled down, we did march by the reviewing stand where Premier (his correct title) Nguyen Cao Ky and US Ambassador, Henry Cabot Lodge stood symbolically side by side.

It was a memory none of us will forget.

This is what was written in the 1966 Fourth Infantry Division Yearbook about that parade and our participation in it:

November 1, 1966

Not only was there fighting in the jungles south of Saigon, but there was the unique terrorist war in the streets of Saigon itself. November first is Vietnamese National Day and Alpha Company, 2/12 Infantry had been selected to march in a parade in Saigon. It would be a change from the ordinary routine of hacking through thick Delta jungle. A day in the Republic of Vietnam's lively but noisy capital.

The troopers washed their jungle fatigues, (*not true, as we were not issued jungle fatigues by then*) cleaned their weapons and jungle boots...

On October 31st, the Ivy troopers were given the day off to

visit Saigon. The following day was the big parade. It would be one to remember.

The Ivy Division's contingent marched through the streets decorated with Vietnamese national colors and jammed with tens of thousands of cheering, waving people. Suddenly there was a strange sound. The whoosh of recoilless rifle rounds overhead. Then the explosions somewhere nearby. It was an all too familiar sound.

Lt. Allyn J. Palmer, the Company Commander turned his head nervously. So did the guidon bearer PFC Larry Savage. But Alpha Company didn't falter or break cadence. They had for all practical purposes witnessed a Vietcong terrorist from recoilless rifle rounds fired into the center of the city in an effort to frighten and intimidate the Vietnamese on the very day they were commemorating the establishment of their Republic.

It was that kind of war."

Whoever wrote that piece took a lot of liberties when writing about our involvement on that day. We were quite a distance from the targeted reviewing stand when the attack took place.

This describes our understanding of our parade experience based on what we were told and saw on that day. There was a great deal of history leading up to that day. In a January, 2020 Alpha's Pride newsletter, I shared the real story of that attack 54 years earlier after an exhaustive investigation of how and why that incident took place.

The History Behind The Parade

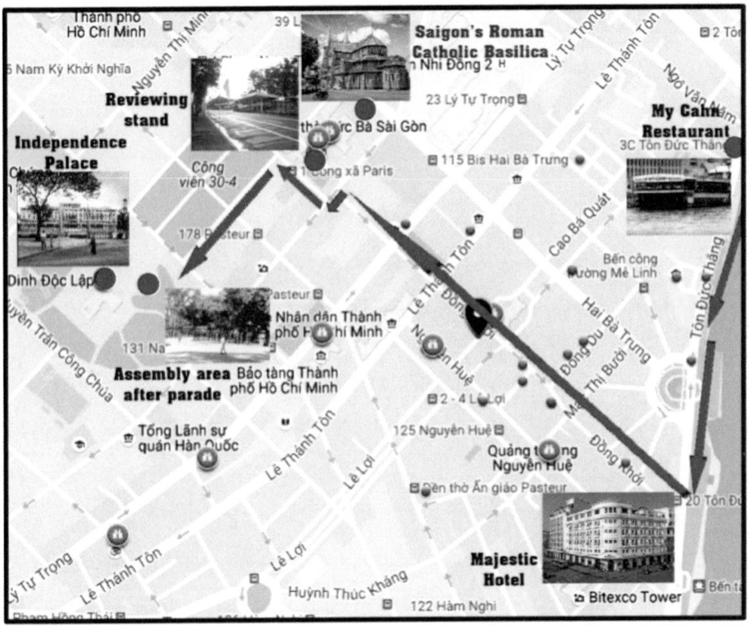

The real story that none of us knew as we marched in that 1966 parade goes back to February 27, 1962. On that day, two renegade South Vietnamese Air Force pilots commandeered two Douglas A-1 Skyraider bombers, flew up the Saigon River, headed up Tu Do Street and hung a left. They then dropped two 500 lb. bombs on the palace, hoping to kill the president of South Vietnam, Ngo Dinh Diem and his hated brother Ngo Dinh Nhu (who could forget his reviled wife, Madame Nhu?) The attack lasted almost an hour as they kept climbing above the 500-foot cloud ceiling then returning to deliver napalm and additional bombs.

They had enough ordinance to destroy the entire palace but were driven off by the palace guards. One bomb penetrated a room in the western wing where Diem was reading, but failed to detonate, leading the president to claim that he had "divine" protection. Madame Nhu suffered minor injuries. Three of the palace staff were killed and thirty others were injured.

The first pilot, Pham Phu Quoc, was shot down and became a folk hero after Diem was assassinated. He was killed later in the war in a raid over North Vietnam and a famous song was written about him ('Epic song of a man named Quoc'). Nguyen Van Cu, the second pilot made it all the way to Cambodia where he received political asylum.

Now the story of that palace going back to 1858 when the French were looking to expand their empire attacked Vietnam (then known as Cochin China). After successfully conquering the southeast Asian nation, they began to set up their governance of the colony. Part of that effort was the building of a magnificent palace that would eventually house 800 people. As a result of the Franco-Prussian War of 1870, the palace was not completed until 1873. The structure was named Norodom Palace after then king of Cambodia, Norodom.

In March of 1945, the Japanese kicked out the French and used the palace to govern their new colony. In September, 1945, the Japanese surrender and the French return to the palace to resume their exploitation of their earlier conquest. As a result of the agreement to end the French-Viet Minh War in 1954, Vietnam was partitioned in half with an agreement to hold elections later to determine which way the country would be governed. Ho Chi Minh, was made leader of the Communist division in the north known as the Democratic Republic of Vietnam.

The southern sector became known as the Republic of Vietnam. Control of the south was handed over to Ngo Dinh Diem. As geography matters, he inherited the magnificent Norodom Palace, now retitled Independence Palace in Saigon.

Ho refused to live in the residence that was given to him in Hanoi. He thought it looked too European. Ho still received state guests there, and eventually built a traditional Vietnamese stilt house and carp pond on the grounds.

Diem had no such qualms about living in the luxurious Norodom Palace after the French were expelled from Vietnam. Upon taking office, Diem quickly developed a reputation for using force rather than democratic means to initiate change. Beginning in 1955, he used ARVN troops to reverse Communist land redistribution in South Vietnam and return landholdings to the previous owners. Fearful of Viet Minh popularity and activity in rural areas, which had increased as a result of Diem's cancellation of the scheduled 1956 elections, Diem uprooted villagers from their lands and moved them to settlements under government or army surveillance. He forcibly drafted many of these peasants into the Army of the Republic of Vietnam, increasing his unpopularity in rural areas even further *(remember that old lady we saw on the way to Bearcat?)* Diem's government was also unpopular because it had an overwhelming Catholic bias and contained several unpopular, key figures who were members of Diem's own family, the Ngo family. Although Catholics made up less than a tenth of the Vietnamese population, Diem himself was Catholic, as were all his other family members in the government. Diem's government engaged in often vicious persecution of Buddhists, who made up the overwhelming majority of Vietnamese citizens, particularly peasants. Diem's brother Ngo Dinh Thuc, the influential Catholic archbishop of Hue, in particular came into conflict with Buddhists. Diem continued his nepotistic trend by installing his youngest brother, Ngo Dinh Nhu, as the leader of the government's secret police organization, the Can Lao. Moreover, because Diem himself was not married, his sister-in-law, Nhu's wildly unpopular, Francophile wife, Madame Nhu, became South Vietnam's de facto first lady. In the years that followed, Madame Nhu would emerge as a notorious figure in Vietnam and on the world stage; arrogant, extravagant, and prone to nasty, on-the-record comments, she created one public relations disaster after another for the U.S. backed Diem government.

Diem and his brother brutally administered the country and in 1959 under the guise of rooting out Communists in the country, they non-discreetly arrested even innocent citizens. It was mainly an effort to suppress discontent. The ruthlessness of this effort became common knowledge and by 1960, the NLF (*National Liberation Front*) was born and our adversary for the coming war was placed on the chess table. To make matters worse, many of those new recruits were formally anti-Communists.

After the assassination attempt of February 27, 1962, Diem and his administration left the rubble of the palace and ordered the palace torn down and replaced. They didn't have far to go as they took up residence at the Gia Long Palace, a French palace a quarter of a mile southeast of the palace. (*Those Frenchmen loved their palaces, non?).*

However, Diem did not have a long-term lease on the palace as a long simmering military coup d'état reached a climax on November 1, 1964. Does that date sound familiar? That was what we were celebrating marching through the streets of Saigon in 1966. That was the day that troops of the 7th ARVN DIV, led by Nguyen Van Thieu, paid an early morning visit to Gia Long Palace to tell him that they had a surprise for him. Diem had other ideas and a battle erupted. It went on all day and that night, Diem and his entourage escaped via an underground passage to Cha Tam Catholic Church in Cholon. They were located on the next day and Diem and Nhu were executed inside an armored personnel carrier. It wasn't pretty, but political assassinations seldom are.

Madame Nhu lost all of her property and holdings and was sent into exile. She ended up on the French Riviera in France, where she died on April 24, 2011.

Nguyen Van Thieu eventually became President of South Vietnam from 1965 through 1975 when the North Vietnamese came crashing through the palace gates at Independence Palace. The palace was quickly renamed Reunification Palace by the victorious North Vietnamese. Today the palace is open to the public each day from 07:30-12:00 then 13:00-16:00. Tickets are purchased at the main gate and visitors will be checked for security reasons.

Which brings us back to our story right after Captain Jon Palmer released us to Saigon on the day before the November 1st parade.

Now you know what the celebration was all about when we marched in the parade, but you don't know how big a tinder box Saigon became while we were touring Tu Do (Liberty) Street. The street was named after the French were driven out in 1955. In 1975, after the Communists took over, it was renamed, 'Dong Khoi' (General Insurrection) street. *What a country!*

The Communists knew that the National Day, 1966 celebration, two years after the fall of Diem was going to be a big deal for the South Vietnamese government. The world would be watching and they were determined to join the festivities and planned accordingly.

While us enlisted men were hooting and hollering on Tu Do Street on the day before the parade, Jim Olafson, our XO, was taking in some of the culture of the city. Unbeknownst to him, (*I know because I asked him*) he was shooting that photo of the palace on the same day of its grand re-opening. They rededicated the newly built Independence Palace on the morning of Monday, October 31, 1966 more than four years after it was bombed. Jim and his entourage just missed the event, arriving in the afternoon.

In the New York Times story of the rededication of the palace, I learned how the Saigon government was really scrambling to organize National Day and the six days of events surrounding it:

"The commemoration of the overthrow of Ngo Dinh Diem were hampered with dozens of mishaps. The overburdened South Vietnamese bureaucracy struggled with the added problems of organizing spectacles.

As an example, someone forgot to bring the keys for the massive gates of Independence Palace to the rededication by Premier Ky and General Thieu. The official party arrived only seconds after two harried workmen had finished cutting through the bolts with hacksaws."

So how did they fill in six days of events to celebrate the overthrow of Diem? They did so by organizing talent festivals, rowing meets, tennis tournaments, receptions, band concerts, and speeches ... many, many, speeches.

So, as we are trickling back into Camp Alpha with fond memories of touring the capital, the South Vietnamese National Police were busy rounding up Vietcong saboteurs. In a letter that I wrote late on November 1, right after returning to Bearcat, I wrote this about the night when we did a practice march along our route. "I noticed that there were many patrol boats cruising the Saigon River that night." It was no wonder, as Saigon was on high alert after Vietnamese intelligence agents in the capital uncovered 900 pounds of TNT secretly stored in the city on this evening by terrorists. "The South Vietnamese National Police tonight announced the capture of 10 leaders of a Vietcong terrorist gang and nearly 900 pounds of TNT. Maj. Nguyen Thien, Director of National Police Operations, said his men arrested the terrorists in Saigon yesterday. Major Thien said his men discovered the TNT in three caches along the Saigon River. He said that, in all, 1,476 blocks of explosive, each about the size of a bar of laundry soap had been uncovered. The blocks, all unmarked, weighed 880 pounds together. Vietnamese sources said the TNT had been wrapped in old inner tubes and buried in mud banks. The Vietcong have usually used 22 to 88 pounds of explosives in most of their bombings in Saigon and seldom more than 100 pounds. The explosives the police took in the last 24 hours would thus have been enough for 10 major attacks. one of the bars of explosives is enough to blow up a jeep." (*Noted on that same newspaper of Nov 1, thirty-eight Americans were killed in Vietnam on that one day. The average American casualties per day for all 1966 were 17 Americans killed. In 1967, the average jumped up to 31 KIAs per day. During the year 1968, when the Tet Offensive took place, the losses mounted to a politically unsustainable 45 American troops losing their lives in Vietnam per day*).

MSG Pete Coxon, a member of the USAF Military Police was part of the security force protecting the parade route on Nov 1. MSG Coxon and his men were set to march at the very end of the parade near our position, near a notorious restaurant called the My Cahn Floating Restaurant. That restaurant was very popular with the middle and upper-class citizens as well as Americans. Unfortunately, it was also a popular NLF target.

On July 25, 1965, the restaurant was bombed by sappers. First, they triggered a bomb within the restaurant and as the panicked patrons rushed to the gangplank, they were greeted by a claymore mine. In all, forty-two people were killed and eighty others were wounded.

Peter Coxon mentioned that his unit was placed at the end of the parade and said there was another Air Police Unit in the front and that was by design. They were to provide security for all the other units in the parade as they were the only groups that were allowed to carry ammunition in the parade. Our company didn't have a bullet amongst us.

The Air Policemen tried to swap duty with others as they were keenly aware of the dangers presented by marching in a parade where a terrorist could easily toss a grenade within their midst and escape unnoticed into the heavy crowds. No one would take up their offer.

We were still too naive to understand how exposed we were to danger.

They arrived, like us, somewhere between 4 and 5 AM. It was still dark and we found a patch of grass to lay on along the river street near the My Cahn Restaurant to await the 8AM startup.

I don't imagine that we were more than four blocks from where the next story unfolded, yet it didn't make much of an impression because none of us could remember it later. I guess we just thought that this was a typical Saigon day. It wasn't.

"After about an hour (*actually the attack began just after 7AM*) we heard loud explosions rip through the city, coming from the direction of the parade stand. That was followed by the sound of emergency vehicles responding to the scene. About five minutes later, the 716th Military Police and QC's (Vietnamese Military Police Corps) came tearing down the street in vehicles and deployed on the floating restaurant. Covering each other, they occupied the restaurant, oblivious to us watching them.

When they finally noticed us, they came over and asked if

we had seen any Viet Cong activity originating from the floating restaurant, as counter-radar mortar had indicated that the mortars fired at the parade stand had come from the floating restaurant. We assured them that nothing had been observed at the restaurant, as we had been there for the last hour.

About an hour later, we again heard loud explosions coming from the direction of the parade stand. We immediately visually checked out the floating restaurant which revealed nothing. Sure enough, a short time later here came the MP's and QCs again. This time they came over to us and asked if we had seen anything as the radar was still indicating that the rounds were being fired from the floating restaurant. We assured them that we had not.

So, what did take place on that day in Saigon? Actually, it was a much more eventful day than we had been told or suspected?

Here is the actual story that was written in newspapers back home about November 1st in Saigon. Foreign Correspondent R.W. Apple, Jr., of both the Wall St. Journal and New York Times described the attacks of November 1st this way:

"Vietcong terrorists disrupted South Vietnam's National Day celebration today, firing recoilless-rifle shells into downtown Saigon for the first time. Other insurgents threw a grenade near the crowded bus terminal in the central market place, causing heavy casualties. A preliminary canvas of Saigon hospitals indicated that at least eight persons had died of terrorist incidents—five Vietnamese men, two Vietnamese women, and an American naval commander."

Commander Richard John Edris, was the man killed instantly in front of the basilica near the reviewing stand during the initial volley of shells on Nov 1, 1966.

He was born in Pottsville, PA, and raised in Ohio. He began

his naval career in the NROTC unit at Ohio State University, with his graduation and commissioning in 1951. As part of the Naval Military Assistance Command Vietnam (NAGMACV) he developed some of the earliest operational plans for the swift boat operations on the Mekong River.

He was buried at Arlington National Cemetery. His wife, Nancy, never remarried and when she died in 2015, forty-nine years later, she was laid to rest alongside the commander.

"…On Longtau River, Saigon's main shipping link to the sea, the Vietcong mined and sank a United States Navy minesweeper early this morning. A Military spokesman said that the crew of the 57-foot vessel had taken heavy losses. The river was closed to shipping until later today.

The attack in Saigon began just after 7AM, about an hour before the scheduled start of a huge military parade. It lasted seven minutes. None of the rounds struck the stands or the parade route, but one exploded within 100 yards of the seats where Premier Nguyen Cao Ky and Lieutenant Gen. Nguyen Van Thieu, the Chief of State, later sat."

All these incidents took place and we knew or were told nothing about the attacks on this day. The only thing that we knew as we boarded the deuce and halves is we marched in a parade that was uneventful while we marched in it. We were told that there was a mortar attack originating from across the Saigon River but we never were unduly concerned about the dangers that we faced on that day. Probably just as well.

Meanwhile, while we were riding back to Bearcat, apparently the Saigon Government was also not concerned, even if they knew more than we did. On the following day this dispatch was published in the morning newspapers:

SAIGON HOLDS DINNER DESPITE AN ATTACK

Saigon, South Vietnam, Nov.1—Outwardly unshaken by the first Vietcong shelling of the center of Saigon, the South Vietnamese Government staged an elaborate reception and dinner for its own dignitaries and foreign guests tonight. The affair was held at the newly rededicated Independence Palace less than a half a mile from where the two barrages took place earlier in the day. Casualty estimates varied. Tonight, military sources scaled down previous estimates and listed seven Vietnamese civilians and an American Navy Commander as dead and 32 Vietnamese and five Americans as wounded. Seven Vietcong suspects were arrested near the shelled parade area.

OPERATION ATTLEBORO

Largest Operation To Date

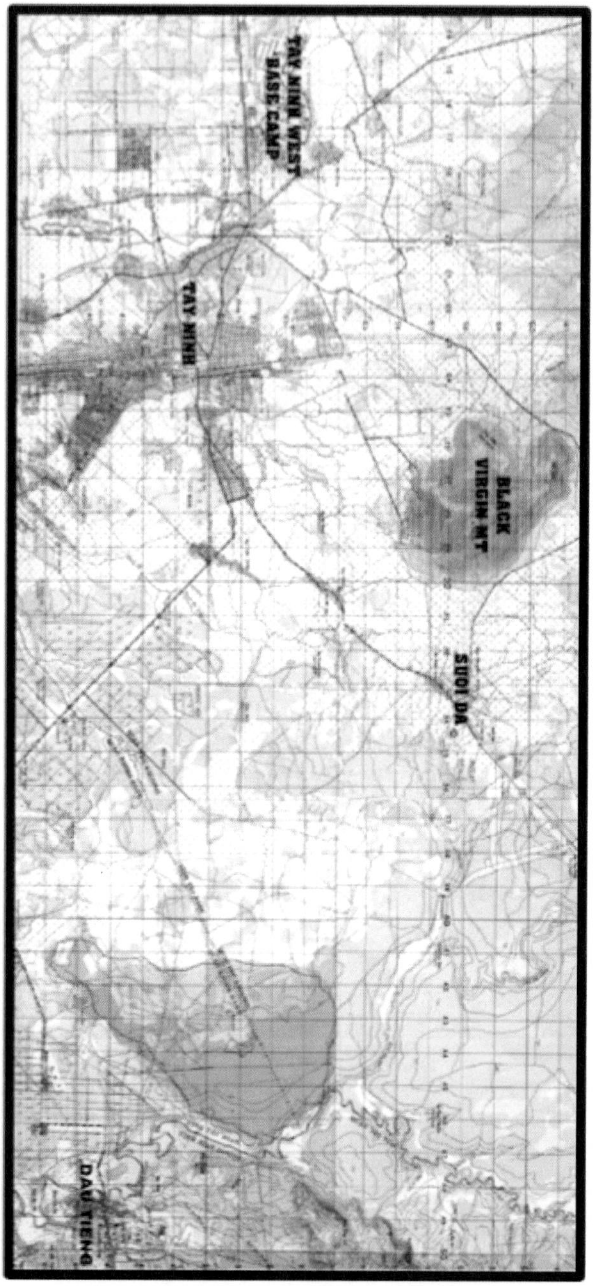
TAY NINH WEST
BASE CAMP
TAY NINH
BLACK
VIRGIN MT
SUOI DA
DAU TIENG

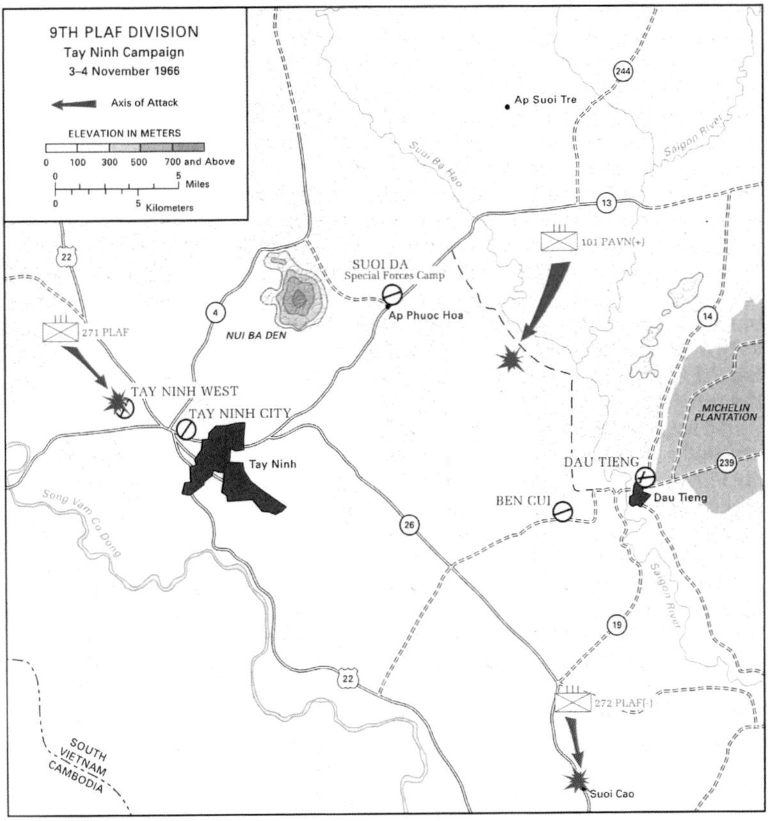

The 196th Light Infantry Brigade (LIB) was part of the Reserve 98th Division during World War I and only trained troops and did not participate. During World War II the brigade defended Hawaii after Pearl Harbor, and again saw no action.

It was deactivated in February of 1946. It was reactivated on September, 1965 and trained at Fort Devens, Massachusetts. Their original mission was to be sent to the Dominican Republic, again avoiding combat during the Vietnam War. They were not so lucky this time.

After training, in a surprise move, it was sent to Vietnam on a troopship and arrived on 14 August 1966. The brigade was sent to Tay Ninh, a Provincial Capital, where it was directed to build a brigade size base camp northwest of the city. That camp became known as Tay Ninh West. Unbeknown to them, this movement was part of the grand scheme to establish a forward base for the much-anticipated operations scheduled for 1967. Dau Tieng would be the second forward base serving in the upcoming offensive.

The 196th LIB brigade was commanded by Brig. Gen. Edward H. DeSaussure. DeSaussure was a superb artillery officer but was lacking in infantry tactics. This became a real problem as the situation escalated quickly during his unit's first operation. He was not prepared to handle the rapid action that his unit faced in the Fall of 1966. It would cost him the command of the 196th Light Infantry Brigade at the end of the operation.

General Westmoreland was worried about a dry season offensive that he expected to begin in War Zone C. That area along the Cambodian border had long served as a staging and a supply depot for Vietcong soldiers threatening the Saigon corridor down south.

To preclude that attack, he ordered the 196th LIB, which was constructing their basecamp, to send two of their battalions into the field to test the enemy's activities. That operation came to be known as Operation ATTLEBORO, named after the Massachusetts city where DeSaussure lived while he was training his troops at Fort Devens, Massachusetts. Naming operations after familiar cities of Commanders was normal at that time.

The first phase played out during the month of September-October, 1966. During this phase, the brigade committed two of his battalions to the operation and left his last back at camp to continue working on the camp. For the most part, those months produced little enemy contact. However, many large stores of enemy supplies and equipment were located and destroyed.

After searching the Tay Ninh area for enemy bases, DeSaussure sent one of his battalions eastward in an area west of the Michelin Plantation that was laden with thick elephant grass and heavy jungle. During this period, the search uncovered a remarkable piece of intelligence that would open up the area for another American operation. During their probe, they came across a hut on the edge of the jungle which was occupied by an old man. After the South Vietnamese interpreters questioned him, they discovered that he was a Vietcong tax collector. Inside the hut was a map showing the location of many of the supply bases hidden north in War Zone C. Time would be critical as the enemy would soon learn of this incident. In addition, they located a major bomb factory hidden in the jungle, not far from the clearing. I will have more to say about that factory in a future chapter.

Westmoreland was correct in predicting a major attack in the area northwest of Saigon. Nguyễn Chí Thanh was a senior Communist party member. As the war scaled up significantly with the introduction of massive America troops, Thanh was sent to South Vietnam. There he served as the Secretary of the Central Department of South Vietnam, and the Political Commissar of the Liberation Army of the South. He was one of the North Vietnamese war hawks who pushed for victories, whatever the cost.

If you served in the jungles of Vietnam and wondered why the enemy was always so close when the action began, it was because of Thanh. He was the leader who promoted the 'grab them by the belt and fight' strategy used by the Japanese in World War II to overcome American fire superiority.

He was present in South Vietnam during the summer of 1966 when the 9th VC Division suffered heavy losses. Regardless of the losses, by the fall of 1966 he was pushing hard for a major strike against American forces. General Võ Nguyên Giáp, commander in chief of the People's Army of Vietnam, advised restraint but did not have the political muscle that Nguyễn Chí Thanh had with the war planners in Hanoi. Thus, the stage was set for a major offensive set to take place in War Zone C.

Thanh and Giap bided their time during Operation ATTLEBORO/phase one, as the Americans uncovered huge caches and supply bases in the area around Tay Ninh and Dau Tieng. During this time, the North Vietnamese four-star generals organized what they hoped would be a demoralizing defeat that would energize anti-war demonstrators in the U.S.

Their target would be the newly arrived 196th LIB, busy building their base camp near Tay Ninh. In addition, they targeted a CIDG unit that was situated in Suoi Da, northeast of Tay Ninh. The mission was assigned to the 9th VC Division their most reliable and experienced force in play then.

Senior Colonel Huang Cam, the Commander of the 9th VC Division, was to carry out a three-pronged attack in the Tay Ninh region.

The attack was to begin on November 3, 1966, when Cam would send his 271st PLAF (Vietcong) Regiment numbering 1,500 soldiers to Tay Ninh West to attack the 196th LIB's reaction force which was busy building the base.

The plan was to annihilate the unit and win a decisive victory. His 272nd PLAF Regiment with two battalions were to link up with the 14th Local Force Battalion, the provincial VC unit for Tay Ninh Province. Their mission was to attack the South Vietnamese territorial outposts at Suoi Cao, 18 miles southeast of Tay Ninh City. This clash was to be mainly a diversionary attack. The final attack force was composed of the remaining battalion of the 272nd VC Regiment and the 101st PAVN (North Vietnamese Regular Unit) which was an inexperienced anti-aircraft and mortar company which had just arrived in South Vietnam. They were to attack the CIDG base at Suoi Da and destroy that force. Before that attack was triggered, the 196th LIB was successfully locating many supply depots and enemy bases in the Dau Tieng area. By October 30, the operation was formally declared a full-size brigade operation. They still had that map which would lose its importance as time passed. With that in mind, on November 1, 1966, American Commanders in the 25th Division loaned them one of their battalions (the 1/27th Infantry Battalion of their 2nd Brigade) to help in the search and to push north. The Commander of the 1/27 Infantry, Major Guy S Meloy, met De Saussure at the Dau Tieng airfield and was briefed on the tactical details of the operation. He strongly disapproved of the tactics of the operation and made that clear to De Saussure. The LIB Commander took note but would not be deterred. Meloy, bowing to his superior's rank, replied with a "Yes, Sir" and returned to his unit to share the plan with his commanders.

Before long, phase two of Operation ATTLEBORO would become the largest operation to date in Vietnam. All elements of the 196th Brigade, 1st Infantry Division, 25th Infantry Division, and two battalions of our brigade of the 4th Infantry Division participated. Numerous ARVN and Regional Forces/ Popular Forces also became involved.

It was not long before General Nguyen Chi Thanh became

aware of what was building up near Dau Tieng and it forced him to change plans on his November 3 attack…

I'm getting a little ahead of myself. While the two commanders were meeting at the air field at Dau Tieng, we were ending our parade march in Saigon and preparing to return to Bearcat on the eve of November 1.

Phuoc Vinh

TAY NINH
DAU TIENG
PHUOC VINH
LAI KHE
BIEN HOI
LONG BINH
BEARCAT
SAIGON

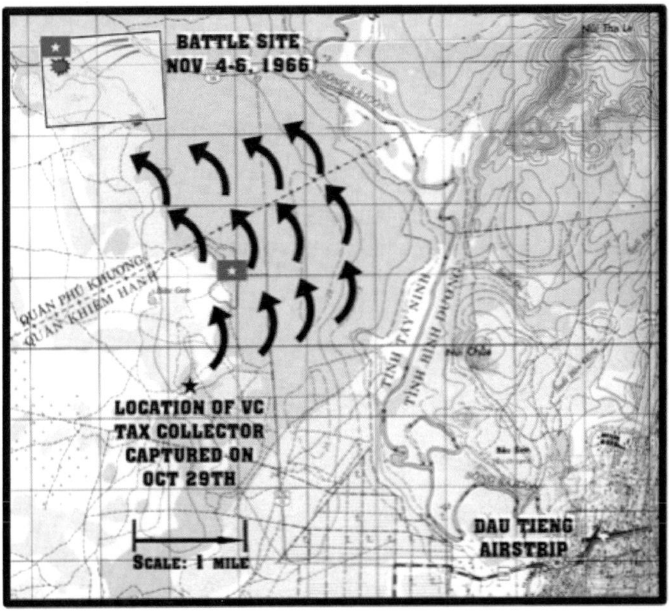
BATTLE SITE
NOV 4-6, 1966
LOCATION OF VC
TAX COLLECTOR
CAPTURED ON
OCT 29TH
SCALE: 1 MILE
DAU TIENG
AIRSTRIP

When we returned to Bearcat after the parade of Nov 1, we learned of a mortar attack on an ammo dump at Long Binh on October 28. That's all we were told.

Long Binh was a massive supply depot, 11 miles in circumference, and was the largest in South Vietnam. It supplied small arms ammunition, mortar, and artillery shells for the American units serving in the Saigon area. It was important and a favorite target for local Vietcong.

This is how the American newspapers reported it at home on the day after the attack:

LONG BINH, South Vietnam—An ammunition storage area exploded here tonight. The explosions rattled the windows in Saigon, 12 miles southwest of the depot…

It was unclear if the explosion was caused by saboteurs who slipped into the base or by the four mortar rounds that were fired into the base that night.

On Nov 2, 1966, the three companies of the 2/12th were transported by vehicles to Long Binh where we conducted our first battalion size search and destroy mission. The three companies entered the jungle north of the base and fanned out looking for whoever caused the damage. I remember that mission for one incident that took place as we searched out the jungle. My company was in a column formation, two platoons abreast. We entered the woodline and worked our way through moderately thick jungle. It was a clear tropical day, with temperatures near the century range with high humidity. After a short while, we heard some small arms fire ring out ahead of the Company HQ group. It did not last long and I received a call on the Company net from the forward position. They indicated that a short firefight broke out with a squad of Vietcong as they came up on a trail. Captain Palmer ordered his two RTOs (myself and Walter Kelley who was carrying the battalion net radio) to follow him up to the front of the company to investigate the incident.

We had an ARVN interpreter and he followed us. When we got to the trail, Palmer was briefed by the lead platoon leader on what had occurred earlier. It seemed that as our men came up to a trail, they noticed a man and woman coming towards them. They were riding in an ox cart and came up to them on the right. As they approached our men, a group of Vietcong came up behind them and began crossing the trail. When the VC noticed the Americans, they quickly turned and fired a volley at us to give us something to think about as they retreated back into the jungle where they came from.

None of our men were hit, but during the action, both the ox and the man's wife were shot. Palmer established security around the area and we walked over to the man and his wife. The man was despondent and tears were streaming down his eyes as we approached with a medic and our interpreter.

That is when I experienced my "Toto, I don't think we're in Kansas anymore" moment. He had his arms around his water buffalo's neck as his wife laid wounded on the ground nearby. She was hit but not hurt badly. I wondered why he was so downtrodden by the loss of the animal, but not by the injury of his wife. I turned to the interpreter to ask him how he seemed to care more for his ox than his wife.

The ARVN soldier didn't seem at all surprised. He looked at me with a smile and explained a little about Vietnamese culture. He said, "That poor man can replace his wife, but probably could never afford another ox to pull his wagon." This was turning into a real-life exercise on the ways of the world for the kid from Massachusetts.

We spent a week on this search and destroy mission, which

for the most part was uneventful. That was not the case for the operation building to a crescendo northwest of Dau Tieng.

Meanwhile near Dau Tieng…Major Guy S Meloy was inserting his 1/27 Infantry battalion into a blocking position when General Nguyen Chi Thanh became aware that much had changed since he devised the November 3 planned attack. He changed the attacks on Tay Ninh West and Suoi Cao to diversionary attacks, using a much smaller force. The intended assault on Suoi Da was completely abandoned. Thanh thought he could achieve a decisive victory against the forces heading for one of his most fortified bases in the area.

He sent the 101st PAVN Regiment, reinforced by an additional battalion, to intercept the American forces operating near Dau Tieng. The mission was to destroy the Operation ATTLEBORO units in turn.

Many books have been written about the ensuing battle and Major Maloy's superb leadership, so I won't go into detail.

I will say that while we were probing the jungle north of Long Binh, Meloy's two available companies (the third was kept at Dau Tieng as a reserve force with another 196th LIB company) were airlifted north of the 196th LIB's probe to serve as a blocking force.

Before it was over, Meloy had taken control of 11 different companies who were involved in the three-day battle. No battalion commander had ever done that before, nor since. The enemy, determined to successfully succeed in Thanh's demand for a decisive victory, sent successive human waves at the surrounded forces.

With Meloy's magnificent control of the situation, Thanh's troops suffered crippling casualties. Still, the enemy would not disengage.

Finally, on the third day of the battle, Meloy was able to

extricate his forces using a brilliant but risky maneuver. Closely coordinating with his artillery support groups, he called in artillery fire directly on his forward units who were alerted the second that the incoming volley left the artillery guns. At that instant, they were ordered to immediately retreat to the rear. They had a few seconds before the shells landed on their old position. When the enemy tried to pursue, they fell into the trap and were cut up.

This battle was the first large scale American/North Vietnamese battle to take place in War Zone C. It led into a multi-brigade operation (the largest up to that time) that chased the enemy survivors of the battle far north into War Zone C (see earlier map).

Brigades of the 1st Infantry Division and later the 25th Infantry Division, the 11th Armored Cavalry Regiment, the 173rd Airborne Brigade, and three South Vietnamese infantry battalions, were all added into the operation.

This is where I need to return to my battalion's participation in the operation…

We returned to Bearcat on November 8 where the 2/12th was immediately put on notice. On the next day, our brigade would be leaving for Phuoc Vinh, the basecamp of the 1st Division's First Brigade.

We packed up all our gear and on the next morning we were trucked to Bien Hoi Airbase. There we boarded C-130 Hercules aircraft where we experienced our first flight in Vietnam. This was to be a different type of experience for us. We took our gear off, placed it in front of us and sat on the floor of the aircraft. In front of us was a sturdy strap securely fastened to the sides of the cabin. We wondered what that was for. We learned quickly upon take off.

The plane taxied to the end of the runway and in short order it lifted off the tarmac. It couldn't have been higher than fifty feet when the nose went up into intense pitch upward which is when we learned the practicality of those straps. We grabbed for those straps and held on for dear life. In Vietnam, it was important that aircraft attained altitude very quickly to evade ground fire from the surrounding jungle. Upon landing, the situation was reversed as we were thrust forward as the plane approached the dirt runway of Phuoc Vinh and dived downward. Again, from a higher trajectory than normal to elude enemy fire from the ground.

We arrived at Phuoc Vinh knowing nothing about why we were being sent there other than a big battle was going on somewhere (they never told us where) and the units that normally would be in Phuoc Vinh were sent to the battle. It would be decades before I learned that the 1/16th Mechanized Infantry battalion, and the 1/28th and 2/28th Infantry Battalions were the units sent to participate in Operation ATTLEBORO.

While at Phuoc Vinh, we were placed under control of the 1st Infantry Division.

Phuoc Vinh contrasted with Bearcat mainly because unlike the sprawling open base camp to the south, Phuoc Vinh base camp was still in its early stages of development. I remember a high berm around the part of the camp where we were sent to pitch our two-man tents and dig foxholes in front of them. My company was set up in orderly rows which would have been a very tempting, can't miss, target for Vietcong mortar men. It never came.

Notable events during our time in Phuoc Vinh:

- We finally received jungle fatigues on Nov 13.
- Lt. Jose Gonzalez, a veteran of tropical combat (he

served in the Congo) advised us to throw away our underwear. The dampness just rots it away in the extreme jungle heat and humidity.

- By then we also realize that watches could produce uncomfortable rashes under the watch. We took to looping the band through a buttonhole of our shirts.
- We experienced our first traversing of rice paddies. We learned to tuck our trousers tightly at the boots to keep leeches out. Upon exiting the paddies, we learned the routine of having our buddies check us out for hitch-hiking leeches. Leeches that were found were burned off with lit cigarettes.
- During rare passes into town, we learned about what was called by the locals "wikkycoke". There was not a great assortment of liquor in Phuoc Vinh but they did have cheap whisky and Coca-Cola. However, for most of the guys, cheap beer was fine.
- During this period, we learned about tropical rain. It rained heavily during this time. The area was built close to the water table so during the heavy downpours our two-man foxholes would fill in quickly with rain water.

Most of our time at the 1st ID's base was spent sending out platoon size patrols to alert us to a possible base attack. I remember only one company sized patrol that became memorable.

We left the base early one morning with orders to patrol northeast of the base. We were to search for enemy activity three kilometers out from the camp. We were accompanied by a local ARVN soldier who would serve as an interpreter should we come across local civilians. We began by passing through the village cemetery to the northeast of Phuoc Vinh.

Along the way, we uncovered a poorly concealed Vietcong tunnel. Bob Livingston and Porter Harvey went down and inspected it, but it was small and inconsequential That was probably the first tunnel we discovered in Vietnam.

We made it out to our target area just as we got a call from the Command Center at Phuoc Vinh. Intelligence had been alerted to a large VC force coming up on us and we were ordered to return to base camp as soon as possible. We were advised if we became engaged, there was no one who be available to reinforce us should we get into a major firefight.

This is Jim Olafson's memories of that patrol:

"I recall that company size operation. We covered the assigned area and dropped off one platoon to establish a night ambush position. When Palmer got the call to return, he called me as my platoon was leading. Palmer said to take a direct oute back to the base camp.

We were moving in a column of twos and entered an area of elephant grass. We had only moved a hundred yards or so before my column on the right yelled freeze. I inquired what the problem was and was told that we were in a mine field. I moved to that column and observed a small circle of U.S. mines. I contacted Jon Palmer and advised him. He asked if we could make our way through them. I gathered up approximately five mines and placed them in a position where I could detonate them using a grenade. We spent maybe 30 or 45 minutes doing this when I told Palmer that this was taking too long and the better option was to back out and go around the area as I had no idea how large this mined area was. Palmer agreed.

Returning back to Phuoc Vinh base camp, I encountered a couple of Big Red One officers at the Officer's Club. In the conversation, I mentioned the mined area and was told that there were a number of unmarked mines around the area. Supposedly an American Engineer unit had gone out and planted the minefields but for some unknown reason no maps were ever made of the mined areas."

We remained patrolling around Phuoc Vinh until November 19th, when we once more boarded C130s and were flown back to Bien Hoi. We then were trucked to Bearcat. When we arrived, we were told to prepare for a move soon to an area to our north where we would be building our own basecamp. We were not told where that location was to be, at least not yet.

We finally were issued our jungle boots, which was a big improvement from our leather boots.

We had only arrived in country in October, just over a month earlier, but our unit had already received allotments for R&R (Rest and Recreation period in another country in Asia) for December and January. It was like eating dessert before finishing our meal for us. Still, who knew what the future held for us. Johnny Martel and I applied for an R&R in mid-January. We would spend a week in Taipei, Taiwan, wherever that was.

As we were preparing to leave for our new area of operation, the move was suddenly put on hold. Operation Attleboro was taking longer than expected and needed to finish before we were sent further north.

I wrote this in a letter home during this period:

"We've been all packed up and ready to move to our new area where our base camp is to be built. We were supposed to go today, November 21st, but we've been told that it will be the 23rd *(actually it was later than that)*. We were told we are going to Dau Tieng, a village 35 miles northwest of Saigon. We're supposed to build our new brigade basecamp on the edge of a rubber plantation."

That pretty much sums up what we knew about Dau Tieng. At that time, we had no idea about what was taking place in

War Zone C or even the big battle tasking place just outside Dau Tieng. We did hear that we were being sent to a region which Vietcong were using as staging areas for attacks on the Saigon district. We were told very little about what to expect other than "there were beaucoup (*many*) enemy soldiers based in that vicinity."

We had no idea that the enemy we would be facing was the notorious 9th VC Division who worked out of War Zone C and nearby War Zone D. We didn't need to be told, so we weren't.

Here is what the army knew about the Dau Tieng and Michelin Plantation that they didn't share with us. This information was acquired in the Fort Nisqually After Action Reports:

- (1) The Michelin Plantation was used by VC Main force units as a safe haven aided by a sympathetic population.
- (2) The *Saigon* River (*running just outside Dau Tieng*) was being used as a main supply route with transfer points located primarily in the vicinity of Ben Cui Plantation, Ben Suc and Thanh An (*located in the Iron Triangle to the south)*
- *(3).* Units located in the area were believed to be the 272nd VC Regiment of the 9th VC Division located at vicinity of XT6340 (*a remote jungle area two and a half miles south of the SE corner of the Michelin Plantation*) and the C-64 Local Force Company operating in the plantation. Operation ATTLEBORO completely disrupted enemy activity in the area.

It was probably just as well that the average GI didn't know any of that intelligence as we prepared to deploy to Dau Tieng.

As days went by, we were informed that we would not be traveling to Dau Tieng until after Thanksgiving, which was celebrated on November 24th. We celebrated Thanksgiving in fine fashion at Bearcat with a traditional Thanksgiving meal. We were all given this handout, which luckily was saved by George Hanna, a third platoon friend of mine.

DAU TIENG

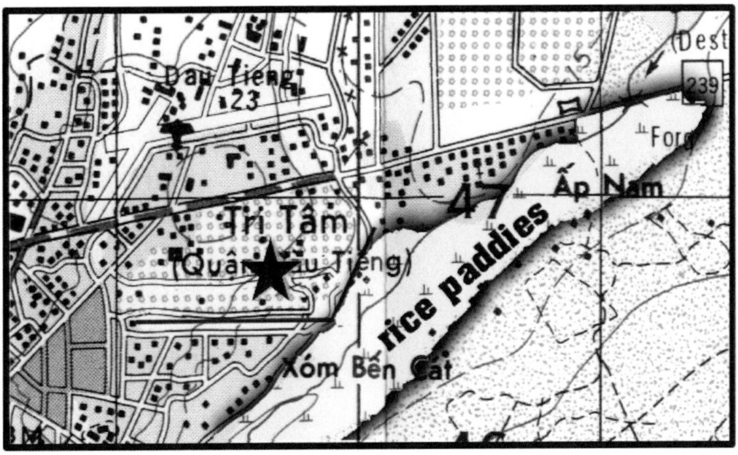

Star indicates A/2/12's location. Straight line shows where the berm was located. There was a clearing of no man's land south of the line with the village in the distance.

On November 25, 1966, the day after Thanksgiving, the 2/12th was transported to Dau Tieng via C123s. This particular aircraft was smaller and needed a much shorter landing strip to land and take off. Dau Tieng had a fairly short runway but it could easily accommodate C123s. The runway was later expanded to accommodate C130s, but that was not until Jan 6, 1967. I don't recall flying out of Dau Tieng on a C130 fixed wing aircraft at any time during my tour there. Most flights we infantrymen took out of Dau Tieng were on either C123s or C7 Caribou airplanes.

As we banked to lineup with the runway at Dau Tieng for landing, we could see the small village of Dau Tieng, just south of our new basecamp. I can only recall one time when they allowed the soldiers based at the new basecamp to visit this small village. My impression of the village was this. "The only way I could differentiate it from the western towns that I saw on TV was that there were no hitching posts for horses outside of the simple businesses."

Some of our men have gone back today to visit Dau Tieng and were surprised to see a modern town with all the utilities typical of an American town today. None of us who served out of Dau Tieng would have dreamed that a possibility back in the 1960s.

Our C123 aircraft landed uneventfully and we disembarked. We noticed another group of soldiers preparing to take our place on the plane on the side of the runway. These were 1st Division troops who had participated in Operation ATTLEBORO and were preparing to return to their base at the conclusion of the operation. Many passed us with that far off look on their face. They seemed really eager to replace us on the C123.

The 2/12th Infantry and 2/22nd Mech were the only 3rd Brigade Infantry battalions available to begin work on the basecamp at first. The 3/22nd became part of Operation Fairfax near Saigon and remained there until Jan 14, 1967. Our base camp was named 'Camp Rainier'. That name was chosen for the iconic mountain, Mount Rainier, situated near Fort Lewis where we all trained for Vietnam.

My company assembled on the side of the runway and soon were marched southeast of the clearing and entered part of the rubber plantation. Before long we were told that we would be set up in the shade of some rubber trees. Talk about prime property! We're in the tropics, where the temperature reaches 100+ regularly during the day, and here we are in the shade of rubber trees. We could not believe our luck. We little understood that it would not matter much, as typically we would be in the 'shade' of the jungle only for a short 3-day period each three weeks. The rest of the time we were on 'Search and Destroy' missions in the jungle. By then our clothes were near falling off us so we needed to return to 'camp' for a two or three-day period of maintenance. Then the cycle repeated itself.

We dropped our gear and looked around. We were set up near what would be the southern perimeter of the future basecamp. The entire company was in the rubber trees of the massive Michelin Rubber Plantation and the company was split in two by a dirt road running east to west within our rubber trees' location. Our perimeter responsibility faced just south of us. There we would establish a berm line with bunkers and an interconnecting trench line. Facing the village, we connected to HQ Co to the west and B/2/12 and C/2/12 to the east.

Initially, we had no perimeter to secure as we dug foxholes and lived in two-man shelter half tents. To illustrate how poorly America was prepared to deploy to a tropical climate, all I need to do is share this about our introductory shelter at Camp Rainier.

After a week or so, 5-man hexagon shaped arctic tents were issued to us to replace our pup tents. These were notoriously ill-suited to the tropics and during the daylight hours, the sides were rolled up to get air flowing through the tents. They were designed for arctic conditions and were unbearably hot during daylight hours. It wouldn't be until mid-January before standard issue tents arrived and were placed on wooden platforms. As I wrote earlier, we didn't spend a lot of time in camp, so it didn't matter much to us pounding the boonies in search of Charlie.

The Michelin Tire company was formed in the 1890s by two French brothers from Clermont-Ferrand, one of the oldest cities in France. From its formation, the company spearheaded many new innovations in tire design. During the year that we arrived in Vietnam, they partnered with Sears to produce the first radial tires. This made Michelin the predominant tire manufacturer in the world. Of course, this led to an increasing demand for raw material. The plantation was important to the company, but was becoming an increasingly hazardous enterprise for the Frenchmen overseeing the operation at the plantation. The NLF was squeezing the operators hard for more and more 'protection' money for non-interference in their operation. This led to the flight of most of the Frenchmen out of Vietnam in 1965, leaving behind only a small contingent of overseers who were there to 'greet' us when we arrived at our camp.

The rubber plantation and the region around it were important to the enemy's war effort. The VC jealously guarded their enclave around and in the plantation. ARVN troops learned that when, on November 27, 1965, (almost a year to the day before we arrived at Dau Tieng) the 7th Regiment of the 5th ARVN Division was completely destroyed by 9th VC Division troops operating in the plantation. ARVN forces avoided the Michelin plantation from that point on until American troops arrived months later.

The Michelin Plantation was vital to both the Saigon government and the NLF. The plantation was six miles wide by nine miles long and was the largest rubber plantation in South Vietnam. This one establishment was responsible for seventy percent of the foreign exchange coming into South Vietnam. The plantation itself contained 18 separate villages which housed many of the workers of the plantation. These villages were used by the Vietcong to collect taxes from the workers and to attain supplies that the Vietcong needed for their hidden jungle bases. Much of the time, the NLF could freely travel from village to village to not only hold indoctrination classes, but to abduct young men to serve in their units.

Michelin was under tremendous pressure to abide by what the local VC demanded from them and also the concerns of the American forces targeting this region for offensive operations. They were in that proverbial squeeze that many businesses in the country felt during the war.

Fortunately for us, by the time we arrived at Dau Tieng the 9th VC division had been bludgeoned during Operation ATTLEBORO and driven far north of the plantation by American and ARVN forces in pursuit.

That situation made for a relatively nonviolent period during the month of December compared to what was to come. To be sure, we did run into small enemy forces at the hidden base camps, mainly protecting food and supplies stored in the hidden bases. However, for the most part, they were uninterested in engaging us for long. Mainly, they hit and ran.

Operation FORT NISQUALLY was the name that was called for our first operation originating out of Camp Rainier. It was named after the historical trading post set up by fur traders and farmers operating under direction of the Hudson's Bay Company. It was built in 1832 and situated near Tacoma, Washington in 1832. I visited the Fort Nisqually Museum in

2005 and they were astonished to learn that our first operation was named FORT NISQUALLY.

Early in December, the 2/12th Commander, Marvin D Fuller left our battalion to serve in Cu Chi as a Brigade Commander. He didn't leave before installing the 12th Infantry Totem Pole which he managed to ship to Vietnam after it had been saved for years after it was given to the regiment by the Governor of Alaska when it served there in 1952. (Alaska was then just an American territory but still governed by a Governor whose name was Ernest Henry Gruening).

The totem pole had soldiers portrayed from all the wars that the 12th Infantry Regiment had engaged in since the War of 1812. Once it was in Vietnam, Fuller added a North Vietnamese soldier to the top of the Totem Pole. Unfortunately, the totem pole was somehow left behind and was lost forever in Vietnam. Such a shame.

With the war expanding as quickly as it was in those days, our new battalion leader was chosen by General William C Westmorland, commander of United States forces during the Vietnam War from 1964 to 1968. Field Grade Officers were at a premium in Vietnam, but Westmoreland didn't have to look far for Fuller's replacement at the 2/12th. He commissioned one of his Staff Officers, Joe Elliott, to take the reins of the 2/12th Infantry. This was not an insignificant event as you will learn later how his leadership played out during the biggest battle of our tour.

Joe joined the Army during World War II when he graduated early from high school at 16 years old. He was too young to fight but his high intelligence scores led to the army sending him to Clemson College for future development. Upon graduation at 18, he was sent to Amhurst College in preparation of attending West Point. He graduated from West Point in June of 1950, just in time for the breakout of the Korean War a few days later. He served as a Second Lieutenant and at one time in his tour he was attached to the Marines during the Battle of Chosin Reservoir. All the participating units received the Presidential Unit Citation for that action.

Many of his West Point class joined him in Korea and suffered so many casualties that the army realized that they had to protect future officer leaders from being killed at such a rate. Joe was sent to Japan as part of that effort.

That experience led to the six-month limit for front line officers in Vietnam. After that period, officers would return to the rear for the remainder of their 12-month tour. Joe spent the next few years stateside after the war and when the Vietnam War heated up, he was sent to Vietnam to serve on Westmoreland's staff. There he served as the Chief of the Operations Section. It was his job to keep Westmoreland briefed on what had happened overnight in Vietnam. Eventually he was sent to us.

All the battalion officers threw a going away party for LTC Fuller. On the next morning. he left the 2/12th to assume command of one of the 25th Infantry Division brigades that was based at Cu Chi.

The Fort Nisqually operation had a coinciding operation in December that was called 'Ponder's Corner' and was primarily a company and battalion size efforts to secure the jungles surrounding our base camp. Operation ATTLEBORO had identified that the enemy was securely entrenched in the area. It contained a number of way stations for enemy troops arriving down the Ho Chi Minh on their way to the Saigon district. The bases were there and it was our initial mission to find and destroy them, which we did with some success.

Here is an excerpt from a letter I wrote home on December 11, 1966 during this period which typifies the early operations around the camp:

"Back at camp now after a short company size operation. While on operation we discovered a VC base camp in the jungle near Dau Tieng. There were four small huts half built, and a deep well, which we blew up. The camp had 12 foxholes in a perimeter and each had a dugout in them for protection against artillery.

The only articles we found left behind were a belt buckle and a toothbrush. We did have one accident during this patrol. One man, Bob Murphy, carrying an M-79 grenade launcher, had one of his thumbs over the barrel of the weapon as he carried it through thick jungle. He passed through an area with dense vines and one of them caught on the trigger of the unlocked weapon and fired off a grenade, taking off one of his thumbs. He was one of the shortest men in our company but also one of the toughest. He never led out a sound after the mishap. He just turned to the soldier nearest him and asked nonchalantly, 'Now what am I going to do without that thumb?"

Actually, quite a bit. After recuperating, he was sent home, and eventually with hard work and ingenuity, rose in the ranks of the Warwick Rhode Island Public Works Department and became a Senior Manager.

As for the discharged grenade, it mostly travelled a few yards and harmlessly blew up an anthill. It highlighted the dangerous occupation that we were engaged in.

I ended that letter with:

"It is now 10 PM. Tomorrow at 6 AM we leave for a five-day Battalion size operation. Intelligence reports sighting a battalion size force of North Vietnamese regulars in the plantation and we're going to go looking for them. Have to get some sleep now. They are waking us up at 3:30 AM for that operation."

That battalion size force was notably part of the 272nd VC Regiment which was thought to be a couple of miles south of the Michelin plantation. Intelligence was rarely timely and in fact, the unit had moved eight miles further north and was camped in a heavily jungled area east of the Michelin Plantation at that time.

From a Saturday, December 17th letter:

"During this Operation, we swept through six villages in the plantation area, all suspected VC sanctuaries. We captured eight VC. In one of the villages, we found a French dispensary. On the wall, written in red was this warning:

"US Officers and Men, your lives are precious to your families, wives, and children. Do not let Washington by itself decide your destiny so that they may achieve the selfish ends of the Imperialists' Warmongers and their stooges...

"...All of us who became eligible had their Combat Infantry Badges presented to them by the Battalion Commander yesterday, the 16th.

"We're due to go back out on operation on the 19th."

Notable about what we learned in operations southeast of the Michelin Plantation was that the villagers in this area were very much Vietcong sympathizers. In many instances, when we surrounded a village and entered it, all we found were women, children, and old men. Occasionally we found a draft dodger, but those were few in number. This particular area, you may be surprised to learn, was depicted in the movie 'PLATOON'. Early in the movie, you will recall there were some scenes showing soldiers burring down some village huts and moving the villagers to other areas. Oliver Stone, who directed and wrote the screenplay based on his experience in Vietnam, joined the army from Saigon where he had been working as a teacher. He ended up in our brigade, and served as an infantryman in Bravo Company, 3/22 Infantry starting in September of 1967, ten months after our initial December sweep of that area.

By September, 1967, it became apparent that the area was a major source of food and supplies for the 9th VC Division and that was not going to change. The decision was made to destroy the villages and move the inhabitants to Government protected villages far from their homes. That move was the one portrayed in the movie. This was a brigade size operation and all three of the maneuver battalions of our brigade participated.

In the 1970 updated version of the topographic maps, all those villages were still marked on the map but marked as destroyed. There were a lot of those over the years.

As noted earlier, after 30 days in a combat zone, infantrymen are presented their most valued award, the Combat Infantryman Badge. Once we were given our CIBs, many of us were allowed to travel to the Dau Tieng village to have photos taken of us with our newly authorized CIBs.

Before we left for Dau Tieng village, we were assembled and addressed by First Sergeant Springer. We formed in a circle around Top and told to sit on the ground.

"You've just received the most precious award given to an infantryman. I think it is a bit premature. So far, every time you've located an enemy base camp, the enemy has decided that what is in the camp is not worth fighting for. That will change and then, and only then, will you have earned that badge. You're just holding them in safekeeping for now. You've yet to earn them." Top knew what he was talking about. We began earning them the next month.

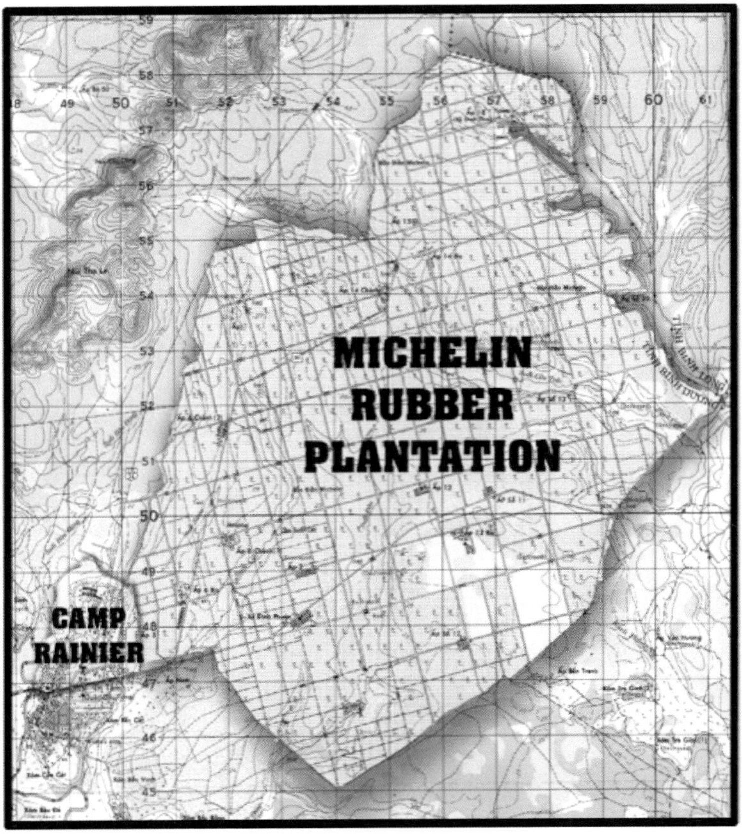

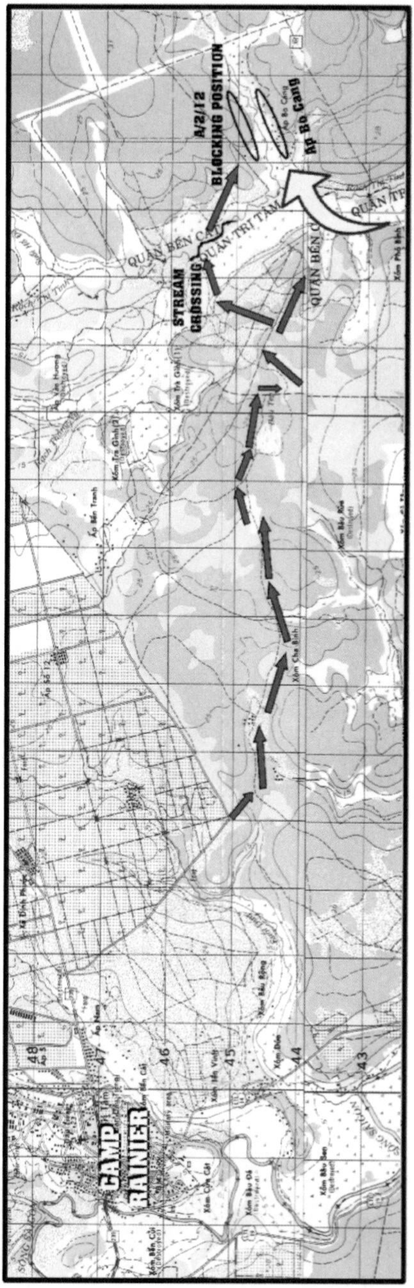
A/2/12
BLOCKING POSITION
Ap Bo Cang
STREAM
CROSSING
CAMP
RAINIER

We had taken two other night time raids on villages in the area before the December 18 -19 invasion on a village southeast of the Michelin Plantation. However, this was the one which had the greatest impact on me and the one I remember the most.

On the night of December 18, 1966, the 2/12th Infantry Battalion snuck out of camp at 9 PM under the cover of darkness. We worked our way southeastward to where there was a number of trails that were located south of the plantation. We chose the one that went due east and traveled three and a half miles along it to reach the area east of the Michelin Plantation. (So much for the warning of staying off the trails in Vietnam).

Once there, the battalion was split up and the three companies of the battalion and the HQ Company moved into their assigned positions to raid the village of Ap Bo Cang, considered a major Vietcong haven. A Company was assigned the responsibility of forming a blocking position northwest of the village. The other companies were sent to other locations to seal off the village and deny the enemy an escape route.

Along our way, my company came across a stream, not too wide and shallow enough to span. The stream ran through light jungle but thick enough to prevent a clear view ahead on this moonless night. I was about chest level into the tributary when all of a sudden, I lost my footing and slid down the mud to an underwater hole that submerged me completely underwater. Lucky for me, Johnny Martel who was behind me, spotted part of my antenna out of the water and reached underwater to pull me back up. In the blackness of that night, I could have been lost for good (there was no way I could get myself out of that hole, between the weight of my equipment and the slippery mud that was unclimbable). It would have been a tough letter to write home to Mom, "We regret to inform you that your son went missing in action during a night operation. He was there at the beginning of the operation but was not when daylight came up. His body was never recovered. Heck, we don't even know when or where he went missing." Luckily, my buddy Johnny saved the day.

After we crossed the stream, the company entered a lightly foliated jungle and we proceeded to our target location adjacent to the village. By then it was 4AM and we spread out, hidden on the edge of the jungle with a clear view of the village. We waited silently for the sun to rise which was scheduled to happen at 6:30 AM on that day.

Somewhere around 6AM, while still dark, we heard the unmistakable sound of chopper blades originating from south of our position. As the choppers came closer to us, there was a feeling of anticipation, not knowing what would come next. They were still out of our sight, but as it flew closer to the village at tree top level, the show began.

Probably a half dozen choppers arrived from the south and turned radically right as it aligned with the western end of the village. First came the spot lights that were turned on, quickly followed by Vietnamese music playing out of a loudspeaker attached to a chopper. I assume that it was patriotic music but it was not like anything I had ever heard before. They weren't playing 'The Ride of the Valkyries' which Coppola used to inspire troops attacking a Vietcong stronghold in his movie 'Apocalypse Now'. This music was chosen to instill terror on the villagers. I know it made my skin crawl.

After a short while of circling the village, the loudspeakers on the choppers turned on a speaker which gave directives to the villagers on what would be coming next. The interpreter with us smiled and told us that it was giving instruction to the villagers to come out of their huts and line up to face questions from ARVN interrogators who were looking for Vietcong being sheltered in their village.

As daylight broke, the Recon Platoon of the 2/12th along with ARVN troops proceeded to round up the villagers for questioning. I don't recall a single shot being fired and after a short period, medics were brought in for a Medcap mission. Seemed to us that there was a whole lot of effort put in to receive so little payback.

For me, that night became one of those milestones of my tour. It felt surreal. We had never been exposed to this type of psychological warfare and frankly, the scene gave us the creeps. We could not imagine what was going through the minds of the villagers.

We continued searching the jungle in that area and my company located an enemy base that contained huts and an extraordinarily large cache of polished rice ("Polished rice refers to rice which has been milled to remove the husk, bran, germ, and varying amounts of the nutrients contained in them, leaving a starch-rich grain", according to the Internet. I think a better way to describe polished rice is what we call 'white rice'.

That rice cache was estimated to be 35 tons. We also found 2,000 pounds of salt. I don't know how those numbers were determined, but that's what they came up with. We began moving the rice to a nearby road and open field to allow it to be transported to the Saigon government. It soon became apparent that this would take too long. We instead set explosives to the captured cache and blew them up. For good measure, we tossed in CS grenades to see if that would contaminate the food. I always wondered how successful that solution was to the problem associated with destroying large stores of enemy food supplies.

This was how Intelligence sources described our first full month at Dau Tieng in after-action reports: "During the month of December, enemy activity was limited to sniper fire and limited mining incidents. C-64 Company was active throughout the plantation collecting taxes, conducting propaganda lectures, and purchasing supplies. The 272nd Regiment was noted to have moved from XT6340 to XT6055 (noted earlier). No offensive type activity was observed or reported."

On Christmas day, most of us were in camp for Christmas dinner.

Here are some final thoughts as I leave the holidays month behind us. For years on TV, they would show Bob Hope Christmas Shows originating in countries around the world. Somehow, we thought that we had at least that to look forward to in Vietnam. Not to be, what folks don't know back home is someone had to assure that the enemy is kept off guard during those periods. That meant patrolling and operations. That took boots on the ground. That meant our feet were in those boots.

It would have been much too dangerous for Bob Hope's troupe to perform for us at Camp Rainier, as it was still a lightly secured base at the time. Instead, Bob played to troops at the Cu Chi base camp, which was much safer.

That's not to say we all missed out on the show. According to the scuttlebutt, there was an allotment of eight men from my company who were sent to Cu Chi to view that Christmas show. Their names were chosen at random by a drawing. They were sworn to secrecy about their selection. They must have taken that oath very seriously as after 20 years of reunions since the year 2000, not a one of those men stepped forward to say, "Yeah, I saw Ann Margaret with Bob Hope in Cu Chi for the 1966 Christmas show." On the other hand, maybe that was just a rumor that was spread to make us feel that at least we had a shot at that opportunity. Maybe no one went.

Our consolation prize was a visit by Martha Raye at Dau Tieng in January. Martha was an old comic movie star who entertained us with comedy and songs. She performed on a stage built in front of the plantation overseers' swimming pool. She had traveled with Bob Hope troupe shows going back to World War II.

This time she came alone except for an elderly guitarist who accompanied her. He played on a Gibson Les Paul guitar attached to an amplifier whenever Martha sang a song.

The most memorable part of her show was the finely polished jokes that were somewhat risqué for the time. She had been performing in front of soldiers far from home for so many years that she knew her audience well.

The guys loved her. She was no Ann Margaret, but she was former Hollywood royalty and she cared enough to leave home and entertain us in an isolated, perilous area when no one else would. She instinctively knew that none of us would be able to see Bob and his troupe at Christmas because of security obligations and wanted to show us that we mattered too.

After the show, she sat on the stage and signed autographs for all the guys who wanted them, and there were many. We never forgot her kindness to us.

No other entertainment troupe performed at Dau Tieng until June of 1968 when Vicki Lawrence, (a star from the Carole Burnette Show), Melody Paterson (who is best remembered from the TV show 'F Troop', where she played the character 'Wrangler Jane)' and Miss Long Beach (whose name would be forever lost to time). Johnny Grant, the perennial host whenever a new star was placed on Hollywood Blvd's Walk of Fame was the host of the shows. Shows at Dau Tieng were few and far between.

EARNING OUR CIBs

The First Is Always Worst

The year 1967 was dubbed the 'The Year of The Offensives' in Vietnam by the Center of Military History.

On July 28, 1965, the pledge was made to commit major troops to the Vietnam conflict in a speech delivered by President Johnson. To attain this, he doubled the draft quotas and later expanded it even more, to establish a major expeditionary force to South Vietnam.

Just before we arrived in Vietnam in October 1966, there were 330,000 troops in Vietnam. (*Remember that figure 184,300 troop total in Vietnam that was quoted to us in March 1966 by Sgt. Harris?)* By January 1, 1967, that deployment total reached 385,000 troops. After almost a year and a half of building up American forces, the pieces were finally in place to go on the attack in Vietnam.

From October 14th to the start of the new year, almost 12 weeks, my company had not suffered a single casualty by enemy fire. That was about to change soon.

The new year rolled into Camp Rainier in a celebratory

fashion at midnight. It began on the berm when, starting on the far side of the camp, a 'mad minute' of intense fire began and extended all along the perimeter in wave fashion. It only ended when the wave reached its origin.

We may not have been so festive had we known that the 9th VC Division had been receiving replacements from North Vietnam. They had recovered from Operation ATTLEBORO and were preparing to target the units who invaded their territory. Prior to January, only rearguard actions, trying to protect basecamps, were available to the enemy. Now they were almost back to full strength and capable of offensive action. Top was right, the day was coming when the enemy would become bolder and take the action to us. They were not the only ones preparing to take the fight to the enemy.

With the American troop strength finally in place for a major thrust into War Zone C, in December, 1966, American war planners were preparing for that eventuality. However, this offensive was postponed due to a fear that a major enemy offensive was preparing an attack on Saigon itself. Three events in Saigon during December prompted a hasty decision to forgo the War Zone C assault.

On Dec 4, the VC Military Region 4/6th Battalion attacked Tan Son Nhut Airbase, damaging eighteen aircraft. That same evening, a satchel charge exploded on the roof of a building in a US compound inside Saigon, wounding several Americans. Five days later, the Vietcong had unsuccessfully attempted to blow up the Binh Loi Bridge, a vital link in Saigon. The enemy was becoming bolder and more aggressive in the Capital District and the attacks were originating from the Iron Triangle. It made sense to attack that region first to ease the pressure on the Capitol. They were also aware that the 9th Infantry Division from Fort Riley, Kansas would be arriving soon and could be used to free up more seasoned soldiers for the big battles later planned in War Zone C.

To deal with the pressure on the Saigon region, Operation CEDAR FALLS was hastily established and put into action.

As I referred to earlier, a good deal of enemy activity was originating from the Iron Triangle, just north of Saigon. To deal with that threat, an American/South Vietnamese Army coalition would attack the enemy hidden in the triangle with thirty thousand troops to root out the enemy's staging area. The 2/22nd Inf was used in a support role for that operation. The 3/22nd was also sent towards Saigon to shore up support for the troops of Operation Fairfax. The 2/12th Infantry was not involved in either operation, instead was used to protect the development of our new base camp. That was how I recall clearly how Dau Tieng was used as a jump off point for an air assault early during Operation CEDAR FALLS.

I learned about the big buildup taking place at the helicopter pad at Dau Tieng on the morning of January 8, 1967. I ran to the edge of the clearing with others where I could see all the troops lining the side of the chopper pads, waiting to board helicopters down to Ben Suc. There I first learned the name Alexander Haig. He was then known as a Korean War Veteran and then Commander of a First ID Infantry battalion. I was told that he was a very highly regarded leader and was destined for great things. They were not wrong. He served as White House Chief of Staff for President Nixon, Supreme Allied Commander Europe under President Ford, and Secretary of State under President Reagan.

Seizing the Iron Triangle

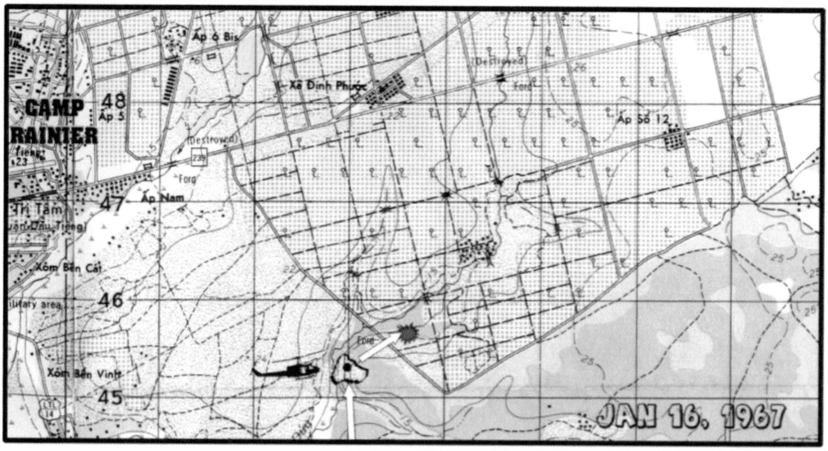

"Just before first light on 8 January, troop carrying UH-1 choppers began landing at the Dau Tieng airstrip, the staging area for the assault on Ben Suc. Shortly after sunrise, men of the lead 1st Infantry Division unit, the1st Battalion, 26th Infantry, led by Lt. Col. Alexander M. Haig, boarded the waiting aircraft. Ten minutes later, sixty helicopters took off, circled Dau Tieng to gain altitude, and formed two "Vee" formations, each with three flights of ten helicopters. After flying directly south from Dau Tieng to give the impression that some place other than Ben Sue was the objective, the helicopters, at a point twelve kilometers west of Ben Suc, dropped to treetop level for a final approach eastward into three landing sites west, north, and east of the village. In a further attempt to gain surprise, no artillery preparation or air strikes preceded the operation.

Traveling at 100 miles per hour, the helicopters crossed the Saigon River and precisely at H-hour, 0800, began to land. Within ninety seconds, all six flights had cleared the landing zones and were on their way back to Dau Tieng. Thirty minutes later, some of them returned to deposit a company south of Ben Suc, effectively surrounding the village.

There was little resistance within the village. The inhabitants were rounded up and relocated to a government-controlled area. The enemy know longer could use the village as a base for attacks on Saigon. It was demolished.

When the operation ended on January 28, 1967, the Iron Triangle was made into a 'Free Fire Zone' and anyone moving in it was considered to be an enemy combatant."

According to the Operation FORT NISQUALLY after-action reports, the 272nd VC Regiment and local force, C-64 were operating in the Michelin Plantation area prior to Operation ATTLEBORO. The C-64 unit was responsible for collecting taxes, conducting propaganda lectures, and purchasing supplies for the Vietcong in the field. Operation ATTLEBORO completely disrupted enemy activity in the area during December, 1966.

Operation FORT NISQUALLY continued up north in January.

Planning those operations was the Brigade Operations (S-3) Officer, assisted by Charles "Ed' Smith, the Assistant S-3 (Operations Officer) for the brigade. Their job was to plan the missions and follow up on those operations.

Ed, as he prefers to be called, came to our brigade in early December 1966. He had served a previous tour in Vietnam from Nov 1962 to Nov 1963 as an advisor to the South Vietnamese 51st Infantry Regiment in Quang Ngai Province, in the Northern II Corps Area.

While serving in his brigade role, he would later request to command an infantry company when the opportunity arose. When Jon Palmer was sent to battalion staff after serving as the A/2/12 Commander, Ed would lead the A/2/12 from April 16, 1967 to July 4, 1967. I came to rely on him considerably when I was assembling the brigade's history over the last twenty years.

Of the January period, Ed revealed that the Brigade Commander, Marshall B. Garth, didn't want us just sitting around while CEDAR FALLS was taking place further south. He instructed us to design both company and battalion search and destroy missions. Three of those missions stood out for us who served in my company the month of January.

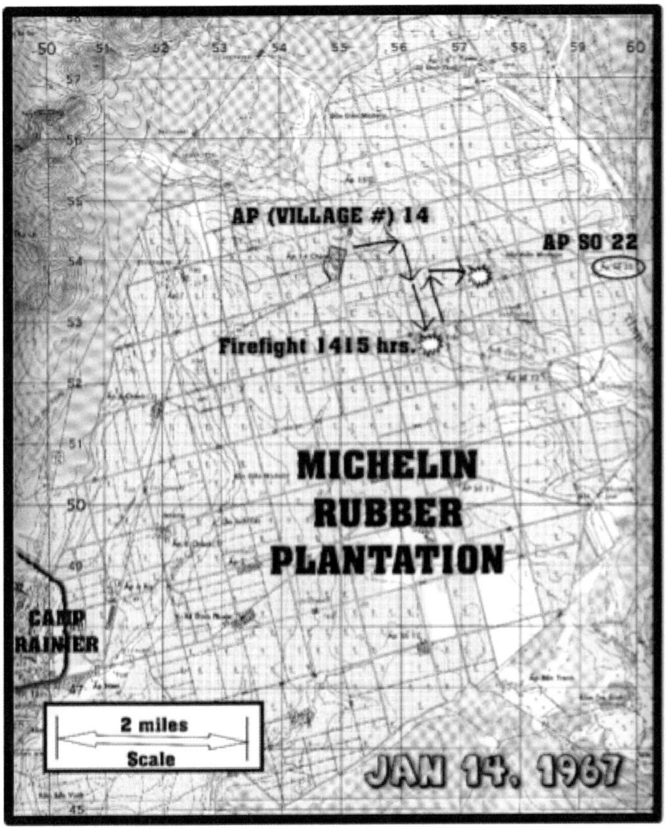
AP (VILLAGE #) 14
AP SO 22
Firefight 1415 hrs.
MICHELIN
RUBBER
PLANTATION
CAMP
RAINIER
2 miles
Scale
JAN 14, 1967

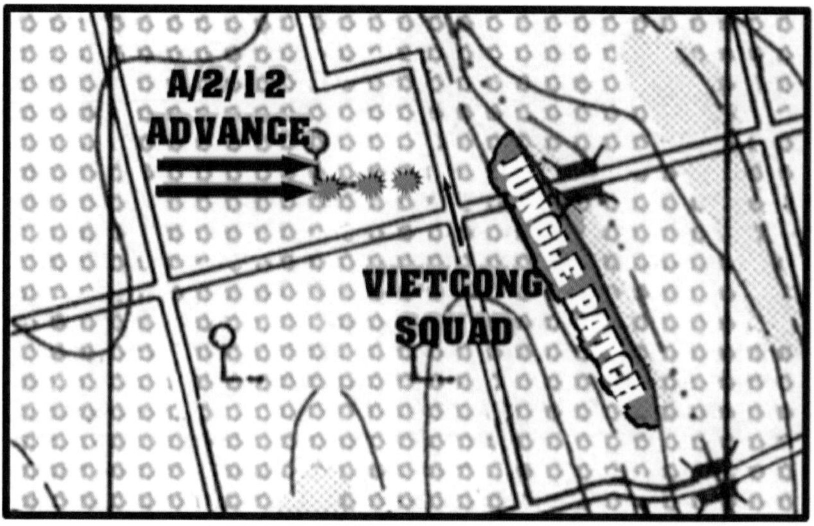

A Company moving east through the Michelin at 1657 hrs.

The first took place on January 14, 1967. The 2/12th Infantry Battalion had been operating in the plantation for a couple of days. On this particular day, the battalion snuck up in the early morning darkness to surround AP 14 and search it for Vietcong. A Company was placed along the northern border of the village, Recon Platoon in the south, B Company was on the west with C in the east.

At first dawn, the trap was triggered and two VC tried to escape the village, one on a bicycle and another on foot. They were driven back into the village.

At 0726 hrs., A Company also captured two other individuals trying to escape north of the village. They were carrying three Chicom machine guns, one 'grease gun' (M3 – 45 Cal submachine gun), and one hand grenade.

Interestingly, it was noted in the day report at 1033 hrs. that Stanley Resor, the Secretary of the Army, was visiting Camp

Rainier this day and following the action at brigade headquarters.

At 1300 hours, the village mission completed with limited success. The battalion was resupplied and dispersed to different locations in the Michelin. A Company was sent south through the rubber trees and had contact with a VC squad at 1415 hrs. The result was one VC had been wounded but managed to get away. After that brief engagement, our mission was changed and we were sent back northward. Once we got into position to begin a movement further east, we pivoted in that direction.

We were headed towards a night defensive position in column formation when 200 meters to our east an unsuspecting VC squad came into view on a plantation road ahead of us. The time was 1650 hrs. The lead platoon, spotting the enemy hit the ground. The rest of the company followed their lead. The platoon leader called back to me on my radio and explained the situation. I shared what I heard with Jon Palmer and he took the headset from me and called the lead platoon. He ordered them to not engage the enemy. He would take care of that squad with an artillery strike. A strike was coordinated and soon a barrage of 155 artillery rounds rained on the rubber trees to the west of the rubber trail that the enemy had been traveling. Once they saw what was happening, they ran to the east where they faded into a jungle patch which would effectively cover their withdrawal from the attack.

From where I was about halfway up the column on the south side, I could hear the rounds travel overhead towards our lead men. The first round was in the rubber and each additional round was falling further and further west into our ranks. The locality filled with smoke and thunderous blasts from the artillery barrage. The men hit the ground and tried to make as small a target as they could.

As it became very apparent that something was going terribly wrong, a 'check fire' was called in to shut off the artillery immediately. It was too late as A Company's lead platoon suffered the company's first fatality in Vietnam.

When the dust finally settled after a few terrifying moments, a cry of "Medic" rang out from the front of the column. It was too late, Joseph Noel, a Weapon's Platoon Squad Leader lay mortally wounded. Dave Cunningham was in Noel's squad. He was carrying a radio for Lt. Robert Huffman, the Weapon's Platoon Leader at the time of the mishap. He and Dave were situated less than twenty-five feet from Joe Noel when the round that killed him exploded, sending shrapnel away from them.

"Originally, I thought that first round was a marking round, but thinking about it, I don't think that was the case. It would have tipped off the enemy ahead. In addition, the second round came much too quickly after the first round. So, what I saw was actually an air burst. It was followed quickly by another shell, which is the one that killed Joe. When that one hit, Lt Huffman and I were blown back and Huffman fell on top of me. Neither one of us were injured.

The rounds seemed to come in groups of three. *(An Army battery of 155s is composed of six guns)*. Before it could be terminated, twelve rounds fell into our midst.

After it subsided, we searched for victims and saw Joe Noel laying lifeless nearby. He was the only one who lost his life in that volley. In fact, he was the only one who was wounded in that tragic mistake."

There was an investigation into the incident and it was determined that an artillery gun was not properly secured and the nose was climbing as the rounds went off. This would result in a round not reaching the proper location that was targeted. It would fall short; thus, it was called a 'short round'. A mandatory count was made of the number of rubber trees that were damaged in the barrage. According to the agreement with the Michelin Plantation owners, the United States was obligated to pay them $600 for each rubber tree that was damaged. The brigade never took that into consideration when artillery was called in. Cost of doing business, so to speak.

For the remainder of our time in Vietnam, the men of A/2/12 would always shudder when the call of "Fire Mission" rang through our ranks. We never truly trusted that young artilleryman on the other side of that mission. As it turned out, this would not be our only man killed by friendly artillery fire, as you will read later. Still, artillery fire was still the infantrymen's best friend in Vietnam when the going got tough. They saved many grunts lives in combat.

After Joe Noel's body was airlifted out of the scene, we proceeded towards that patch of jungle to our east. Inside that jungle we set up a defensive night position.

As was the custom, Johnny Martel and I dug our own foxholes and those for the company commander and first sergeant. It was a very somber moment for all of us as little was said as we prepared the position in the jungle. When we were through, darkness was slowly enveloping our positions. Right behind us was First Sergeant Springer who was leaning backwards against a four-foot high bolder. He was staring at the ground, not saying a word. Johnny and I walked up to him to get his opinion on losing our man to a tragic accident.

"How are you doing, First Sgt?" He didn't say anything for a while. Finally, still gazing down, he said to no one in particular, "The first is always the worst, always," as if to try to convince himself. Then we noticed his eyes tearing up. We had never seen this kind of emotion from Top who we figured was the toughest man in the company. Joe Noel was one of his boys that he had nurtured and prepared so diligently for nine months at Fort Lewis and to lose Joe so senselessly was hard on him. It was difficult enough on him to lose men through enemy action, but this tragic accident was incomprehensible.

To be sure, this incident shook all of us up. The seriousness of what we were doing in Vietnam and the possible consequences hit us like a ton of bricks. Top was right, some of us would have our lives taken away from us in Vietnam and that was just the nature of war. To make matters worse, it was not just the enemy who could hurt us. Friendly accidents also were a possibility. Ours was a dangerous business with serious consequences.

On the official battalion spreadsheet showing the names of those killed in action, the cause of death was marked "Misadventure". Many years later, the words "Friendly Fire" was added.

In Joe's hometown, Hope, Rhode Island, a square was named in his honor at the junction of the two main village roads. It was put up by a local VFW post and dedicated on Veterans' Day, 1967. The association would later visit both this site and his gravesite during our 2011 annual reunion that was celebrated in that area.

Tom Noel, Joe's brother, has been an associate member of Alpha Association since its founding. In January, 2017, all the New England association members were invited to attend a Catholic memorial mass on the fiftieth anniversary of Joe Noel's tragic loss. Around ten of us were able to attend that Saturday evening service.

The mass was celebrated at Tom's church in West Warwick, RI. Before the mass began, Tom approached me at my bench to ask if I would like to participate in the mass with a platoon friend of Joe's named Al Peckham. He asked us to go to the rear of the church during the liturgy and pick up the water and the wine offering and bring it to the front where we would hand it to the priest who was celebrating the mass. We both agreed.

We were cued by Tom when it was time to proceed to the back of the church. We immediately and proceeded to the back where a table held vials of water and wine that we would need to bring forward.

Once there, we fetched the bottles off the table, turned and waited for the signal from the celebrant indicating that he was ready to receive the vessels.

While waiting, I had a few seconds to survey the layout of the church. For a brief moment, in my mind, the columns supporting the church became rubber trees. The placement was almost exactly what I recalled that day forty years earlier. To the front of me, there was an aisle where the priest would be waiting to receive the vessels. In my mind, that aisle running perpendicular in the front of the church turned into the rubber tree dirt road where the VC squad, unsuspecting, walked in front of the company.

The moment was brief, but not forgotten.

I was nudged out of my flashback by a poke from Al Peckham, "Hey, the priest just called us up." I composed myself and we both walked to the front of the church where the priest awaited our arrival. The trees disappeared, replaced by these beautiful columns that supported the picturesque church. The front aisle transformed into the aisle where the good father received us with a smile.

It never leaves you. The first one 'is' the worst.

Return Of The Dragon

Monday, January 16, 1967

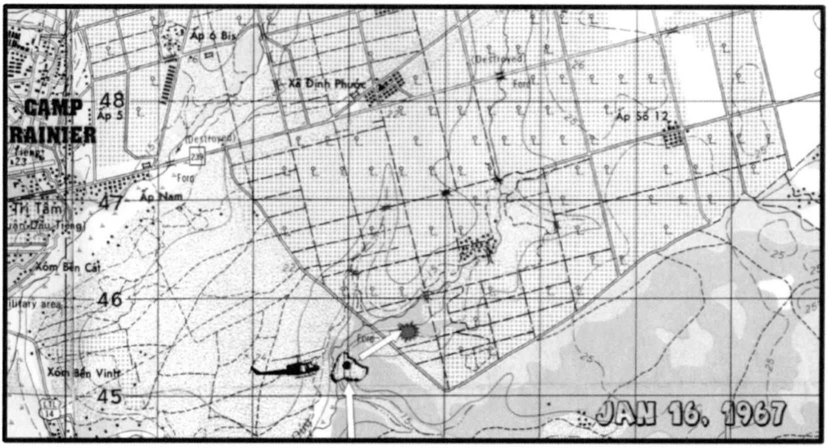

On this day, the 2/22nd was on an operation south of the Michelin plantation and the 2/12th was working its way up to the southern perimeter of the plantation. Just as my battalion was approaching a clearing south of the plantation, Alpha Company was alerted by radio that there were two men who were scheduled to go on Rest and Recuperation and needed to return to the basecamp immediately. Those soldiers were Johnny Martel and myself. We knew it was coming up soon, but had no idea what day was set for us to leave. We learned that we needed to be at Camp Alpha, near Saigon on the 17th for our January 18th departure for R&R in Taipei, Taiwan.

The company was told to drop the two individuals at the upcoming clearing. Once there, they were to wait for the arrival of a Huey which would be sent from basecamp to pick us up. Once back at camp, Johnny and I were to report back to our company area at Camp Rainier to pick up the paper work for the R&R.

The battalion found the clearing they would be passing through and Johnny and I were left behind with a radio and smoke grenades. The time was about 1100 hrs. We waited patiently for our ride back to camp, but it was slow in coming. At 1125 hrs. we heard the sounds of a battle taking place in the direction where the 2/12th had been heading. The action was prolonged and still going on when the radio came to life and a chopper pilot contacted us and requested that we 'pop smoke'. (This is a common practice that is used to identify friendly troops on the ground before risking a landing in enemy positions). We threw out a purple smoke grenade. He spotted it and called out "purple". We acknowledged the color and he approached our position in the clearing.

When he set down, we boarded the chopper quickly and we took off to the east and turned quickly to the west to return us to the basecamp. We couldn't have been 50 feet in the air when the pilot turned to us and asked if we had used a CS (tear gas) grenade. Johnny and I looked at each other and shrugged our shoulders. We said that we were not aware of the use of tear gas in the area, but there was some kind of battle taking place to our northwest and that may be where tear gas was used. Johnny and I talked about that incident and neither one of us had picked up the scent of the gas, but who knows? After a couple of weeks in the jungle, we would become immune to unpleasant scents of any kind. This is what the after-action report documented about the enemy contact on that day: "At 1540 hrs. Recon 2/12th engaged 3 VC coord XT538455. Results 2 VC KIA. The battalion reported that the VC were employing riot control agents during the operation. Due to the favorable winds, no one was hampered and no US casualties were suffered."

In the summary report of Operation Nisqually, they wrote this about that event: "On January 16 to the south of the plantation a platoon sized unit was engaged. The VC platoon unit immediately attempted to break contact by using a chemical tear agent. This was the first occasion that VC units have employed a chemical agent in this manner."

I have no idea why there is a disparity in the numbers of enemy engaged, but I do know that from that day on we carried our gas masks out in the field. This was okay with me as the mask holder and mask made a nifty pillow when wrapped up with the towel that we wore around our necks to soak up the sweat around our face in the jungle.

Sweating enough to impede your vision? No problem. All you need do is reach down to dry yourself with the towel that later be used as a pillow case for your gas mask pillow. I can't imagine what it smelled like, but by then we were getting used to the stench of jungle warfare.

Not much thought was given to that event until a fateful day almost forty years later, during an Alpha Association reunion. On that day, each of our leaders had an opportunity to give an account about what they knew in Vietnam and what the rest of us did not. *Actually, over the course of the twenty years of researching our story, I came to learn that there was a lot of information that didn't get down to company level. In fact, even battalion commanders were not always fully informed on intelligence or strategy matters.*

John Concannon, served as a platoon leader and later executive officer for A/2/12 during our training at Fort Lewis. As I wrote earlier, he was moved up to battalion level just before we left for Vietnam. There he served as the Battalion S-2 Intelligence Officer. He told us a story at that reunion that no one was aware of until that gathering.

John was near the front of the battalion movement when they came up on some unaware Vietcong who came into view. The point men of Recon Platoon engaged the enemy. When the action began, John was up close enough that he could make out where the Vietcong were laying down a base of fire on the attacking force. To pinpoint where the enemy was situated, he was ordered to 'pop smoke' on the enemy's position. He was close enough to toss a smoke grenade at them so he personally reached down for a smoke grenade and in the excitement of the moment, he accidentally grabbed a CS (tear gas) grenade by mistake and threw that at the puzzled Vietcong. Hey, it happens. There are a lot worse mistakes made within the fog of war.

The fact that neither side had gas masks forced all the combatants to leave and withdraw to the rear. When the gas cleared, Recon Platoon moved back up and discovered that during the initial action, the enemy suffered two soldiers killed in the action. They were left behind by the Vietcong and retrieved by the platoon.

Once back at basecamp, the battalion officers were debriefed by their brigade counterparts. No one actually saw the tear gas incident, so John, figured, "Why upset the applecart?" It was an honest mistake and who really needed to know the truth? So, the event was reported as a Vietcong initiated incident. From that moment on, all soldiers sent to the field would be required to carry gas masks.

There were no other incidents of CS attacks by the Vietcong reported that I could find going ahead more than a year after the incident in brigade journals. Still, that gas mask sure was handy as a pillow.

Tuesday, January 17, 1967

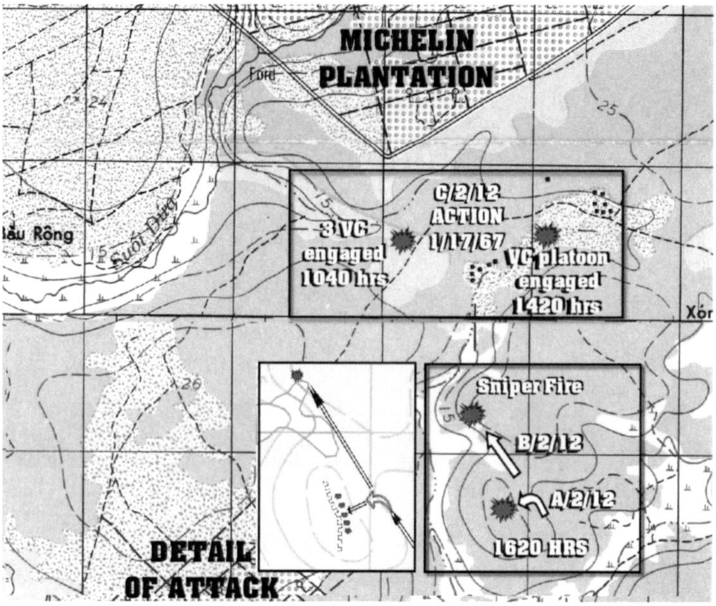

Map showing the 2/12th action of January 17, 1967, just south of the plantation

Units of all three battalions of the brigade were locating an increasing number of enemy base camps in the southern regions of the jungle south of the Michelin Plantation. Contact was picking up also with some units even reporting receiving mortar fire from the 9th VC Division troops. The Vietcong were now emboldened by additional troops arriving in their midst from North Vietnam. They became much more aggressive. Enemy bunker lines that were readily conceded to our advancing troops in December were now being contested in January. The division's troop were broken into small units and spread out in our Area of Operation.

The 2/12th battalion was operating south of the Michelin at two separate locations. C Company was located to the Michelin's southern border while the remainder of the battalion was airlifted three kilometers south with the intention of sweeping north.

Charlie Company was the first to run into a group of three VC at 1040 hrs. on this day. After a short firefight, the VC disappeared into the jungle, leaving one of their dead behind. The dead VC was armed with an AK47 rifle. That weapon and his web gear was recovered and would later be sent back to camp.

The company resumed its march east and almost four hours later, at 1420 hrs., came up on an enemy bunker system manned by an estimated VC platoon. After a small exchange of small arms fire, the company disengaged and artillery and airstrikes were called in to demolish the enemy base. C Company only suffered one wounded man from this action. After the strikes were terminated, the company surveyed the complex to determine how many Vietcong were killed. They did not find a single dead enemy and the camp was left empty of any gear, food, or weapons.

The day went on with both 2/12th groups continuing to search their areas. The main body of the battalion was sweeping north towards the plantation, a couple of kilometers south of the C Company sweep.

B Company was in the lead and in column formation when they came up on a clearing to their west. They did not enter the clearing at that time but continued on northwest to see if they could locate any basecamps that skirted these clearings, as sometimes was the case.

They moved up until they came up to the clearing in the northeast corner of the field. As soon as they began to enter the clearing at that point, they came under heavy sniper fire from the tree line ahead. The fire became intense. The decision was made to move A Company, then just behind the lead unit, around the lead company and travel northward to protect B Company's flank and to see if they could also engage the enemy force.

Second Platoon was in the lead for Alpha Company that day and instantly took a left to begin maneuvering around the B Company men, who were then in the down position trying to assess the degree of danger facing them at the clearing. As 2nd Platoon began that maneuver, they unexpectedly walked directly into an enemy bunker complex. The VC entrenched there immediately opened up with a great show of force, including machine gun fire and grenade attacks. A number of our company soldiers were injured during the initial volley.

This is how the men of Second Platoon recalled the action on that day:

Larry Walter

"We were on a battalion sized search and destroy operation on that date. B Company was in the lead with our company directly behind them. It was like any other day for the unit. There was little evidence of enemy activity in the area, but the countryside was still dangerous and enemy forces could be hidden in the jungle and preparing to attack our unit at any moment.

"Bravo Company was in the lead of the battalion when some of their lead men came under fire from enemy forces across a clearing north of us.

We all crouched down to determine which way this engagement was going to progress.

We were the second company in the line of march and we were soon ordered to move left around B Company and work our way north to engage with the attackers.

My platoon, the Second Platoon, was the lead platoon for

the company on that day and instantly took a left to begin going around the B/2/12 men, who were now in the down position trying to assess the degree of danger facing them at the clearing. I was in the left column with my men as we headed west. Not too long after we began that movement, we unexpectedly walked directly in front of an enemy bunker complex on the top of a small hill. We scrutinized the area to try to determine if it was manned. All of a sudden, the bunker line opened up on us with everything they had. A number of our men became casualties during the opening salvo.

The Vietcong were well prepared for an attack. We did not realize it until later, but the VC had trimmed all the brush and leaves leading into their base camp to a level just about knee height. When the firefight broke out and we hit the ground, we were exposed directly into the enemy's cleared field of fire. The noise was deafening as volley after volley of enemy fire was directed towards us. My squad, in the left column, was pinned down and dealing with casualties as we tried to figure out how to disengage successfully with our injured men.

The Vietcong, manning their bunker line, made that retreat as difficult as they could. These soldiers were well protected and disciplined."

Peter Filous

"I was the leadman going into the bunker complex. I was shot through my left hand in the opening volley and I was told later that it was probably a round from an M1 rifle. The bullet had penetrated the barrel guard of my weapon and came in from the left before hitting my hand. I immediately went into shock and passed out. When I came to, I realized that I had been pulled back from the front and Ed Belisle was trying to remove my bandage from my pouch to dress my wound…"

I need to acknowledge Ed for his action on that day which was never rewarded by any medal. That was not unusual as commanders never had the time to properly reward their men in those days. Ed was an usual warrior and I need to describe him in detail.

Ed was drafted out of Holyoke, Massachusetts, just a few miles north of Springfield, MA. Basketball was invented in Springfield and four years later, in 1895, volleyball was invented in Holyoke, eight miles north. Go figure.

Just before Ed was drafted, he ran into trouble with the law. Once he was in the army, Holyoke law enforcement stopped pursuing him and he settled into life in the army. He always wondered if this was a temporary pause or if they were awaiting his return from Vietnam. Upon his release from his two-year draft tour, he immediately joined the navy. He found a home there and made a career out of that branch of the service for the next twenty years. The Holyoke authorities must have figured that he paid his debt to society by then and never tracked him down in the 1980s when he was once more a civilian.

Twenty years had passed since the end of WWII and Ed was still very much distrustful of Germans when he arrived in our company in 1966. After basic training, he was placed in the same platoon as Peter Filous. It was then that he learned that Peter was a recent German immigrant who opted to allow himself to be drafted and remain in America. From that point on, Ed gave Peter a hard time and they never reconciled their differences…until it really mattered.

When Peter was shot and in a coma in the front of the 2nd platoon position, Ed was stirred by the moment and pulled what we later called his 'John Wayne' moment. Ed knew who was in trouble but it made no difference to him, he jumped to the challenge and started to crawl his way up the column to where Peter laid motionless in front of the platoon. With complete disregard of his own safety, he threw Peter on his back, crouched down, and with his back to the escape route, fired his weapon continually at the enemy to get Peter to the rear.

Weeks later, when asked about his heroism during the battle, Ed downplayed it. He told everyone, 'The 'Kraut' was in a jam and to me he was just another 2nd platoon soldier. It was nothing special.

When Ed Belisle was in Henry Osowiecki's squad at Fort Lewis, Oz always complained about Ed ruining the platoon's scores because he was so undisciplined and unkempt. Still, Henry knew that in a firefight he recognized that he would probably rather have Ed by his side than almost anyone else. Ed was a genuine fearless fighter…

We searched for Ed Belisle for years, and finally in 2003 we found his son who told us that Ed had died of cancer only the month before. It would have been a wonderful find for Alpha Association. He never knew how highly regarded he was by the men who served with him in Vietnam."

Peter Filous

"Once in the rear, Sgt. Harris, Second Platoon Sergeant, pulled alongside of me and asked for my ammunition. He was very upset to learn that I had some old M14 ammo pouch and they were packed so tightly with M16 ammo that he could not pull out the much-needed ammunition. He told me that he would kick my butt when he saw me again. (That never happened as it was never mentioned when we next saw each other 35 years later during the 2002 association reunion)."

Back at the front, Larry Walter thought that everyone else but himself had retreated from the battle site. It was a lonely moment for Larry as he contemplated his next move.

On the right column was Norm Smith's squad. At first, his column was not attacked like the left column but that allowed him to determine where the enemy fire was coming from; the bunker line was right in front of him which had not spotted his arrival."

Norm Smith

"Second Platoon was the lead platoon on that day. Peter Filous was in the front of the column to the left. I was in the right column, just behind the point man and served as the compass man.

As we moved to the west, we came up to what looked like a formidable enemy base complex, complete with bunkers.

Our column carefully moved forward looking for evidence of enemy inhabitants. Suddenly there was machine gunfire originating from the bunker line directly in front of the right column. The target of this hail of bullets was, surprisingly, the left-hand column. Soon the men in the right-hand column began laying down some suppressing fire to take the pressure off the men on the left, who had taken casualties from the initial gunfire. Five men were hit by the initial hail of gunfire and were pinned down by the incessant volley. Larry Walter moved up the left column to help the fallen men. When he arrived at the head of the column, I tossed hand grenades near the bunker to allow the wounded men to be moved to the rear.

The men slowly maneuvered back as best they could. It

wasn't easy, as the enemy machine gun salvo continued to make any withdrawal perilous.

Larry and I remained in the front and were now alone. We tried to occupy the gunmen who were relentless firing their weapons. We gave them something to think about as we engaged them and waited for an opportunity to withdraw. Finally, after what seemed like a long time, there was a break and slowly Larry began moving back. I crawled back right behind tossing grenades at the enemy as I went."

After we and the rest of the company retreated a safe distance, the bunker complex was destroyed by an air strike."

Larry Walter

"I recall that there was an air strike into the bunker complex that had attacked us so fiercely as we approached it. We were close enough to the air strike that the spent shell casings from the jet's 40 MM weapons were falling amongst us.

We reentered the base after the airstrike and we did not find a single enemy spent cartridge or for that matter any hardware at the site of the engagement. They were destroyed by the air strike or carried away by the enemy."

Peter Filous

"The clearing that was used to evacuate me from the battle was a short distance from the battlefield. I was flown to a MASH unit where I was initially treated. Eventually I was sent to a hospital in the Philippines. It was originally determined by the medical people that I would lose a finger. I fought to prevent that and was successful. Ultimately, some bone tissue and tendon material from my hip was used to make repairs to my hand.

In May, I was sent to Japan for more surgery. I, in due course, ended up at Walter Reed Hospital, where I remained until September when I was finally released, eight months after the battle.

It was a difficult period to be returning home to, characterized by a disenchanted public that thought little of the sacrifices endured by the servicemen who were sent to Vietnam. I kept to myself and buried myself in my work. Most people that I worked with didn't even know that I had served in Vietnam."

That was a common refrain of most of us who served in the infantry in Vietnam.

During that battle alone, A/2/12 suffered ten casualties, but no deaths. Those ten wounded were the first casualties for our company (outside of the Joe Noel death the week before) since our arrival. That was pretty amazing, looking back at that period. We had been in country for a quarter of our year length tour, and were relatively unscathed by enemy action.

Unfortunately, this battle would serve as a trendsetter for us. The action would escalate considerably eight days later when the action would move to the badlands of the jungles west of Camp Rainier. It was there on January 27, 1967 that Medic Donald Evans would display extraordinary heroism leading to his being awarded our nation's highest tribute, the Medal of Honor.

Prelude To The Battle

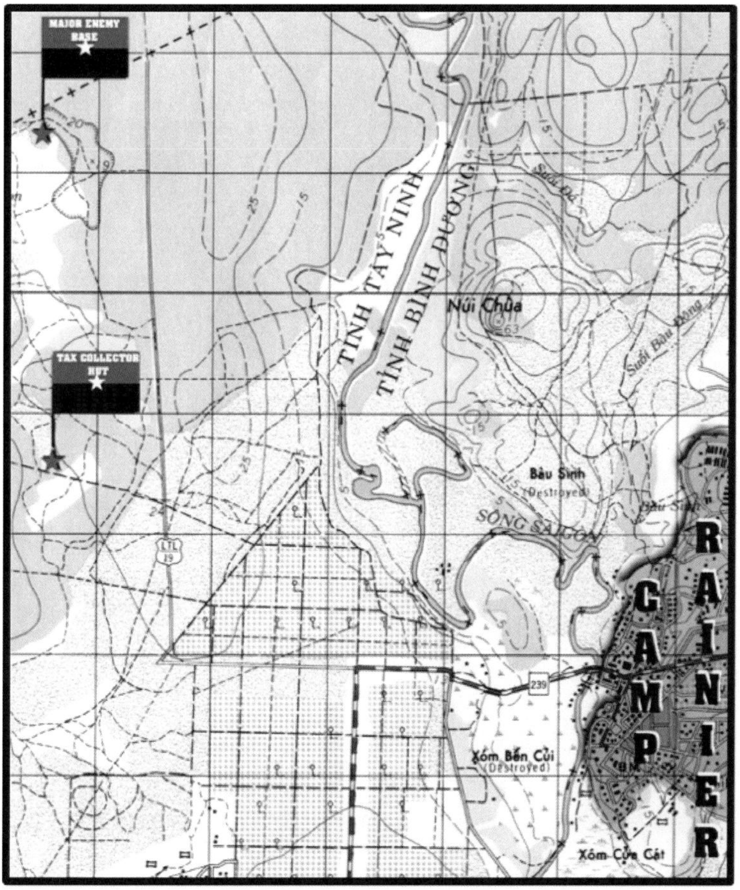

If you look at the map above, you will see the location of the hut where the tax collector was captured during Operation ATTLEBORO and the enemy base camp that was located during the same operation.

On January 27, 1967, A/2/12 Infantry stumbled into that same fortified camp. We had no idea of the history of that camp before we were sent into the base. However, one distinction of this January 27 battle stood out for us. One of our own men would be awarded the Medal of Honor for his gallant action on that day. He would become the first Medic and the first 4th Infantry Division soldier to receive the MOH for action in Vietnam. His name is Donald Evans.

After the January 17 battle, the next few days were mostly routine, spending most of our time on operations around our base camp. There were no incidents that took place which would indicate that the enemy was prepared to draw a line in the sand and fiercely defend one of their bases. That changed on January 27, when A Company would locate a base that was an important piece of the 9th VC Division's war strategy.

January 18

Our battalion remained in the field for one more day after the January 17th battle. Moving north from their earlier position, A Company discovered a trench system, 50 meters long built on an azimuth of 90 degrees and 100 meters along an azimuth of 200 degrees. Upon closer examination of the location, they discovered two bunkers, eight additional foxholes, one hut, one water well, a small pack of documents, one 20 lb. claymore mine, a small amount of small arms ammunition, and assorted clothing.

That base was hidden in the jungle less than a hundred meters from the rubber plantation. It was close enough for the enemy to sneak into the plantation, collect taxes, gather provisions, indoctrinate the local inhabitants, and melt back into the wilderness.

January 19

The 2/12th battalion was airlifted back to basecamp by Hueys for a rest and equipment maintenance period.

The 3/22nd battalion units continued operation southwest of base camp, hopscotching the area via choppers to confuse enemy scouts. Surprisingly, the 3/22nd on this day finally received the rest of their battalion, who were lent to other units elsewhere. This was the first day since Camp Rainier was established on November 28th that the entire 3/22nd Battalion was together again.

January 20

After spending one day in camp, Recon Platoon of the 2/12th and the A/2/12 Company were sent out at 1300 hrs. on a Bushmaster Operation southwest of camp in another rubber plantation area.

A/3/22nd was sent on another Bushmaster operation north of camp near the Sawback Mountain area.

Both units continued to find deserted base camps, but except for a few sightings of small Vietcong clusters, they experienced little enemy interaction.

January 22

On this day, 2/22nd units were reunited at Camp Rainier, just as the 3/22nd was two days earlier. Some of their units were sent south to participate in Operation CEDAR FALLS in early January.

January 23

Both Bushmaster units were recalled back to camp from their operations on this day.

C/1/10th Cavalry, (Buffalo Soldiers) which was a 4th ID Reconnaissance Squadron, was working with us during this period of operation FORT NISQUALLY. On this day, they discovered and destroyed 6,000 lbs. of rice a couple of kilometers from the clearing where A/2/12 was inserted on Jan 20.

January 24

The only incidents reported on this day was the fact that C/1/10 had two mine incidents.

January 25

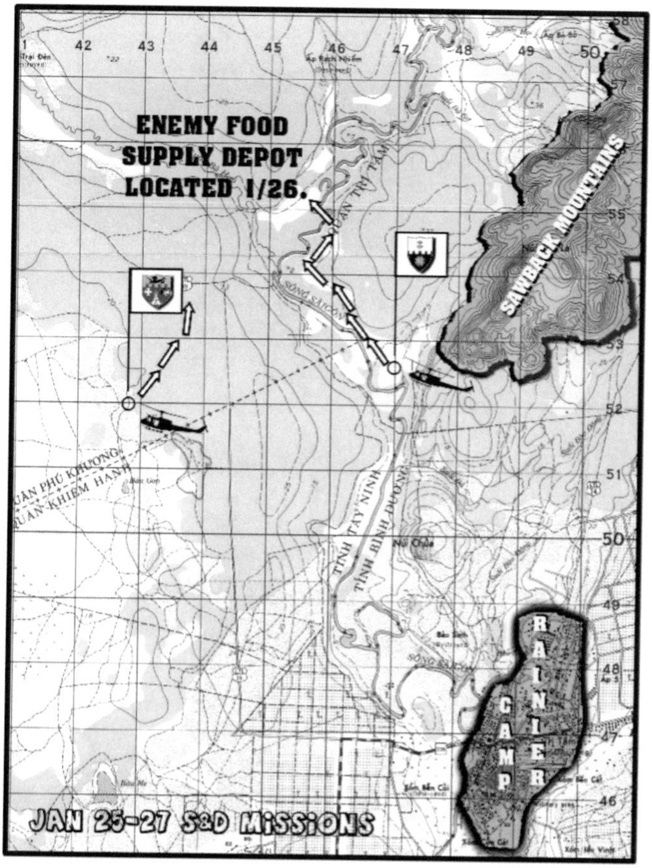

Captain Palmer was elsewhere during this operation and Executive Officer Jim Olafson led the Company.

Both the 2/12th and 3/22nd were airlifted northwest of camp on search and destroy missions. The units landed in clearings and sent their companies in separate directions to search for bases, which typically were built just inside clearings and not far from trails. The 2/12th would be searching on the west side of the Saigon River (a valuable means of transportation for the

Vietcong heading north and south) with the 3/22nd searching on the eastern side. It did not take long before both ran into deserted enemy base camps. The Vietcong were on the move and not yet inclined to defend bases in these sanctuary areas. On this day, the 2/22nd was performing local patrols around Camp Rainier.

January 26

The 3/22nd searched northwest of their drop zone searching the jungle along the clearings along the way. Eventually they discovered an enemy supply depot which contained 11,000 lbs. of polished rice. As that destruction of the rice took place, a small group of VC fired on the troops and the 3/22nd unit disengaged and a heavy artillery attack took place on the base. When they returned after the barrage, they discovered that the base was quite a bit larger than they thought. There were bunkers and huts within the base, which may have had more VC stationed there. Only a rear-guard force engaged the 3/22nd, which allowed others to escape.

It was becoming clear to those who were keeping score, that the enemy was well supplied with food and equipment and this area was their travel destination to resupply troops on the move.

While this action was taking place, the 2/12th was being sent right into a major base camp site that was discovered during Operation ATTLEBORO. The base consisted of a number of concrete bunkers, which were rare in the Dau Tieng area. This base was very near to the location where the Battle of Dau Tieng took place.

Somehow, we missed it and there is no mention of locating it in any of the journals that were written about this period. In

July we would return to this area, when I had three weeks left in my tour, and we did locate those concrete bunkers, which were still partially left standing. Heavy air strikes would eventually destroy that base location that summer, but I have little doubt that it was rebuilt later in the war.

Every infantryman that served in Vietnam would agree that their tours contained many days of monotonous slogs through jungle terrain interrupted by sudden outbreaks of extreme violence. It always seemed that the enemy controlled the tempo of the war, except when they were surprised by our visits. Sometimes they fought; many times they moved on to return to the same location after we left the area.

The soldiers of the 9th VC Division were trained to inflict maximum casualties on their enemy when the odds of their success were at a maximum. They were also instructed to never allow their enemy to assess the casualty count or the condition of their men after a battle. That being the case, most of the time American soldiers rarely felt the exhilaration of determining the extent of damage they did to the VC. Their dead and injured were dragged away, leaving little behind for infantrymen to score. That made it difficult for us to determine if our sacrifices were worth the cost to our units. We only knew what our casualty count was, and it was growing by the week.

I do know that on January 27, 1967, A/2/12 engaged in a battle that would cost our company dearly. Military historians would share the stories of sacrifice and extraordinary bravery, but nothing about the final accomplishment. This was unlike past American wars when a battle would be shown as part of a military strategy and how it affected the strategy of the war.

Westmoreland's "war of attrition" strategy (kill many more

of them than they kill of us) left little for the grunt to celebrate because many times the enemy left behind little for acclamation. It left the American foot soldier with a feeling of bewilderment. We all wanted to do our part, but to what end? For many, it was a war of survival as countdown days to returning to the 'World" became the measuring stick to our 'progress'. With that as a goal, was it any wonder that so many of us returned to America and suffered with survivor guilt. Those of us who survived this affliction only came to terms with our guilt by realizing the fact that our survival was just meant to be.

THE MEDAL OF HONOR BATTLE

Dual Camp Complex Located

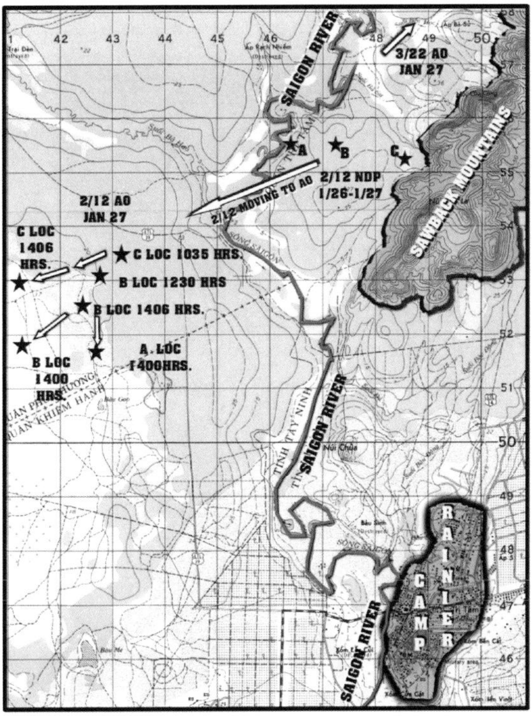

This map illustrates where the 2/12th and 3/22nd battalions were situated according to the different reports that I used to document the situation on January 27th, 1967. The 3/22nd after finding that supply base not far from the Sawback Mountains on the 26th was sent further north on the 27th. The 2/22nd was sent back to Camp Rainier for maintenance and resupply.

The 2/12th was ordered across the Saigon River on the 26th and set up a NDP (Night Defensive Position) just south of the 3/22nd. In the morning, they were once more sent back across the river to search out that area.

The stars on the title page map represent the called in locations for the three companies leading up to 1400 hrs. on that afternoon.

From morning until 1400 hrs., the 3/22nd was able to locate a couple of base camps that were left unprotected in the north. The 2/12th had less luck and up to 1400 hrs. reported no enemy sightings or discovered base camps.

That all changed at 1427 hrs. when C Company reported observing three VC in the clearing and engaged them. The VC snuck into a nearby jungle patch and disappeared.

At 1447 hrs., twenty minutes later, B Company engaged an unknown number of VC and reported they were in pursuit of them. Soon after, B Company requested a dustoff for two WIAs in need of evacuation at 1510 hrs. That incident took place near that sinkhole or land depression. It is shown on this map like an inverted contour line on this map showing the numbers 20 and 19.

At 1448 hrs., A Company reported that they were engaged with five or six VC and were receiving small arms fire and grenades from the enemy. That enemy group escaped without producing any casualties on A Company. A Company then was directed to move east towards the tree line when a bubble helicopter carrying Colonel Garth, the Brigade Commander, appeared overhead. He stopped the movement to the eastern jungle and instead of continuing their march east, they were directed to turn left into the jungle. Lt. Jose Gonzalez, the Second Platoon Leader, forewarned his men, "Prepare yourselves. We are about to make enemy contact."

At 1551 hrs., Second Platoon, leading the company, began that movement into the jungle. It barely entered the woodline at XT435513 when they came under attack from entrenched enemy Vietcong who attacked the A Company soldiers with small arms and machine gun fire and rifle grenades.

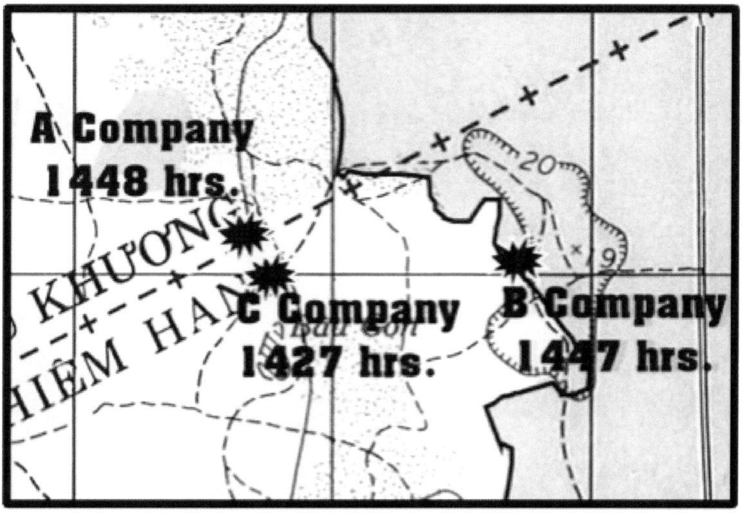

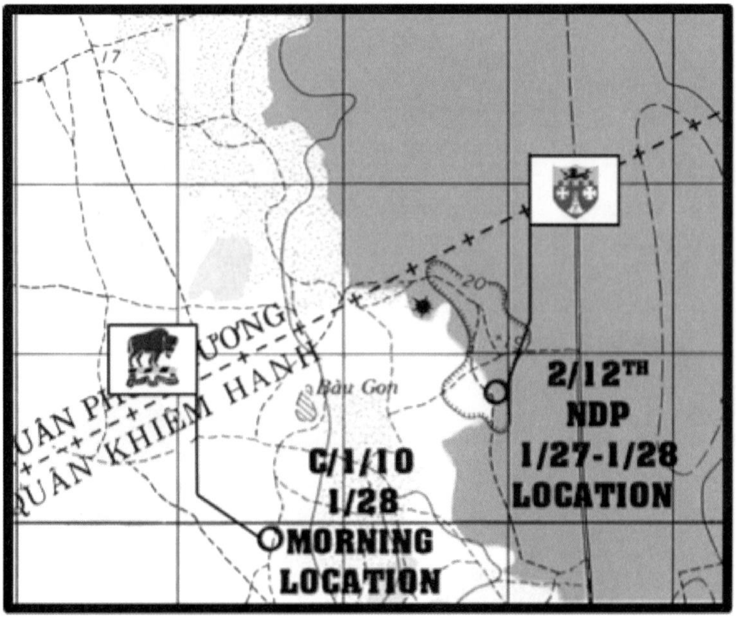

This was to be the first of two separate assaults against the enemy's base camp. When Second Platoon led the company into the battle, they quickly suffered mounting casualties. Medic Donald Evans, then attached to the Third Platoon, rushed forward to the action to treat the Second Platoon wounded. He even dragged some of them back to the clearing.

After the loss of a number of Second Platoon members to enemy fire the order was given to move back to the clearing to reorganize and treat the wounded. After a short while, the order was given to A Company to return to the base camp and to once more attack the enemy position. That short interruption was all the enemy needed to fortify their positions and move up reinforcements. When the Third Platoon was sent back into the jungle, Don Evans was still treating the Second Platoon wounded. He left them in the care of the Second Platoon men and reattached to his Third Platoon unit and headed back into the action. The losses quickly mounted before the order was given to move back to the clearing.

These are the stories of the men who participated in that significant battle.

Porter Harvey, a Third Platoon Squad Leader Had This to Say About What Led Up to That Battle:

"We began that day searching for enemy through the brushwood area that features vast areas of high grass surrounded by jungle. Not the type of terrain to please an infantryman.

I thought the turf that we walked through on that day was prairie like, with small brush and twigs scattered around the clearings. The pace set that day made me think that the Company was being pushed hard to reach some undeclared objective. We spent the day in open fields moving through one clearing to the next.

At one point, late in the afternoon, we were ordered to move into the jungle to the east. The Company marched to the east in columns, with Second Platoon in the lead. They were followed by Third Platoon with the First Platoon and Weapons Platoon taking up the rear."

Henry Osowiecki Was a Squad Leader in the Second Platoon On That Day:

"We were working through the clearings all day long. The pace was demanding and we were looking forward to the end of the day when we would have a chance to rest.

"As our platoon pushed through the clearings, we noticed a command helicopter flying tree top level over the area. Suddenly that chopper raced past us at low altitude. Shortly afterward, a command was given to the Company to stop and hang a left and enter the tree line to the north. I just knew that we would be making contact with the enemy. I was not mistaken.

We swung our column to the left and slowly approached the jungle. It was shortly after we entered the jungle when we received a hail of gunfire and grenades. Bunker and entrenched enemy were waiting for us and we were taking casualties at an alarming rate."

Porter Harvey, Added This to the Story:

"When Second Platoon came under fire, John Faidley, from my platoon, the Third, passed by me moving up the line to help the Second Platoon that had taken casualties and were trying to get them evacuated from the battle line. I asked him why he was moving up to the front before his platoon was "committed." John told him, "That's where the bad guys are, aren't they?" It would be the last time that I saw John alive. No one who knew John would be surprised to read this story. John was one of those soldiers whose instincts would always lead him to where the action was taking place.

"John worked his way to the front where he became a casualty himself and was unable to move. A number of men tried to retrieve him from the front, but the enemy was using him as bait and were picking off those who tried to save him."

It should be noted that the brigade was suffering a disproportionate number of casualties whenever men tried to save their buddies who were pinned down in the front of a battle. This exasperated Colonel Garth, the Brigade Commander.

Ed Smith, then on brigade staff, explains Garth's frustrations this way: "Garth had a meeting with the brigade staff officers to tell them that he would no longer approve any medals for any soldiers who went to save their buddies who were pinned down at the point position. We all looked at each other in disbelief. How could we share that with the men in the line units? The message never was passed down the line." More about Colonel Garth's attitude towards medals shortly.

Henry Osowiecki

"An immediate call for a medic rang through the jungle. The second platoon medic, a recent arrival in country only a week earlier, froze and refused to move into the withering fire that was emerging from the front. The medic just kept saying to bring the wounded to him. This is not written to judge his behavior one way or other, just to lay the groundwork for what was to happen later. I doubt that there were many of us who did not hesitate or freeze a bit when an order was given that represented something that was counter to our survival instincts.

"This is where daring was needed and Doc Evans, the Third Platoon Medic, responded to the challenge. Before his platoon was committed to the battle, Donald Evans moved up into the fray. His heroics that day have become an inspiration to the men of Alpha Company and dare I say, to all the medics of the 4th ID who followed.

"Finally, the Second Platoon was able to disengage and return to the clearing. A few minutes passed before a second push into the enemy base camp was made. This time it was Third Platoon's turn. The platoon moved past the men of the second platoon who were busy administering first aid to their wounded."

Porter Harvey

"Slowly we entered the jungle and it was not long before enemy fire erupted from the bunker line. They seemed even more prepared for us than they were when Second Platoon arrived earlier. My platoon was picking up casualties even greater than the original assault.

"I remember two incidents that come to mind from that battle. One involved a man who was shot through the throat and had to have an emergency tracheotomy performed on him during the battle to allow him to breathe. (He survived this and at the 2003 reunion he was able to meet the family of Donald Evans. Many more survivors of that battle were there to celebrate the gallantry of Donald Evans on that day.)

"A second incident took place when I moved up the line looking for wounded and cover for them while they awaited extraction. One man, Bobby Benson, who survived and is in our association today, had been shot in the groin area by the time that I reached him. He was bleeding pretty profusely by the time that I crawled to him. He asked me to check to see if he was shot in his 'family jewels'. I was surprised by the request, but I checked anyhow. I told him that the bullet was in his thigh. He said, "Good, I thought I was in trouble there for a while." So here we are hunkered down in a pitched battle with an entrenched enemy and he was relieved even though we were still under heavy fire. Eventually I was able to get him to the rear where he was treated. He too survived the battle, was sent to a hospital to recover, and once more was sent back to our company."

In Vietnam, unless a soldier received a wound which would lead him to being unable to function in battle, he was sent back to his unit after he recovered. If a man received three wounds, he would qualify to be sent to the rear unless he chose to remain on the front. This man did in fact return to us after he recovered. Three months later he was wounded once more. He survived Vietnam and is today living in upstate New York, not far from where he was drafted 55 years earlier.

One of the men who was shot during this battle was Larry Barton. He was the Third Platoon Leader's RTO and one of those who recovered and rejoined us after his recovery. He was the man who replaced me when I was chosen to carry the Company Commander's radio in Vietnam and was a friend from my 3rd Platoon training days. You will read more about him later.

Bob Livingston, a Third Platoon Squad Leader, Added These Recollections of that Battle Here:

"I remember how hot it was on that day and I don't think we were all low on water at the time of the battle. I remember going forward with the 3rd Platoon. When we got close enough to the bunkers, I told Brian Cosman to shoot directly into an aperture of a small bunker. He did and that silenced that gun. I saw Porter Harvey throw a hand grenade and I heard it bounce off something and didn't go off.

"I saw Donald Evans moving about and I shouted at him, 'Where's your rifle?' He never answered me. He just went forward and went about his business of treating the wounded.

"At one point during the battle, I saw Sgt Jim Harris carrying a wounded man on his shoulder and bringing him to the rear and out of harm's way. I remember R.L. Smith, Gary Hipp, and I going forward to get Armand Aufiere who was wounded and up a small trail. When we arrived, we saw him just lying there lifeless behind a small log on the trail. I remember the VC shooting at Smith and I as we stood up to get Armand away. The rounds from that burst severed a small tree right between us. We then saw the hole in Aufiere's head and gray matter that was exposed on the left part of his forehead. One of the Vietcong was close enough that I could see that he intended to grab Aufiere's rifle as we pulled him away. I used my front site blade to snag his rifle sling and pulled it away with us. Rounds were hitting all around us during this withdrawal.

"After getting Aufiere back, we reentered the battle and it was then that I saw Doc Evans coming down a trail with his arms flapping like a rag doll. He ran right into me and started to collapse. I used his pack straps to drag him along. I knew this was very bad because it appeared that the muscles between his elbow and bicep had been shot or blown away. I remember yelling to another guy to give me a hand but he just could not or would not get up. A medic came up and we rolled Doc over, I yelled in Donald's ear to stay with us and he took a couple of big gulps of air and was gone.

"During the battle, I remember Sgt Armstrong being shot in the throat and the medic trying to keep his airway open.

"Eventually we were pulled back and the attack was taken up by artillery and air strikes.

"What I concluded on that day was that it was a terrible waste of good men injured and killed for such a shitty piece of ground. (*More on that shortly.*)

"It was such a sad dirty war. I can call these memories up on the screen of my mind almost at random and sometimes they come unbidden."

While this battle was taking place, B Company, situated in the jungle southeast of the engagement, was ordered to redirect their movement to the east and to move to the battlesight to flank the enemy on the right. They didn't get there in time and A Company was withdrawn from the battle and air strikes were ordered on the enemy base camp. A Company withdrew at 1650 hrs., an hour after Second Platoon entered the enemy base camp.

All the wounded were retrieved with the exception of John Faidley. No one knew if he was alive or not, but they did know that he was so close to the enemy that every time anyone tried to retrieve him, they too became a casualty. The heartbreaking decision was made to leave without him. His rescue was futile and was only adding to the casualty list. Not knowing if he was still alive didn't help. Those that went to the front saw no movement from John when they approached him. It was a tough call, but the right one in my estimation.

At 1700 hrs. 3rd Brigade HQ was informed that A Company suffered 2 KIAs and 10 WIAs from the earlier battle. Sadly, it would grow to 3 KIAs on the next day when A Company returned to the battle site and found John Faidley's lifeless body.

At 1802 hrs. it was reported that A Company sent an element back to the battle sight after the air strikes ended. They were attacked by snipers as they approached and drove them away then remained in the area to observe the earlier battle location for the evening.

At 1950 hrs. the 2/12th battalion reported this to Brigade Headquarters:

"Evaluation of activity—A multiple bunker system was discovered at XT435513 FAC pilot reported that he spotted a series of trenches, plus secondary explosions that took place where bombs were dropped. Estimate that a minimum of a Company of enemy manning the base. B & C had arrived at the NDP with A Company enroute."

At 2020 hrs. the 2/12th reported that A Company had arrived and they had set up a battalion sized NDP in the jungle near the clearing at XT439508. They were then approximately 600 meters southeast of where the battle had taken place earlier.

At 2035 hrs. the 2/12th requested three more air strikes,

At 2110 hrs. one platoon of the C/1/10 Cavalry (Buffalo Soldiers) was ordered to move at first light to XT 425499. Their mission was to support the 2/12th troops in the morning to determine the scope of the camp.

At 2135 hrs. an additional airstrike was placed on the enemy base and an additional airstrike was planned for daybreak.

The Day After The Battle

January 28, 1967

0826 hrs. 2/12th reported that the reinforcements, 1st Platoon, C/2/12, had arrived west of their position and were preparing to support a movement towards the Jan 27 attack location. When they were in position, all units moved forward to the battle location.

Arriving at the base, they secured the area and searched the base.

At 0930 hrs. B Company reported locating 20 bunkers and a tunnel complex, each about 100 meters apart. They indicated that the air strikes had damaged the tunnel system. Two burned weapons were discovered, but nothing remained of their owners. One additional enemy soldier fatality was located and deserted by his comrades. That was unusual.

At 1016 hrs., B Company reported that they had found a freshly dug grave and would be digging it up.

At 1038 hrs. B Company indicated that after digging up

that grave, "they found nothing but a hole." Perhaps something important was buried there by the enemy and was dug up and taken with them as they evacuated the base?

At 1124 hrs. "A Company reports that they found the MIA reported yesterday, MIA status changed to KIA". This was John Faidley's body.

At 1246hrs. B Company located another part of the base camp 50 meters east of the original battel coordinates were. Here is the entry for that time period: "B/2/12 found a base-camp at XT433514. Within the base camp was a mess hut 20 feet X 15 feet, a living area and tunnels. In addition, they found 20 thatched roof huts around the entire perimeter of the base. They saw many E and F Silhouettes (*E and Type F target are familiar silhouettes used for target practice. The former represents a standing man, the latter, a man in prone position).*

A dummy training rifle was also discovered. Numerous papers, including training manuals, and personal articles were captured. Destroyed the huts and all personal articles and forwarding papers on."

At 1307 hrs. A & B remained in the area and C company was sent to XT434498, near a jungle area south of the battle-ground.

At 1515 hrs. helicopters arrived for resupply of 2/12th Infantry.

At 1625 hrs all 2/12th troops headed to join C Company who remained where they were sent earlier and arrive at 1700 hrs. C/1/10 unit is released from 2/12th control and headed back to Camp Rainier.

For those of us on the ground, it was just another base discovered which at first was thought to be guarded by a squad, later determined to be at least a company during the action. Both estimates were wrong. It was bigger and much more important, as evident by the After-Action Reports that were released on June 6th. That was long after most the combatants who served in A/2/12 during the battle were transferred to other units. They never learned of the significance of that battle for the security of Tay Ninh/Dau Tieng area of operation.

Here is what was written about that battle in the AAR:

> Operation Fort Nisqually After Action Report, dated 6 June, 1967
> C. Enemy Locations:
>
> Item (5) Operations to the south of Ben Cui Plantation located a munitions factory XT435513. An examination of the factory indicated at least mines, small arms ammunition, and grenades were manufactured there. Mining incidents along Route 239 dwindled to nothing after the factory was destroyed. It was approximately one month later before significant mining incidents again began occurring along Route 239.
>
> Item (6) The large training complex located on Jan 27 XT435513 was identified as a large training area for the 380th Training Bn. of the 9th VC Division. A camp operated by the same unit was located in the same area during Operation ATTLEBORO."

You will note in Item 5 that the destruction of the factory base resulted in safeguarding route 239 from mines for a month. Route 239 was an essential road leading southwest of Dau Tieng where it would run into Hwy 22, a major north/south road leading to Cu Chi and Saigon. Forces from Dau Tieng and Tay Ninh used this road daily for resupply convoys. Finding that munitions factory and eliminating the threat of mining incidents was indeed a big deal. I would bet you any amount of money that the men who discovered that base had no idea what impact they had in securing the main supply route for both Tay Ninh and Dau Tieng.

Item 6's notation was most interesting to me. "*A camp operated by the same unit was located in the same area during Operation ATTLEBORO*". Oh really? How come when the Second Platoon arrived just south of the camp why were they not warned about what was found there less than eight weeks earlier? What did Colonel Garth and his pilot see from above that was not relayed to the people on the ground? We'll never know the answer.

When the company returned to camp a couple of days after the battle, Henry Osowiecki was able to document the names of most of the men who became casualties on January 27th.

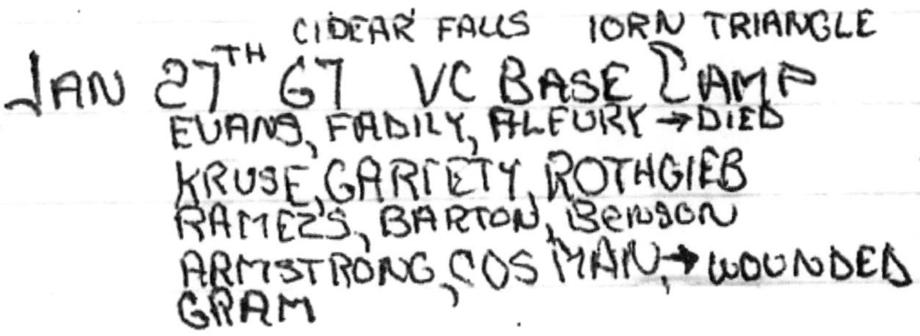

CIDEAR FALLS IORN TRIANGLE
JAN 27TH 67 VC BASE CAMP
EVANS, FADILY, ALFURY → DIED
KRUSE, GARRETY, ROTHGIEB
RAMEZS, BARTON, BENSON
ARMSTRONG, COSMAN, → WOUNDED
GRAM

This is intriguing to me for a couple of reasons. First of all, note that he wrote that the battle took place as part of Operation Cedar Falls. He also notes that it happened in the Iron Triangle. Operation Cedar Falls had ended on January 26th, the day before the battle. It was part of Operation Fort Nisqually, a 4th ID brigade operation. Also, it did not take place in the Iron Triangle. We fought that battle more than 30 miles north of the Iron Triangle. I remember that in January we received a lot of misinformation. There were spies everywhere, so today I can only assume that we never knew specifically where we were going or what our plans were to prevent us from spilling the beans. Do you know what? They were probably right.

Henry identified 3 KIA and 9 WIA from our Company who became casualties from that battle. His diary was close. The official tally recorded by brigade records was 3 KIA and 10 WIA from the January 27 engagement. Henry missed one.

As of that day, nobody from A Company who participated in that battle had an inkling of the significance of that battle. Many will not know the full story of the battle until they read it here.

The Aftermath

The battalion returned to base camp on January 29th. I had just arrived back at Camp Rainier from R&R as the battle was taking place. One of the guys from the Headquarter Section of A Company knew we were coming back so he met us as our chopper touched down at the helipad. When we jumped off the Huey, he ran over to us to tell us that A Company was in battle and was taking heavy casualties. We ran back to our area of the base and climbed into the communications bunker to monitor the radios. By that time, 3rd platoon was being committed to the battle and things were going from bad to worse. We could hear through the radio speakers the sounds of the explosions, gunfire, and the frantic communications emerging from the battle. We knew something big was happening, unlike any of our earlier battles.

When the company returned from the field a couple of days later, we met them as they trudged down the company street nestled in the rubber trees at camp. There was little of the chatter that was typical as men returned to camp for a couple of days of rest. They seemed in shock.

I followed the third platoon men as they silently traveled to their platoon tent. Some threw gear around their area in anger. This was not the same group that I left only ten days earlier south of the Michelin Plantation.

I visited my friends from the Third Platoon, the platoon that suffered the three KIAs and I could tell that they needed to vent. I let them express their frustrations and listened to their stories one by one. The emotions displayed were quite remarkable. Some were despondent, some were angry, while others wondered how they had gotten into such a mess. It was the third time that they were reminded that war was, in fact, a very dangerous proposition.

It brought me back to 1SG Springer's Vietnam leave speech. "I want to bring all of you back home, but I am not going to be able to. Some of you will lose your lives. Some will be crippled. Just the way it is." Or the other pronouncement, "It's not very hard to get a man to go to war. Getting him to go a second time is far more difficult." In fact, we did have men from my company return to Vietnam for a second and sometimes, even a third tour. Most went home and tried to forget the horrors of what they saw.

However, even more than those declarations, was Springer's warning, "Someday we are going to come up on a base which the enemy decides he will ferociously defend. That is when you will know the true meaning of the Combat Infantryman's Badge. That is when you will be required to earn the honor of wearing it."

I spent a half an hour or so listening to the stories of my

friends when Lt. Jim Olafson, who led the company on that day entered the tent. That's when I left and returned to the HQ Section tent where they were going over what they experienced or learned from the battle.

Jim Olafson presented himself to get the more complete story of Donald Evan's heroics during the battle. As the commander on that day, he felt the need to properly determine how Donald should be recognized for his action on that day. After listening to the combatants, Jim felt compelled to submit Donald Evans for a Silver Star for Gallantry. It seemed a sure thing based on what he recorded from the guys engaged in that battle. Donald deserved that recognition and Lt. Olafson was determined to succeed in acquiring it for Donald.

After the debriefing, Jim returned to the Company HQ tent. He sat down and began writing the letters to the families of the three slain A Company soldiers. As Executive Officer, it was his responsibility to extend the condolences of the company to the kin of those who lost their lives while serving in his unit. It was a difficult but necessary procedure that tugged at his heart. For the soldiers, it was an accepted part of the deal. For the families, it was different.

He finished the letters, each very personalized, and began to organize the details of Donald Evans' heroics with a recommendation that he be awarded a Silver Star for Gallantry, posthumously. The normal procedure was to send the recommendation up the hierarchal line first to company level, then to the battalion, and on to the brigade for approval. If it was rejected at any level, the recommendation was disapproved.

Company and battalion's approvals were quick in coming but then the recommendation was held up at the brigade level. As I recorded earlier, Colonel Garth was not one to indiscriminately authorize medals.

He turned down Jim Olafson's Silver Star request cold, but he did say that he would approve a Bronze Star instead. Perhaps this was because of his earlier opinion that he was losing too many men who were dying trying to save their buddies. Jim, not knowing the earlier directive that was never passed down to company level, was incensed at Garth's decision and this only made him more determined to resubmit another recommendation with more information to increase the chances of approval.

He returned to the men who saw Evans' heroics and reinterviewed them. At the end of those interviews, Jim thought to himself, "You know, I think old Garth was right. Evans did not deserve a Silver Star. He deserved the Medal of Honor!" With that in mind, Jim began the slow process of recommending Donald Evans for our country's highest and most prestigious military decoration that can be awarded to recognize American soldiers in combat.

He then wrote a recommendation for the awarding of the medal of Honor for Donald Evans with the additional details.

He began the process by seeking the approval of Captain Jon Palmer, the A/2/12 Commander. Obtaining that consent, the proposal was forwarded to battalion where it was carefully considered. From that point on, Jim never heard any more about the request until he left Vietnam. The good news was it was not disapproved outright.

I, like most in the men of my company at the time of the battle, were aware that Doc' Evans was recommended for the MOH but didn't have much confidence that the effort would be successful.

I left Vietnam in late August, 1967, and was discharged from the army on December 12, 1967.

Jim left Vietnam in September, 1967, and remained in the

army. Having not heard anything about the progress of the MOH effort, he put the recommendation in the back of his mind.

The wheels move slowly in the army and sometimes those who served in combat never get the updates on medals until long after they return to civilian life. Case in point, when our unit was awarded the Presidential Unit Citation in October 1968, few of us draftees were aware of it.

During early July, 1968, 17 months after the January 27, 1967 battle, I bought a *Boston Record American* newspaper (now defunct). On the back page was a copy of a picture showing Donald's family in Washington DC. They had been invited to a ceremony when the Medal of Honor would be presented to Donald's family. I didn't know much about his family, but I did know Bonnie through pictures that Donald showed me while Donald and I were training together. I knew then that Jim Olafson's efforts were rewarded and Donald Evans was indeed awarded the Medal of Honor for his extraordinary gallant action during that action in Vietnam. I was elated.

Here is the General Orders which authorized the MOH award for Donald Evans:

"General Orders

34

HEADQUARTERS

DEPARTMENT OF THE ARMY

Washington D.C., 5 July 1968

AWARD OF THE MEDAL OF HONOR

CITATION: For conspicuous gallantry and intrepidity in action at the risk of his life above and beyond the call of duty. He left his position of relative safety with his platoon which had not yet been committed to battle to answer the calls for medical aid from the wounded men of another platoon which was heavily engaged with the enemy force. Dashing across 100 meters of open area through a withering hail of enemy fire and exploding grenades, he administered lifesaving treatment to one individual and continued to expose himself to the deadly enemy fire as he moved to treat each of the other wounded men and to offer them encouragement. Realizing that the wounds of one man required immediate attention, SP/4 Evans dragged the injured soldier back across the dangerous fire-swept area to a secure position from which he could be further evacuated. Miraculously escaping the enemy fusillade, SP/4 Evans returned to the forward location. As he continued the treatment of the wounded, he was struck by fragments from an enemy grenade. Despite his serious and painful injury, he succeeded in evacuating another wounded comrade, rejoined his platoon as it was committed to battle, and was soon treating other wounded soldiers. As he evacuated another wounded man across the fire covered field, he was severely wounded. Continuing to refuse medical attention and ignoring advice to remain behind, he managed with his waning strength to move yet another wounded comrade across the dangerous open area to safety. Disregarding his painful wounds and seriously weakened from profuse bleeding, he continued his lifesaving medical aid and was killed while treating another wounded comrade. SP/4 Evans' extraordinary valor, dedication, and indomitable spirit saved the lives of several of his fellow soldiers, served as an inspiration to the men of his company, were instrumental in the success of their mission, and reflect great credit upon himself and the Armed Forces of his country.

GO 34

By order of the Secretary of the Army:
W. C. Westmoreland,
General, United States Army,
Chief of Staff
Official: KENNETH G. WICKHAM,
Major General, United States Army,
The Adjutant General.

Many of our men never learned of this award until decades later when they learned it from us in our association newsletters.

Jim Olafson who did all the heavy lifting so that Donald was properly recognized for his bravery never received notification of the award. Jim learned of the award only through an article in *THE ARMY TIMES*. The General Order which authorized the medal for Donald was dated on July 5, 1968. He learned of it when he received the next issue of Army Times, two weeks later.

That citation that accompanied that MOH presentation essentially reflected the words of the men who were at the battle of January 27, 1967, and chronicled by Jim Olafson after the battle. That Medal of Honor award had the distinction of being the first awarded to a Medic during the Vietnam War and the very first soldier to receive it for the Fourth Infantry Division.

It took less six months before Donald Evans would be additionally honored at military bases around the world. The first to do so was Camp Foster, a Marine base on Okinawa. They named their Branch Medical Clinic (BMC) after Donald Evans (BMC Evans). That clinic is an extension of the U.S. Naval Hospital on Okinawa.

In January, 1969, Bonnie, Elsie, Donald Evans Sr. were

flown to Okinawa to attend the ceremony naming the clinic BMC Evans to honor Donald.

It should be noted that from that day on, Elsie Evans was a tireless advocate for veterans. She was tenacious in dealing with officials whenever veterans were involved. It would not be an overstatement to say that she devoted the rest of her life to whenever veterans were involved. It would not be an overstatement to say that she devoted the rest of her life to veteran causes. It was said that at the Covina city hall, whenever she arrived to challenge officials on veterans' issues, the common refrain was. "Uh-oh, here comes Elsie."

Her crowning achievement was the establishment of the Donald Evans Jr. Memorial Courtyard behind the Covina city hall. The plaza was dedicated on Nov 11, 1997. The plaza paid tribute to Donald and the other 19 Covina area men who lost their lives during the Vietnam War.

Fort Sam Houston is the fort where Donald and all the Vietnam era medics of the U.S. army train for duty. On June 1, 1972, the auditorium was named in honor of Donald and every Army Medic since then has received their graduation certificates on the stage of that assembly hall.

A larger, more modern hospital was proposed early in the 1980s at Fort Carson, Colorado. After construction, a number of names were considered in the naming of the facility. The most popular choice was to name the medical center after Donald Evans.

The official dedication took place on 5 June 1986 and Donald's family and Jim Olafson, who promoted Donald's Medal of Honor, were invited to attend.

This is a story I heard from Richard Evans about the days leading up to the dedication.

"Prior to the dedication, Don's mother was shown a copy

of the painting. She noticed something that nobody else had picked up on. The painting showed Donald wearing his Medal of Honor medal, which would have been ludicrous as Donald was presented the medal posthumously. She insisted that the painting be altered and the medal be removed. As usual, Elsie was placated and the medal was eliminated from the painting before the dedication."

On the day of the dedication, the family met Jim Olafson for the first time, but the schedule was so tight that they never really got to spend much time with each other. It was not until our 2002 annual reunion, 16 years after the dedication, when Jim and Richard met once more and were able to share stories with one another in detail.

There were other dedications to Donald over the years, but these earlier honors were the most important.

A Moment In Time

On January 27, 1967, Donald was engaged in a battle where he displayed extraordinary heroism. The battle took place almost 19 months after President Johnson notified the American people in a televised speech on July 28, 1965 that he was sending whatever troops would be needed to deal with the Vietcong threat in Vietnam. That was a turning point in the war. We were about to reach another milestone in the war; the year to crush the NLF with massive forces.

American troop strength in Vietnam reached 400,000 during the week of January 27, 1967. A total of 123 Americans were killed and 716 were wounded during this one week. To that date, January 27, 1967, 6,978 Americans were killed and 39,977 were wounded in Vietnam. However, the allies were in defensive mode up until that date. That was all about to change as the allies were set to go on the offense and attack the enemy's strongholds all over South Vietnam. The troops were finally in place to move to a new and riskier phase.

U.S. troop numbers peaked in 1968 with President Johnson approval of increasing maximum number of U.S. troops in Vietnam at 549,500. The year was the most expensive in the Vietnam War with the American spending \$77.4 billion (\$576 billion in 2021 dollars) on the war. Even more sobering was the American death total that reached 11,363 deaths by the end of 1967.

American forces and their allies up and down South Vietnam began in earnest the 'Year of the Offensive' that was designed to make the North Vietnamese pay a staggering price in lives for their efforts to gain control in the south. The effort was successful in that respect, but meant little to convince the Communist to give up their effort.

WAR ZONE C

Operation Gadsden

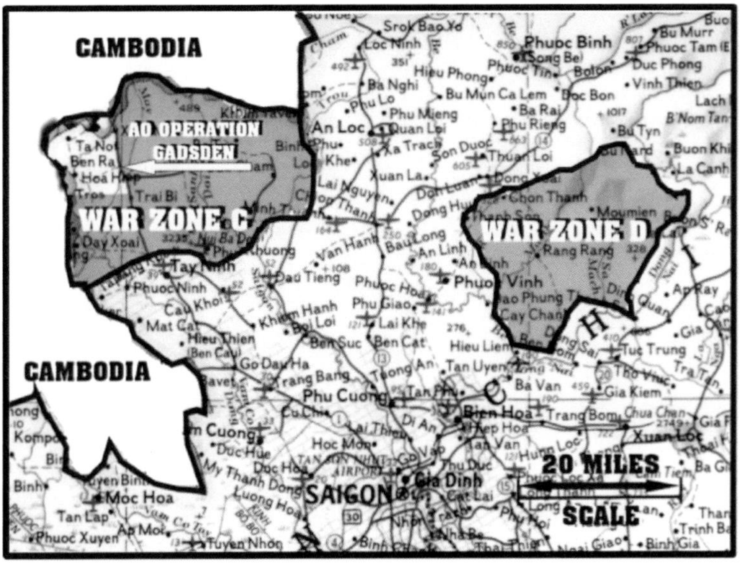

CAMBODIA
Operation Gadsden
2-21 Feb, 1967
FSB Lee
LTL
20
QL
22
Logo
Trai Bi
Xom
Giua
Tay Ninh
West
Tay Ninh
City

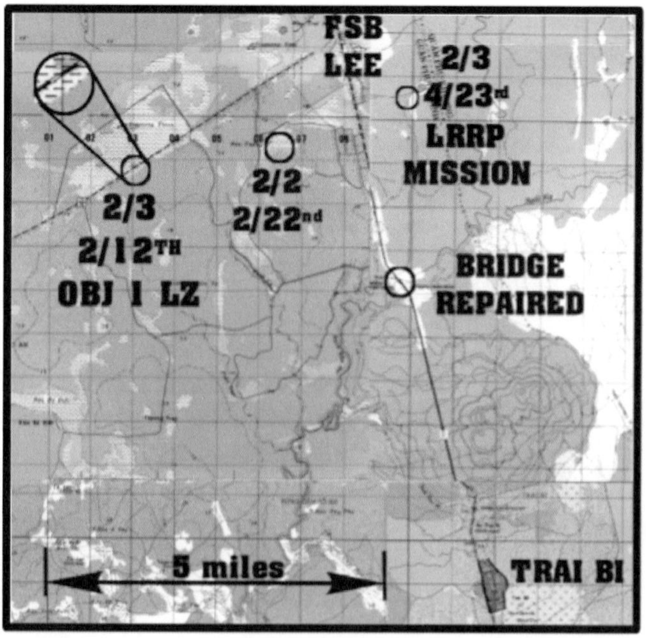
FSB
LEE
2/3
4/23rd
LRRP
MISSION
2/2
2/22nd
2/3
2/12TH
OBJ 1 LZ
BRIDGE
REPAIRED
5 miles
TRAI BI

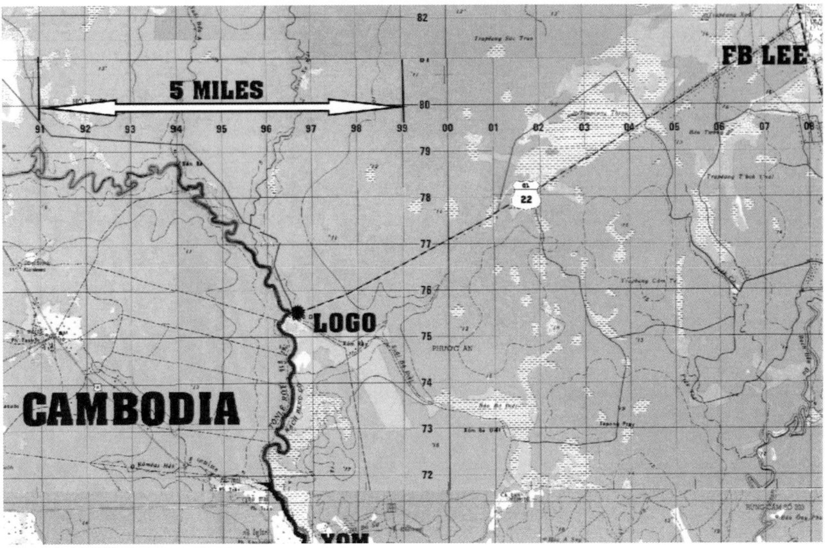

February 2–21, 1967

There was little time to grieve for our Company as we were now set to leave our security assignment and become part of the big push into War Zone C. In the previous two weeks, A/2/12 Infantry had suffered 16 casualties and three deaths in our 160-man Company. It was time to brush ourselves off and move forward, a little wiser for the effort.

Operation Junction City, the big one, had been in the planning stages for many months. Except for the pressure on Saigon, it would have happened earlier. To set the table for the Junction City maneuver, a diversionary operation named Operation Gadsden was put in place. The 3rd Bde. of the 4th ID, under operational control of the 25th Division, was to be sent into northwestern Tay Ninh Province between Highway 22 and the Cambodian border.

The premise for the attacks into the western border area of

War Zone C and inside War Zone D were to misdirect the enemy's attention with a couple of deceptive moves.

Gadsden began on February 2, 1967 and ended on February 22, 1967. It was conducted by one brigade of the 25th Infantry Division, the 196th Light Infantry Brigade, and our Third Brigade of the 4th Infantry Division. Operation Tucson was conducted by two brigades of the 1st Infantry Division and deployed to the Long Nguyen Secret Zone in the western border of War Zone D. The Long Nguyen Zone was a well-known enemy sanctuary base and was well defended. Tucson began on February 14 and ended on February 18 with little enemy contact.

Lo Go and Xom Giua were deserted border villages and were major supply centers for the Viet Cong forces where shipments from Cambodia were transferred to local units. It was believed that the area of operation contained extensive supply and ammunition caches, communications storage areas, hospital facilities, base camps, and major training complexes. In addition, personnel and supply routes to and from Cambodia were expected to be found. All of that proved to be true.

We were going to interdict those supply routes of the enemy during the Tet Holiday, February 8-12, 1967, which was always a dangerous period in South Vietnam.

We were told to expect to be challenged by units of the 271st VC Regiment of the 9th VC Division who used this AO as a haven and resupply base.

Prior to actually being sent into War Zone C, we were told for the first time during our Vietnam tour that we were being deployed into a "free fire zone", where every person we encountered in the jungle was considered enemy and subject to attack.

The 271st Regiment was then a 2,400-member unit that

was thought to be operating in the area between Logo and Hwy 22 to the east.

The operation was under command of Colonel Marshal B. Garth, Commander of the 3rd Brigade of the 4th Infantry Division. The units involved in the operation were:

2/12th Infantry—LTC Joe E Williams.

2/22nd Infantry, Mechanized—LTC Ralph W Julian

3/22nd Infantry—LTC John A. Bender (20 Feb-21 Feb)

4/9th Infantry, 1st Bde, 25th Division—LTC Robert A. Hyatt (12 Feb-17 Feb)

4/23rd Infantry Mech,1st Bde, 25th Division—LTC Walworth Williams (31 Jan-18 Feb)

2/77th Field Artillery—LTC Walter R. Rehm

3rd Support Bn (Prov)—LTC Joseph A. Westbury

Co D, 27th Engineers—CPT Johann R. Kohler

Additional units arrived in the area as more and more bases were located.

Operation Gadsden Brigade Reports

Jan 31

The 4/23rd Mechanized Infantry, 1st Bde, 25th Div. was attached to our brigade.

Feb 1

The operation began for the brigade with a motorized movement to Trai Bi by the 2/22nd Mechanized Infantry and the 2/77th Artillery Battalions on this day.

The movement was uneventful.

Feb 2

The 2/22nd was sent forward from Trai Bi to secure Objective 1 which would be used to insert the 2/12th into our new AO for the operation. Twenty-eight air sorties were flown across the axis of advance and objective area that day. When the advance reached a river which had been destroyed along their route, the engineers were choppered in to repair the crossing site.

Along the route, the APCs were slowed down by a number of mines that were placed along the road and a few sniper incidents.

They ended three kilometers short of the site where they needed to secure a landing zone for the 2/12th troops. This resulted in the 2/22nd not making it to their objective before nightfall.

While 2/22nd was on their movement north, the 2/12th was airlifted into Trai Bi on C-123s aircraft at 0830 hrs. The

3rd Brigade of the 4th ID Command Group arrived with us and set up a Forward Commander Center in Trai Bi.

The 2/12th immediately began preparing for a helilift to Objective 1 (XT030785) later that day. As the LZ was not secured, our battalion spent the night at Trai Bi. It seemed that command had a tendency to overestimate the ability to negotiate the jungle and the enemy resistance to arrive at a predetermined site in time to secure a landing zone. This would lead to a disastrous outcome during the next month.

What I recall most from our arrival at Trai Bi was how the base was secured by a Special Forces A Team, numbering ten men and a small contingent of CIDG forces before our arrival. The commander of the Special Forces group met us at the landing strip and I was impressed by their boldness. They seemed fearless to this young draftee.

Feb 3

This day the 2/22nd was able to secure Obj 1 and the 2/12th flew into the clearing on choppers beginning at 0915hrs. Our mission was to conduct search and destroy missions and secure FSB Base Lee, just north of the landing zone.

As you can see, from that LZ we were now within five miles of Logo, our first target and XOM GIUA, our second target was just three miles south of Logo. However, there were many enemy base camps in the area to protect those two targets.

As soon as the 2/12th arrived at the LZ, the 2/22nd headed southwest towards Logo. At 1303hrs. after traveling about a mile, they came under fire from automatic and small arms fire from 5-7 VC set up along the road. Seven of their troops were injured but not killed before the enemy was driven off. One VC was killed and they captured his AK47 assault rifle. In addition, they collected his arsenal of four Chicom grenades, 47 magazines (this man was armed to the teeth if you can believe these records) 205 rounds, and two cartridge belts.

At 1705hrs, the battalion discovered a large cache of items and food supplies. By 1800hrs. one VC was killed on the river. 2340hrs. VC probe their night position with Automatic Weapons, and on it went.

The enemy was definitely in our area of operation.

The terrain in our AO varied from low flat terrain and cultivated fields to scrub brush and forested area with double and triple canopy. The latter offered excellent concealment and poor observation. Avenues of approach were generally limited to developed trails and cleared areas. Obstacles included streams, dense forest, and in places heavy mud in paddy areas near the river. Being that it was the dry season, the movement of foot troops and tracked vehicles was much less difficult than what was encountered in War Zone C during Operation Attleboro. This was due primarily to the drying out of the flooded areas since the rainy season ended. During the entire operation, the entire area AO experienced only .01 inch of rain.

The whole operation consisted of search and destroy missions that located many bases dispersed all over the area. There were few large bases, but most of the bases were platoon or smaller in size. Many were defended by trench systems and strong bunkers that were built to withstand artillery and air attacks. Sometimes they defended them ferociously, other times they would not or leave a rear-guard force behind that allowed the main body to escape.

This is a story about one of those bases that was defended tenaciously. It was a typical experience during this operation.

The story starts with a letter home that I wrote to my brother-in-law (I never would have worried my mom with details I shared with in-laws).

February 7th, 1967

"...Since the beginning of Operation Gadsden, we've been guarding an artillery support base (FSB LEE). We are not too far from a place called Trai Bi (actually, eight miles). There was a major enemy base 500 meters from where I am now but it was destroyed by a B-52 strike before we got here. So far, my company has not made any enemy contact. Other units have not had the same experience. Units from the 2/22nd Mechanized Infantry ran into two enemy base camps yesterday and suffered 49 casualties."

While writing this book, that last sentence seemed like an exaggeration. That would be unusual for me. Writing home, I always said I never needed to exaggerate or embellish stories as the truth was so remarkable in itself.

I reviewed the after-action reports from that day and sure enough, there it was.

"6 Feb (D+4)

- (a) 2/12 inf: Continued to work out of Obj 1. At 0945 hrs. the Bn. (-) departed for S&D vic XT0473. At 1230H Co A destroyed 200 lbs. of salt at XT047733. (I wondered about the salt. By the time the operation was over we had located 20 tons of salt in our AO. Why so much salt, I wondered. Then it hit me, they were using salt as a food preservative, just like it had been used throughout history).
- (b) 2/22 Inf: At 0730H Companies A, B, and C

> departed laager area to conduct a sweep from the laager site to coord WT981741. At 1100H, a resupply ship landing in the area received fire with neg. casualties. Artillery was employed and the firing ceased. At approximately 1145H coord WT984741, Co C received AW fire from 2 VC resulting in 4 US WIA and 1 VC KIA, and the capture of 1 AK47 assault rifle, 3 hand grenades, 1 gas mask, and one set of web gear. At approx. 1230H, Co C reported receiving sniper fire and grenades from an estimated VC size squad resulting in 2 WIA At 1250H (WT983743), Co A received several rifle grenades resulting in 6 WIA. At 1445H Co C fired upon and killed one VC in a tree. At 1550H the first of 21 Tactical sorties were flown against the entrenched VC. By 1700H all elements were withdrawn. A total of 47 US WIA were evacuated. 4 US KIA had to be left in the area of contact. Co C estimated the enemy strength to be a VC platoon in bunkers and trenches (WT983742)

4/23rd Mechanized (25th Division Unit attached to us for Gadsden Operation) is located 4 KM east of Lee continuing an attack on fortified positions."

This was a brief attack as the Vietcong never committed to defend that base and left behind two RPG rounds, two gas masks, assorted clothing, two lbs. of tobacco, three AK47 magazines, one cleaning rod for a 57 Recoilless rifle, 10 pounds of rice, 15 Chicom grenades, thirty 60MM mortar rounds, 15 60mm fuses and a Chicom type 56 carbine. This was a very typical booty we acquired while coming thought the jungle of our AO.

I asked Ed Smith, who was then the Brigade Assistant S-3 during Gadsden, what his recollections from that 2/22nd attack on this day and this is what he wrote back.

"Brigade Command was established at Trai Bi, close enough to the AO so that we could quickly respond to the action taking place less than ten miles north of us.

"That was a tough day for the 2/22nd Mech troops. They were engaged with a determined band of VC who fought from trench and bunker positions. Eventually we had to bring in air strikes to deal with them. Word was spread to the mech troops that the air strike was going to be danger close and were told to get inside their APCs. One group got into the armored carrier but did not close the hydraulically operated rear ramp. When the air strike dropped its load, some of the men in that APC were wounded unnecessarily. As Assistant S-3, it was one of my duties to inform the 25th Division Command Center about our casualty count. They were not pleased to hear that we had suffered so many casualties on the 3rd day we spent in War Zone C."

My company did receive a couple of days off, from the 15th to the 17th, to return to Dau Tieng for 'maintenance'. Actually, it was not all time spent at the NCO and Enlisted Men's Club drinking beer and swapping stories. The nights were open (if you didn't have to go out on an ambush patrol or man the berm bunkers) but the rest of the time the men were put on sandbag duty around our new tents that had arrived during the operation. We even had wooden floors in them. That eliminated our completely inappropriate arctic tents.

By then we were spending so little time in Dau Tieng that the new tents were appreciated, but our real base was out in the jungle. Generally, we would spend two to three weeks out in the jungle before returning for a short two or three days to get our equipment and weapons repaired.

After that short time at Dau Tieng, we were airlifted back near FSB Lee on the 17th. We were ordered to begin building bunkers just inside the jungle on the west side of Highway 22 just south of FSB LEE

This was very unusual for us grunts in Vietnam during that year. Foot soldiers were very mobile in Vietnam and had little time to" dig in". Our priority was to sweep through an area and move on to another area…sometimes returning to the same area over and over, but rarely sticking around before newer intelligence sent us elsewhere. This was not one of those times. So, what was different?

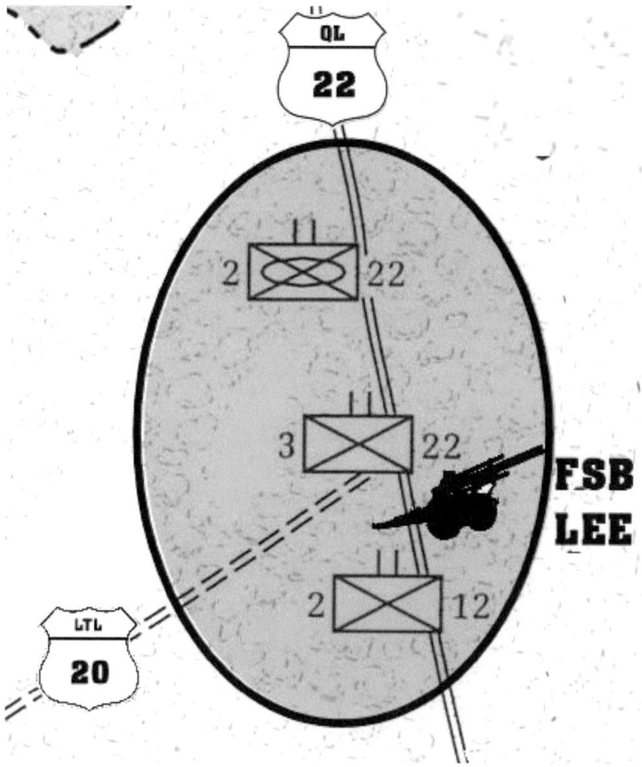

The enemy was being set up in a trap and we were going to be part of an exercise to prevent their escape.

The 4/9th and 4/23rd infantry units that participated in Operation Gadsden left the operation by February 18th and were replaced by 3/22nd companies. Now our entire brigade had all three maneuver battalions committed for what would be coming up as our next mission, Operation Junction City.

Summary of Operation Gadsden February 2-21, 1967

I retrieved this from the Operation Gadsden 25th Division After Action Report dated March 10, 1967:

> "Intelligence gathered from Operation Gadsden confirmed that our opponents included elements of the 9th Vietcong Division (271st and 272nd VC Regiments). In addition, the 70th Guard Regt., the 680th Training Rgt, and numerous elements of COSVN HQ, to include several medical units, were situated in our area of operation.
>
> The AO contained several extensive supply and ammunition caches, communications storage areas, hospital facilities, base camps, and major training complexes. It was expected that the VC would have this area fortified against ground and air mobile assaults. The extensive facilities and the well-developed trails and roads indicated that this was an important VC base area for training and logistical activities. Agent reports, POWs, Hoi Chanh (Vietcong who defect to the allies under the "Open Arms" program), Visual Reconnaissance, Long Range Recon Patrols, and U.S. Special Forces Reports were utilized in the development of the enemy situation during the operation."
>
> Total Personnel Losses: 161 VC KIA (Body Count), 215 VC KIA (possible), 2 VC POW's. Allied losses, 29 KIA.
>
> Total Material Captured: 6 AK-47's, 1 Rocket Launcher AP-

TNG, 5 M79s. 1 CHI COM (Chinese Communist) SMG, (sub machine gun), 1 CHI COM type 57 Machine Gun, 5 CHICOM carbines, 1 US carbine, 3 Mouser rifles, 1 M-14 rifle, 1 rifle (type unknown), 1 shotgun, 1 mortar sight, 2 12.7mn Anti-Aircraft tripods, 1. HMG (heavy machine gun) tripod, 2 Anti-Tank mines, 11 grenades, seven 57mm HET rds, 9 claymores, 4 CBU's, (cluster bomb units, used primarily against armor), 2. RPG-2 RDS, one 75mm rd, one 55-gal drum of CS-1 (tear gas) , 7,850 rds. of SA ammo, 7 1bs of TNT, 1 TA-312, 3 CHICOM radios, 8 hand generators, 1 PRC-25 radio, 1 PRC-10 radio, 1 PRC-6 radio, 1 Microphone, 1 headset, 3 power cables, 75 lbs. of radio parts, 25 radio manuals,1000 lbs. of commo wire,(telephone wire), 1 box of commo parts, 25 radio antennas, 24 CHICOM field phones, 13 CHICOM hand phones, 10 boxes of radio tubes, 2 battery packs, 1 CHICOM transmitter, 1 CHICOM generators, one 12 volt generation, one 12 volt regulator, 7 voltmeters, 3 voltage testers, 1 diesel engine, 2000 flashlight batteries, 300 notebooks, 19 bicycles, 40 bicycle. tires, 70 sheets of tin, 25 gal of calcium carbide, 50 gal of tar, 55 gal of kerosene, 558 1bs of documents, 1 typewriter, 200 shovels, 65 lbs. of clothing, 20 lbs. of maps, 45 medical books, 10 lbs. of medical supplies, misc. field equipment, misc. books, 1 sewing machine, 2 rice polishing machines, 1200 1bs. of sugar, 20 cans of milk, 19 cases of condensed milk, 1000 1bs of tea, 175 gallons cooking oil, 75 lbs. of soup, 171.05 tons of rice, 2 cases of crackers, 740 1bs of peanuts, 120 lbs. of dried fish, 20 gal of soy sauce 30 cans of tomatoes, 20 lbs. of peas, misc. food rations, 5.5 tons of salt, 2000 lbs. of soap and 30 boxes of soap.

Total material destroyed: 28 Sampans, 543 VC structures, 5 footbridges, 4 oxcarts, 588 underground fortifications, 7 trenches, 3 tunnels, 22 boat docks, 17 punji pits, 1 training site, 1 smelter, 1 rifle range, 35 foxholes, 5 Anti-Tank mines, three 500 lb. bombs, two 250 lb. bombs, one 62mm mortar round, two 105mm rounds, 10 rifle grenades, one 4.2" mortar round, 15 lbs. Composition B, (C-4

Plastic Explosive) ten 2.75" rockets, two 57mm rounds, 540 rds. SA ammo, 37 CBU' (cluster bomb units), 64 1bs of TNT, 3 Claymores mines, twenty-six 60 mm mortar rds, 117 hand grenades, one 81mm mortar rd, 1 Anti-Tank mine, 2 booby traps, 55 gals of gasoline, 55 gals of diesel fuel, 270 gals of kerosene, 225 gals of motor oil, 128 bottles of alcohol, 1 bottle of acid, 150 gals of carbide, 50 gals of tar, 200 gals of mash, (used to make distilled liquor), 26 fuzes with igniters, fifteen 60mm fuzes, 50 rolls of chicken wire, 12 flashlights, 1 drum of CS-1, 1 outboard motor, 25 tons of scrap metal, 1000 ft of commo wire, 60 lbs. of clothing, 500 lbs. of resin, 37 bicycles, 1 rice polishing machine, 120 1bs of gauze, 58 VC uniforms, 2 VC gas masks, 2, sewing machines, 100 cartons of cigarettes, five 12.7mm AA tripods, 2 tons of soy beans, 500 cans of tuna fish, 100 lbs. of soap, 600 bars of soap, 35 gal. of hot sauce, 6 cases of canned milk, 3,040 lbs. of misc. food, 20 gals of cooking oil, 1120 lbs. of dried fish, 422 lbs. of peas, 375 lbs. of peanuts and, 215.25 tons of rice."

In one of the final recommendations in the AAR they mention that more consideration needed to be taken into account when determining when troops could realistically reach a destination. They were not considering the difficulty of navigating the heavy jungle by ground troops. This would be a critical error which would prove disastrous three weeks later.

Intelligence gatherings from Operations identified the enemy units that we squared off against during Operation Gadsden. Operation AAR identified them as:

"Elements of the 9th VC Division including the 271st and 272nd Regiments, the 70th Guard Regiment, the 680th Training Regiment and numerous elements of COSVN, to include several medical units subordinate to COSVN."

Summing up what we learned of our area of operation they indicated that:

"The AO contained extensive supply and ammunition caches, communications storage areas, hospital facilities, base camps, and major training complexes.

Yet, for the most part, the enemy determined that it was not yet time to consolidate their forces for a major battle. They just melted away into the jungle as we approached with some crossing into Cambodia. It wasn't time…yet.

OPERATION JUNCTION CITY

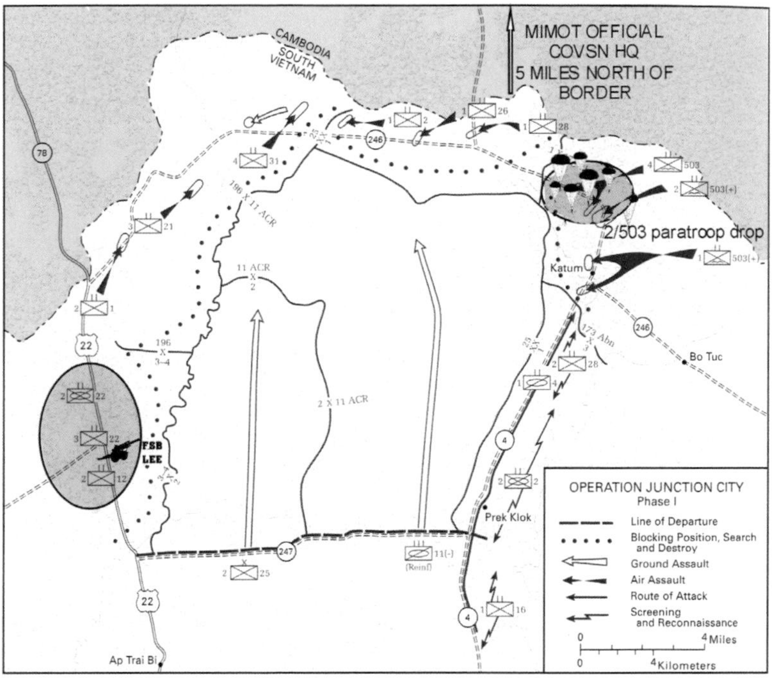

All the pieces were in place. GEN William Westmoreland and North Vietnamese Senior General Nguyễn Chí Thanh were both itching for large scale battles that might bring their competing goals of the war to a quick conclusion.

Westmoreland needed the addition of American troops that had reached over 490,000 by this time. Thanh was also eager for large unit battles that would produce high numbers of American casualties, which he predicted would erode the will of the American public. During 1966, Communist battle deaths had risen to some 5,000 a month. Operation Attleboro, lasting from Sep 14, 1966—Nov 24, 1966 had depleted his 9th VC Division so much so that he was resorting to recruiting replacements from the local populace. His troops were dismayed by the great number of losses without comparable allied deaths. He needed big victories to restore the morale of his forces in South Vietnam. By the time that Westmoreland kicked off the largest operation of the entire war, Thanh had his forces back in adequate numbers to play the same big unit battles game that Westmoreland would be presenting to him in War Zone C.

Operation Junction City was the largest operation of the Vietnam War, initially utilized 22 combat battalions of troops (equivalent to two army divisions), as the attacking force. This gigantic force was supported by 17 artillery battalions and over 4,000 Air Force sorties. An aerial Armada consisting of 249 Army helicopters were flown in this assault. 30,000 troops would participate in this operation. This would be the largest U.S. Army aerial assault since Operation Market Garden in World War II.

The goal of the campaign was to find and destroy the forces using War Zone C as their sanctuary. Even more important, the goal was to find and destroy the COSVN headquarters, believed to be located in War Zone C. COSVN HQ was not designed to be a stationary entity. It was dynamic and tended to move at ease around War Zone C and Cambodia with its protective force and medical and administrative units. Later it was learned that its official HQ was situated in a rubber plantation near a small village in Cambodia called Mimot. Mimot is located five miles north of the South Vietnam border, well out of the target zone of Junction City. Still, what we called COSVN, was rarely in Mimot and was mobile and moved around War Zone C with impunity.

The major invasion of the War Zone C's enemy strongholds, began on D-Day, February 22, 1967. Units of the 3rd Brigade of the 4th Infantry Division were already in position when the operation began. We set up blocking positions near FSB Lee and ran operations in the surrounding areas. In addition to our three battalions, other battalions moved into War Zone C to form a northern archlike blocking position in the shape of a horseshoe.

In the far north, an experiment was launched by the 2/503rd Battalion of the 173rd Airborne Brigade. They convinced the war planners that they could get their unit into position in the northeast corner of the horseshoe quicker if they parachuted in off C-130 aircraft. They argued that they could be organized and in position in minutes after the drop, where it would take 120 helicopter loads and would take them hours to get them ready to move on. It was a tough sell, but so many troops were needing chopper lifts to get into position the parachute drop looked appealing. It didn't hurt that General Westmoreland was a former Commander in the 173rd Airborne Brigade.

They were right, but it would be a challenging affair. They needed to bring their supplies and vehicles with them and the army had never tried a heavy drop in combat. Most of the supplies were pushed out of airplanes in earlier wars. Paratroop procedures had been worked on since the Korean War and they thought that they had figured out a safe way to get their heavy equipment and artillery pieces to the ground by parachute without damaging any of it. It worked. The greatest challenge was getting two 9,600 pound vehicles down to the ground via parachutes. No problem for the 2nd. 503rd troops who adopted the "Avis" car rental emblem and attached that logo to the front of their helmets. Avis advertising campaign in those days referred to Avis' perennial second place finish to Hertz in the car rental business, Avis' motto was "We try harder."

When the first C130 plane arrived over the drop zone, which was a large dried-up rice paddy, the first to jump were the leaders of the unit. Brigadier Gen. John R. Deane Jr, the 173rd Brigade Commander, jumped first, followed by LTC Bob Sigholtz, the 2nd/503rd Battalion Commander.

The drop was successfully completed and no equipment or men were lost in the jump. They successfully landed all 880 members of their unit and equipment on target without sustaining a single loss of lives or equipment. It took them less than an hour to complete this feat, prepare their blocking force base, and send out patrols around their new AO. That was quite an accomplishment for a unit that adopted the Avis campaign slogan.

If you look at the map at the beginning of the chapter, you will see that the westernmost boundary of the horseshow was generally Hwy. 22, paralleling the Cambodian border. On the northern border was Hwy. 246 and the eastern section ran along Hwy. 4. All the maneuver battalions were flown into position by choppers (except for our units, which were already in position at the end of Operation Gadsden and the paratroopers in the northeast AO).

Once the area of operation was sealed up with the border bases and patrols, two attacking forces drove into the horseshoe to flush out the Vietcong and force them into escape routes into their Cambodian safe havens. This attacking force consisted of the Second Brigade of the 25th Division and the 11th Armored Cavalry Regiment. The 25th Division unit was flown up from Cu Chi on C123s and landed at Trai Bi. They were reinforced with an additional mechanized battalion. They attacked northward on the western segment into the horseshoe. The 11th Armored Cavalry with their complement forces of 3,600 men drove up through the eastern sector.

Major Attacks In War Zone C

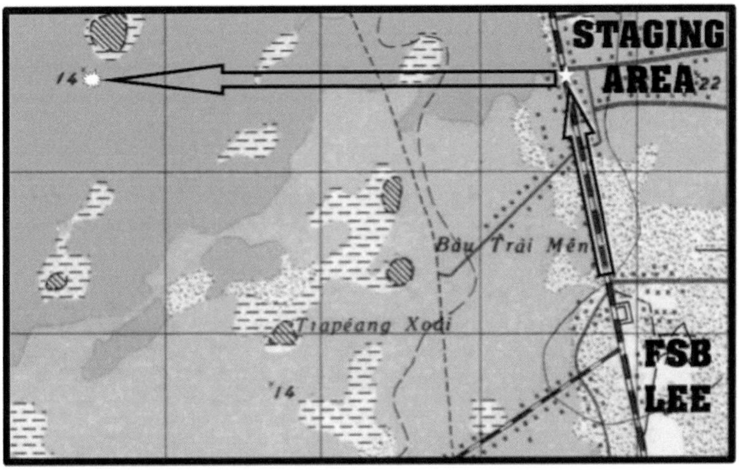

FB/2/12 activity on February 25, 1967

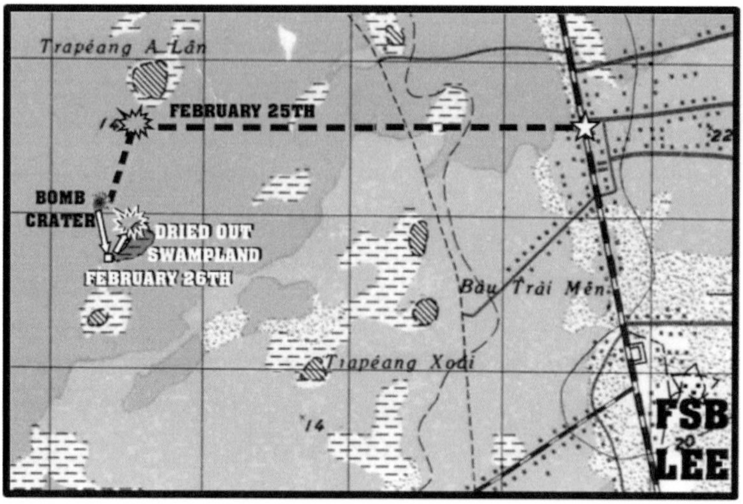

As was typical (probably throughout the Vietnam War), the enemy usually chose when and where to mass for a major assault. He was seldom inclined to do so during this initial Search and Destroy campaign in War Zone C.

Some American units experienced large scale attacks, but not many. One of those battles took place when B/2/12 was out on operation northwest of FSB LEE. I knew about B Company getting pinned own in a fight earlier, as I wrote home about the battle. A/2/12 got into a similar situation a couple of days later when a smaller Vietcong group would not break contact with us and may have been buying time to get a larger VC force in play. They left soon after artillery barrages arrived. For B/2/12, the enemy was more determined.

I didn't know many details about that B/2/12 engagement but decided to investigate when I accidentally discovered that Bob Gold had received a Bronze Star for Gallantry for that day's action. Bob graduated from basic training with A/2/12 in

1966. He was later sent to HQ Company where he was trained to be an FO for the 2/12th's 4.2 Mortar Platoon.

He was awarded the medal posthumously at the same time as some of the Battle of Suoi Tre men received their awards. Come to find out, his widow never knew that. I sent her the announcement and her local VFW followed up on it and had a ceremony there where his wife received the medal in his place.

Now here is the story of that battle and how Bob became involved.

Just as Operation Junction City kicked off on Feb 22, 1967, Bob was sent to B/2/12 to cover for someone who was away on R&R.

Story with commentary from Casey Kramarczyk, and Clark Hamm, both B/2/12 veteran of the battle:

The 2/12th Battalion was at FSB LEE from the 22nd until the 25th. On that day. without much going on, they began sending out company size patrols in the area.

Casey Kramarczyk

"On the 24th, we were told that B Company was heading north of FSB LEE on a search and destroy operation. We were trucked north to a staging area by road. The staging area was secured by APCs, some of which were mounted with twin 50s on them.

"It was moonless that night and pitch black. All of a sudden, we heard fire in the jungle. I could hear communication coming over the radio from a nearby Long-Range Recon Patrol. I heard that one was killed and one was wounded from the LRRP. We were told to mount up to go rescue them, but luckily, that never became necessary. The patrol moved to another clearing and they were choppered out with their wounded and dead man.

"The next morning, we received orders to enter the jungle and head west to where the LRRP had their firefight. We didn't have a problem finding where the action took place as there still were things around smoldering from it.

"I surveyed the area and found three empty bunkers nearby, with an adjoining trench line. Looking further, I found a number of old punji stakes. I figured that these guys weren't rookies.

"The enemy had been here a while. Looking to my right, I caught sight of a huge tree. I investigated the tree and on the back side I saw a homemade claymore mine packed with TNT and obviously intended to be used against mech units coming through. It was not armed with a detonator. I don't know why they didn't take it with them. I destroyed it in one of the bunkers.

"After more searching, I happened to notice a wire hanging 20 feet in the air from a tree near a path. It was blue, which I already knew meant that it came from the French army that fought there in the 1950s. This worried me as it indicated that a seasoned unit was in the area and had communications.

"We followed the wire, staying off the trail. I was on the flank at the time, near the overhead wire. When the wire reached down low enough where I could get to it, I told the men to pass me by and I climbed a tree to cut the wire. As soon as I cut the wire, a firefight broke out (*seen on the earlier map*). I got really nervous as I figured that the GIs might panic and start spraying the trees. Luckily, I managed to get down unhurt and work myself up to the front where I was when I stopped to cut the wire.

"I got into the fight up front on the flank and before long the man on my right was hit. It seemed to me that there wasn't a lot of fire coming from up front so I yelled for everyone to toss a grenade on my order. They did so and it became quiet all along the front.

We got the Medic to treat our man and I headed to the front of the main column to find out what was going on there. I was saddened to learn that Leon Eckhart, the point man, was killed in the initial fire ..."

Clark Hamm

"I don't recall much of that first day other than the fact that my best buddy was walking point and was killed when an ambush was triggered. He was shot in the head and the enemy tried to drag off his body. We laid down some heavy lead and they abandoned the effort. The man's name was Leon Eckhart and he lived around 25 miles north of where I lived in Pennsylvania.

"On February 25, 2007, 40 years to the day after the battle, I did a Memorial service for Leon in his hometown American Legion Post in Lehighton PA. There were 90 people present in the hall, including his two sisters and his brother. I met his former classmates and to this day still keep in contact with his family."

Casey Kramarczyk, confirmed the activity: "I remember that a VC was trying to take his gear and ammunition from Eckhart but was driven away by fire from the front but not before he made off with Eckhart's M16. When I checked him out, I recalled that Eckhart's girlfriend had sent him a gold chain and a gold St. Christopher's medal. We were playing cards when it arrived and he showed it to all of us. The VC took that also.

"Eckhart was at a trail intersection when he was killed. Investigating around, I discovered that the unit, later determined to be a VC platoon, that hit us was busy pulling down the telephone wire, correctly reasoning that we would follow it back to their basecamp. I figured it

would be simple to figure out how far away their camp was by stretching out the coiled wire that was left behind when they broke contact.

"We followed the direction that the VC were coming from and before too long came to an intersection of two trails. It was obvious that the men who trailed us headed east and that was where their base camp was located.

"Captain Leon Mayer was commanding the company at this time, replacing Captain Robert Kavanaugh, who left the month before. My feelings at this time were, "We know that the enemy is at the end of this trail. Let's go in and duke it out before they can call in reinforcements." Captain Mayer had other ideas. He would send us to the west.

"Not far from the junction, we came across a bomb crater where CPT Mayer ordered us to dig in for the night. I thought being this close to the old base camp was unwise. They would be waiting for us the next morning with a much larger force than we would have faced on the first day.

"While digging in, the two VC who were trailing us opened up full automatic while our backs were to them *(according to the AAR, that happened at 1700 hrs.)*. Billy Coggeshall, one of our medics, was killed at that time and a number of our men were wounded. (Coggeshall was awarded the Silver Star, posthumously, for his valor during this battle.)

"The VC escaped and it got quiet. We used the clearing to our south to evacuate the wounded and dead and then went into our defensive position for the night. In the distance we heard the Vietcong preparing to engage us the next day."

February 26, 1967, The Day of the Big Battle

"On the next morning I was sent out with a machine gun crew to check out the area to the east where we knew the base camp was situated. I got to the site and found some freshly dug foxholes near the camp. I reported back to CPT Mayer what I saw. He said nothing."

Casey did not recollect what I learned from after action reports about that morning prior to the big battle. Before heading east, the company moved out to the southeast heading to the southern perimeter of the dry swamp clearing. At 1115 hrs., according to the records, they made contact with eight VC. No casualties were listed for the incident. There was no location reported on the document, but the next incident had a location and it took place at 1135 hrs., so it was near that location. I marked the 1135 hrs. location with a square at XT 043827 icon. A three gallon can filled with documents was found there. B Company began moving north, searching the western woodline of that dry swamp bed.

At 1400 hrs.: Found base camp at XT 045829, several bunkers, huts, and a trench system were found. All huts were destroyed.

At 1535 hrs.: Made contact with 20-40 VC at XT 046830.

Were they ever wrong with that assessment of enemy strength at the time! It was a reinforced Company! The Vietcong had set up a U-shaped ambush that B Company walked into just north of the VC Base camp.

Casey Kramarczyk

"We finally entered the deserted base camp near the swamp clearing from the south at 1400 hrs. First, we came across some foxholes that were interconnected by a trail. Before long, we found a very old trench line that probably went back to the French War. I learned later that this was their old base camp position that was destroyed a year earlier and they just moved their new camp back a bit. Luckily, none of these positions were booby trapped.

In the short distance, I could see an elevated observation post hut. This was in a somewhat open area in the jungle and that hut offered an ideal view of our approach. We stopped for about a half an hour. During this time, CPT Mayer had artillery (and probably mortar fire) brought into the position.

It was during this time that Bob Gold, standing near me, hit me up for a cigarette. He knew that I smoked Luckies and he thought they were too strong for his taste. He took it none the less and said, "You know Case', I'm getting used to these things."

After the prepping was completed, we began to move out past the trench line. (By then it was 1535 hrs.). Within a few minutes all hell broke loose. We got hit from the front, but the most intense action came from the rear. Weapons Platoon, which didn't come out in the field too often with us, was taking most of the fire. The enemy was using heavy machine guns and a heavy mortar attack.

They tried to take me out with a grenade but I didn't notice it until later. It noticed a Chicom grenade lying nearby under some leaves later and lucky for me, it was a dud and never went off.

They had us in a U-shaped ambush with only the side facing the swamp open.

The defensive artillery fire was coming in all around us as they moved it closer and closer to our lines. They got the artillery pretty close and then we were told to move back in the same direction that we entered the base camp. A buddy of mine and I started pulling back and before we made it back to the trench line, I heard someone calling out to me, "Casey, over here."

It was a medic, who I can't quite identify right now. He was treating Bob Gold who was propped up against a tree facing towards the trench line to our rear.

I went over to him and saw that he had wounds and was bleeding badly. He was in a bad way. It told the medic that we needed to pullback and soon as the artillery would be moved into this position shortly. The medic told me he wasn't leaving Gold. I told him, "Of course not. We're going to have to work together to move him." I looked down at Bob and told him that we needed to move him or we would be killed by the artillery that was going to be raining in soon. I explained that it was going to be painful, but we had no choice.

I bent down and he grabbed me by my sleeve and pulled me down to tell me something. He could just barely speak but I believe he said, "My wife, tell her I love her." I told him there was no time for that right now. All of us had to get away.

I told Dennis, who had moved back with men to grab Bob's shoulders, "I'll grab his feet and walk backwards towards the trench line. I'll still be able to cover us as we move back."

We made it to the trench line which was filled with B/2/12 men looking for cover and injured soldiers laid out helter skelter. I told the men to make room for us. Dennis placed Bob in the trench, and we squeezed into the ditch with them.

Shortly after that, the artillery began walking in towards our position from where we were earlier. One of the rounds hit a nearby tree and rained shrapnel on us. It put a piece of shrapnel in my left arm and blew out my hearing. I tried to pull out the hot shrapnel and it burned my trigger finger. Things were only getting worse.

I carried a lot of magazines loaded with 18 rounds (*Typical for us in those days. We never filled the magazine up to its 20-round capacity. They tended to jam easily when filled completely. Most guys loaded 19 but Casey said he wasn't taking chances and loaded 18*). My Platoon Sergeant came looking for ammunition and I reluctantly shared some ammunition. I never fired automatic. I wanted to see what I was killing (which was difficult in the jungle) ..."

Meanwhile at 1600 hrs. B/3/22 (*Oliver Stone's Old Company before he got there six months later*) was attached to the 2/12th and ordered to leave their position and link up with B/2/12 at the ambush site. They would not arrive there until 1754 hrs., almost two hours later.

Casey Kramarczyk: "Bob Gold died in that trench before the battle ended. The artillery and small arms fire went on for quite a while but eventually took its toll on the VC and the enemy fire gradually subsided. By the time B/3/22 arrived, the battle basically had ended. I don't believe that they did us any good by that time.

"It was then time to collect the wounded and dead, collect their gear, and prepare them for medevac. I noticed some of the guys were walking around the jungle in circles with their arms high in the sky. They had dropped their gear, but held onto their weapons. They were sent out on medevac as victims of shell shock. I'm not so sure that all of them were. I suspect some had seen enough and just wanted out anyway they could.

"The group that suffered the most casualties on that day were the officers, their radio operators (CPT Mayer's RTO was killed), and NCOs."

Six men were killed and 24 were wounded, not counting the men who Casey described walking around in circles.

"Lt Haxton, my platoon leader, ordered me to take a head-count of the company after the battle and I distinctly remember the figure that I gave him ... 37 men. That was all that was left after the last two days of action. We were already undermanned when we began Operation Junction City as replacements for our wounded were slow or non-existent before the battle. By the end of the day on the 26th, we numbered less than a platoon in the company."

When I read Bob Gold's citation to Casey, he told me that Bob should have received a Silver Star for his action on that day, not a Bronze Star.

Ed Smith was with Brigade S-3 at the time of the ambush and I asked him what he recalled from the ambush episode.

Ed Smith

"I recalled that battle for a couple of reasons. First Colonel Garth was still upset about how the brigade was picking up what he considered needless casualties. Point men were being shot and their buddies were being picked off as they moved up to retrieve them. By the book, the best way to help your buddy is to take out the source of the fire that took him down in the first place. Of course, when your buddy is on that point position, sometimes the book gets thrown out.

"The second thing that I recall was the intensity of the artillery fire that was sent hurtling to that border battle. The fire was nonstop. During the battle the 2/77th artillery Forward Observer was relentless in calling in fire missions from three different batteries around their position. He was awarded the Bronze Star for Valor for his action on that day."

Summary Thoughts From Me On That Battle.

I was able to connect up with Billy Coggeshall's brother, who still lived in Marshfield, Massachusetts. They were never told anything about how he died. They appreciated learning that Billy died with honor serving the men that he loved in B/2/12. I don't believe that his family ever learned that he was awarded a Silver Star as when I visited his grave 45 years later, his gravestone only recorded that he was awarded a Purple Heart. He was awarded the Silver Star less than two months after the battle. Sad.

On February 26, 2013, exactly 44 years after he was killed, Carleen Gold did indeed receive Bob Gold's Bronze Star medal. The decoration was presented to her at the American Legion Hall in Bob Gold's hometown.

Here is the inscription that was written on the Award:

HEADQUARTERS
25TH INFANTRY DIVISION
GENERAL ORDERS 3 April 1967
NUMBER 1161
Award of the Bronze Star Medal for Heroism
TC 320. The following AWARD is announced.
GOLD, ROBERT J. US52654531 SGT E-5
HHC, 2d Bn, 12" Inf, 3d Bde, 4" Inf Div

Awarded: Bronze Star Medal with "V" Device (Posthumously) Date action: 26 February 1967

Theater: Republic of Vietnam

Reason: For heroism in connection with military operations against a hostile force: Sergeant Gold distinguished himself on 26 February 1967, in the Republic of Vietnam. Sergeant Gold's company was moving through the extremely thick jungle area, when an enemy base camp was discovered. During the advance on the enemy base camp, Sergeant Gold was with the lead platoon moving into the area when the company came under intense enemy machinegun fire. The company immediately took up defensive positions, but Sergeant Gold found himself pinned down on one side of the trail, while his radio operator was pinned down on the other side. The company received additional enemy automatic weapons fire from the rear and left flank. Realizing the importance of radio contact in order to bring effective artillery fire on the enemy, Sergeant Gold determined to reach his radio operator's location despite enemy automatic weapons fire. He called to the radioman to remain in place as he then attempted to cross the road; however, he was seriously wounded in the left shoulder and hip. Once again, without regard for his painful wounds and the intense enemy fire. Sergeant Gold desperately tried to reach his radio operator's position, but was mortally wounded in the attempt. This outstanding display of aggressiveness, devotion to duty, and personal bravery is in keeping with the highest standards of the military service and reflects great credit upon himself, his unit, and the United States Army.

Authority: By direction of the President under the provisions of Executive Order 11046, dated 24 August 1962, and USARV message

16695, 1 July 1966.
FOR THE COMMANDER: JASPER J. WILSON
Colonel, GS
Chief of Staff
Robert S. Young

LTC, AGC
Adjutant General

As I worked on the history of our units who fought the 9th VC Infantry Division, I came to realize that many more soldiers should have received medals for their bravery than actually did. I learned through leaders that with almost continual operations, and little preparation for them, Company Commanders were many times overwhelmed. They were forced to look ahead more than behind. Maybe it has always been so.

Border Defense

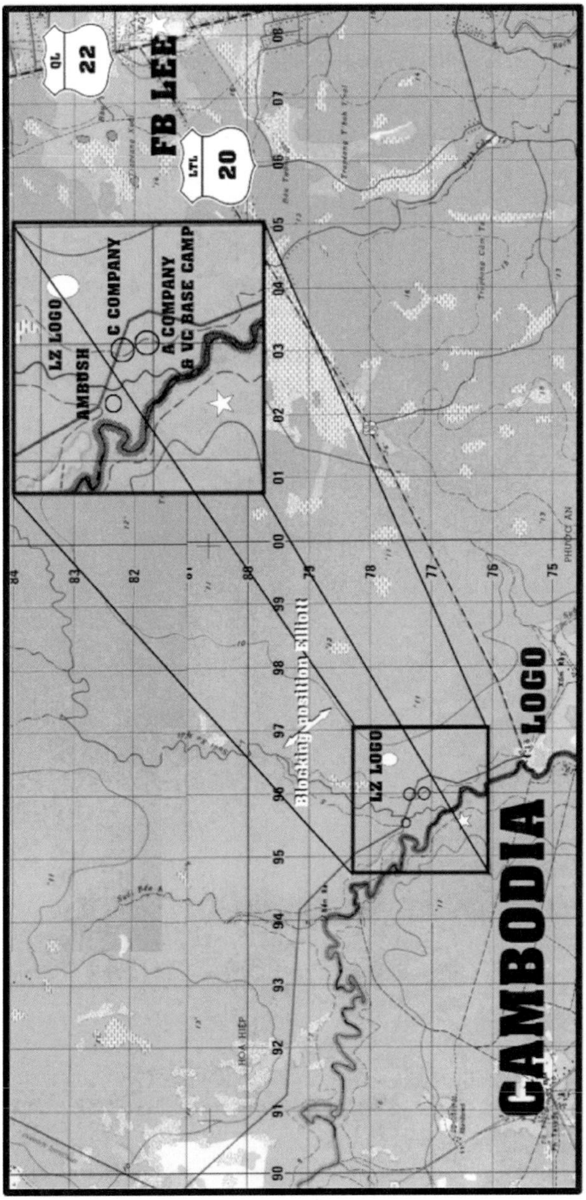

While B Company was recovering from the ambush attack of the 26th at FSB LEE, it was up to HQ, Recon Platoon, A and C Company to take up the slack and continue the 2/12th mission. While other units were driving into the heart of War Zone C, we were there along the border to prevent them from escaping into Cambodia.

LOGO village had been abandoned by the VC during the Gadsden Operation. Still the area was strategic to the Vietcong and many bases had been located in the area, especially near the border.

March 7, 1967

While B Company remained behind, the rest of the battalion was flown by choppers to a clearing cut in the jungle called LZ Logo. Their mission was to establish a blocking position just northeast of the clearing called ELLIOTT, named after battalion commander, Joe Elliott. A, C, HQ Companies and Recon Platoon closed in on that position and set in for the night by 1830 hrs.

March 8, 1967

At 0730 hrs. the 2/12th units moved out of the blocking position and were sent back to LZ LOGO where they would establish a battalion base camp. Once in the clearing, the units began to dig in. A Company was assigned to the northernmost perimeter.

At 1050 hrs., C Company was sent out to scout out the area southeast of the clearing. By 1120 hrs. they reported locating an enemy base camp (*located in the lower circle in the blow up that is labeled A Company NDP*). They located a 55 gallon barrel of CS gas (tear gas) and made plans to destroy it before leaving the area. Then they began to find additional items which included six pigs soon after discovering eight foxholes and punji pits in the area.

While C Company searched the camp, they soon began receiving enemy fire from an estimated squad of VC, resulting in two WIAs.

C Company pulled back and the basecamp area was attacked with air strikes and artillery. When that took place, A Company was sent up from the battalion laager to reinforce C Company.

When both companies returned to the enemy base after the shelling, they found a large basecamp containing bunkers and a trench system and two large thatch structures, each 20X40X15' that contained an estimated 24 tons of unpolished rice. Four 55 gallon drums of diesel fuel was also found, which was convenient, because it was used to burn the rice. Actually, eye witnesses to the burning said that the rice really was not destroyed, but turned into puffed rice.

By then it was 1808 hrs. Daylight was burning away. Both companies set up a night defensive position near the base camp so it could continue the search through the camp in daylight and protect what they had discovered.

An ambush patrol was sent out with nine men which included Joe Kirkup, an artillery FO.

Joe was one of those soldiers who always volunteered for ambush patrols and dangerous missions. This was one of those missions. That ambush patrol was set up near a jungle trail and a dry river bed. Both were excellent approach avenues to the base camp that was found should the enemy be inclined to take it back.

The ambush patrol was guided into position by the Second Platoon Leader, Lt. Jones, and a squad of men led by Larry Walter. It was almost dark and they wanted to be certain that they were properly in position. After the ambush site was placed in position, the second platoon men returned to Alpha Company's NDP by following the dry river bed back south.

At 2045 hrs., a thunderous firefight broke out, emanating from where the ambush patrol was set in place. It wasn't long that it was realized that our ambush patrol had moved and had passed in front of one of C Company's listening post placed in front of their company. In the ensuing action, seven of the nine ambush members had become casualties and once the situation returned to normal, they began moving down that dry bed to return to A Company's night location. Two men never made it back that night. Still out on the ambush location were Joe Kirkup and PFC Tyrone Rambo, a recent arrival in A Company.

Later we learned that the squad leader that led the patrol didn't like the position they were in and decided to later move them. They ended up passing near a C Co. listening post.

I contacted John Napper, who was the C Company Commander during this period to ask for his recollection from that night. This is what he said:

"My recollection is that my LP reported noise/movement to their front. When they popped a Claymore, the A Co ambush patrol answered with small arms fire. The C Co LP countered immediately with its own small arms fire. I received no report of my main body receiving any incoming fire. My main body of the company did not fire, because they knew the friendly LP was to their front. The actual firing of weapons on both sides was very brief. Radio traffic during the lull in firing made it evident almost immediately that we had a friendly fire incident. Reports of contact to me from the LP and reports of contact from the patrol to A Co Commander, Al Palmer, were almost instantaneous. Too close to be coincidence.

I never really knew the details of A Co casualties. I vividly remember firing illumination for the rest of the night in an effort to locate the missing soldier(s) from A Co. At the time, we only knew of one missing. His name was given to us over the radio; and my guys in the LP spent the rest of the night calling out to him. When he never answered, we thought, perhaps, he had been killed in the exchange of fire. Near first light, he answered the calls. My LP suffered no injuries during the incident; no doubt due to the fact that they were dug into prepared positions. However, one of the guys on the LP had a round enter his steel helmet without penetrating the liner. The bullet entered the front of the helmet and followed the contour of the space between the helmet and liner before finally exiting to the rear, and below his ear. We later supplied written documentation of the event so he could take the helmet home as a souvenir rather than have it confiscated."

John Napper

Larry Walter recounted what he remembered from that night:

"We heard a a hell of a firefight taking place where the ambush patrol was set up earlier. Soon after the survivors began arriving down the riverbed. None were seriously wounded, but each had been hurt in the exchange with C Company. We asked if anyone was left behind and that's when we found out that Kirkup and Rambo were still at the ambush site. Lt. Jones told me to get a couple of men as we were going to go retrieve those ambush men left behind. He told us to bring only a few magazines and grenades. They were going to fire continuous illumination for us as we made our way back to the stranded soldiers.

Making our way down to the riverbed, we were walking so low to the ground that anyone who saw us from a distance would have thought that we were a family of turtles making our way along. The shadows produced by the illumination coming down, back and forth, hanging from parachutes, gave an eerie feel to the occasion.

We moved up the gully and as we got near to where we thought we had left the patrol, Lt. Jones kept calling out to Kirkup and Rambo. I told him to pipe down as he would get us killed. This didn't stop him.

Jones kept yelling out that we were coming to get them, and finally Joe Kirkup heard him. Joe answered back, "Leave us down here. There is no reason for getting anyone else hurt. We have grenades with the pins pulled and if we are still alive in the morning, come and get us, but if not, don't get anyone else killed." That is when the two stranded men learned that Charlie Company men were the ones who had opened up on them.

That response by Joe Kirkup I always considered to be the bravest act I saw in Vietnam. We were deep in enemy territory and here he was telling us, don't worry about them.

Here is Joe Kirkup's remembrances from that evening and how he felt about being called a hero for that night:

"We were all standing up and moving when we heard something hit the ground near us. Nothing happened and someone said it was a pig. A couple seconds later there was an explosion. Rambo and I emptied our magazines into the dark. I threw a grenade and heard it hit a tree and hoped it didn't come back. I loaded another mag and then it went quiet. Rambo was carrying a grenade launcher. Every time he tried to reload, we got more incoming.

"After it got quiet, Rambo and I were back-to-back against a small tree. He whispered to me, "Are you scared?"

"I was afraid he would panic so I said, 'No.'

"He replied, 'You're a damn liar.'

"Now that is courage.

"We heard the guys who came to get us call my name. After I warned them off it was quiet for a moment. Then some gook sounding voice called out quietly trying to imitate them. It sounded more like "gurga, gurga." By that point I had discovered the blood running down my arm on to my trigger hand. Contrary to what people told you about my bravery, I'm three parts stupid and two parts chicken. 'Jumpy' is a gross understatement.

"The American voices were easy to identify. Rambo and I got to our feet. Rambo had pulled the pin on a white phosphorous grenade and couldn't find it. He had whispered, 'We'll all burn together.' That young man had more courage in his baby finger than my entire shaking body.

"Now we were moving in a file out of the area. A guy, a 2nd Lt. (*2nd Lt. Jones, the second platoon leader*), I think, had taken the grenade from Rambo and was holding the spoon down. He was the last one in the file directly behind me. He threw the grenade as far as he could into the jungle. Suddenly there was a flash and a muted explosion. I spun around and pointed my rifle at the silhouette in the flames (*Lt. Jones*). I think it was the wet blood that saved him. My hand slipped on the grip just long enough to realize my mistake and point the gun upward. Someone took my rifle away and we continued on.

"In the morning we were taken by chopper to some place where our wounds were treated. Some medic was working on me being guided by a doctor (I guess) who was working on someone else. The only pain killer was a cloth in my mouth. I was lying on my side on a cot when the 'doctor' growled at my medic, 'Damn, you're making a mess. Just sew it up.' Interestingly, what came out appeared to be a piece of broken glass. Many years later I was going through security at Atlanta airport. They had gotten new high tech, triangle shaped metal detectors. The thing started blinking red and beeping at my shoulder. They made me take off my shirt then said it was nothing and that the new units had not been calibrated yet. None of the subsequent medical imaging since then has showed anything."

With that episode put to bed, everyone went to sleep, resting for the next day's mission. Then at 0145 hrs. we heard the familiar thuds of mortar rounds pounding earth. Over at LZ LOGO where we had dug in earlier in the day came under an 82mm mortar/150 mortar round attack. This was personal as all the rounds hit in the Alpha Company area where we had dug in earlier in the day. No one was hurt at the clearing, but they determined where the rounds originated. That's where we would be heading when the sun came up. It would end up being a fruitful adventure for the 2/12th.

March 9, 1967

Marching to the border, the 2/2th walked right into a major base camp at 0855 hrs. The camp contained three bunkers with overhead cover. They were interconnected by a 300-meter trench system. Food found there was thought to be prepared two days earlier. There were cooking utensils at the site and 400 to 500 lbs. of salt were discovered and destroyed.

Crossing into Cambodia, they discovered another base camp at 1245 hrs. Within that camp was buried 300 pounds of printing type and type setting equipment.

After a number of trips to the National Archives researching records, I learned that there is an archive of military photos that are also collected at the site. It's on a different floor so I took the elevator up and saw the office where the collection is kept.

The photo collection is rather extensive, but the 2/12th collection for the Vietnam period was very limited. I found only one photo pertaining to the 2/12th period of 1966-1967. That photo showed Bob Livingston, who was then attached to the 2/12th Intelligence, climbing in that hole where the VC had hidden the type.

On the back of the photo was printed this:

> SC 639267 CU CHI, VIETNAM SGT Robert Livingston examines type boxes used by the Viet Cong for sorting printing type, captured by members of Co "A", 2nd Ba, 12th, Inf, 4th Inf Div.
> 9 March 1967
> Photo by PFC Jack Mraz
> 125th Sig Bn (Inf Div)
> 1258-0841/AGA-67
> UNCLASSIFIED by USAPA, 25 May 1967

For the next few days, the brigade located only a few empty enemy base camps and there were no significant enemy encounters. It seemed as if the 9th VC Division had lost interest in defending jungle bases that they were powerless to defend against our artillery and airstrikes.

Operation Junction City phase one was a disappointing exercise for the operation planners. For the most part, the VC would fight rear action stalling tactics with small groups while main groups melted away into the jungles of War Zone C. As for the mythical COSVN Headquarter unit, there was little evidence that they were even in the horseshoe trap, or along the border. They were like a ghost force, nowhere to be seen.

For phase 2 of Operation Junction City, they would send more troops into the horseshoe and search the area more thoroughly. The feeling was that sooner or later, the 9th VC Division would have enough troops and mount a major attack with a large unit assault. They were right as there would be three major attacks during Junction City II. The moment had come where they realized that sacrificing their men piecemeal would not help their strategy to end the war quickly. The 9th VC Division needed to up the stakes and hopefully come up with a major victory that would raise the morale of their troops and destroy the American public's hope for a quick victory. Westmoreland and his war planners would accommodate their wish and planned for their own major victory.

Last Visit Home

This photo taken by Johnny Martel with my camera is my most precious snapshot taken during my time in Vietnam. It was shot at Camp Rainier in December of 1966 while we were still using arctic tents for shelter. It shows Clint Smith, left, me in the center and Tom Nickerson on the right. Both Clint and Tom were attached to the Headquarter Group of A Company. We got to be good friends once I started carrying the CO's radio and also became part of the Company HQ. In the back you may be able to see a three-image frame carrying the photos of our girlfriends at the time.

Both Clint and Tom were very confident young men. They each had a keen sense of humor. Of course, like most young men put together in a group, they enjoyed the playful teasing that usually was the norm in the enlisted men ranks.

Smith and Nickerson served in the field as the company commander's bodyguards. Over the course of time, with the climbing casualty rate, that attachment became a luxury that the brigade could ill afford to continue.

Whenever the brigade was in the field, the base camp was usually secured by a small group of defenders, usually the Weapon Platoons of the brigade's units. These units were supplemented by brigade and battalion staff security and HQ men and rear personnel who provided technical, supply, and administrative services for the units. Camp Rainier became a very tempting target.

As a result, Clint and Tom were told before Operation Junction City began that they would remain at Camp Rainier to help with security.

When they found out they were delighted. Camp security was understood to be a much safer proposition for the typical grunt compared to those who had to plow through the jungle searching for Charlie. They kidded us as we went out on Junction City. Unfortunately, in Vietnam, some places were safer than others, but there was really no hiding from danger.

So, as we headed out to the airstrip on trucks to begin phase one of Junction City, there were Tom and Clint on the side of the road waving us off with big smiles. Little did they know that their days were numbered and their demise would happen on a routine assignment at Camp Rainier; taking out the garbage.

March 15, 1967

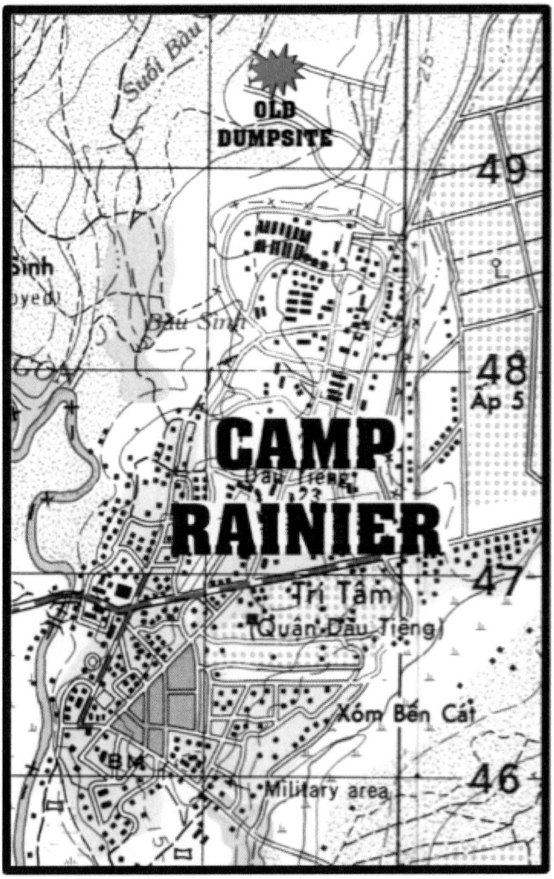

The battalion was in transit from their blocking position along the Cambodian border back to base camp when a tragic mistake led to the loss of Clint and Tom. We learned about the deaths of the pair when we returned to camp but did not know the details until decades after the incident.

FSG Springer travelled only once to an Alpha Association reunion and that was in 2001, the year after the association was formed. That is when Top admitted to us that Smith and Nick should never have been killed. On the morning that they were killed, Springer assigned the pair to collect the garbage barrels in the company area and bring them to the dump to be emptied. This was a routine for them. Unbeknown to them, Springer had received word on the day before that the dump had been moved and forgot to tell the duo about the change. American base dumps were frequently combed through by Vietcong for valuable items that the they could use. Security was the only measure that kept them at bay. Apparently, the Vietcong were aware that the dump had been abandoned. When the ¾ ton truck approached the old dumping ground, they hid and sprung an ambush.

The loss of Clint Smith and Tom Nickerson weighed heavily on him over the years. Few knew the details of that morning and Top could have gone to his grave with the secret. However, I think he felt an obligation to the men he led in Vietnam to tell them the truth. He screwed up. It happens in life all the time, but not usually resulting in such dire consequences.

Soon after that reunion, we learned much more about what took place that morning from Ken Eising who was back at Camp Rainier on the morning of the ambush.

Ken was on base security detail when the incident took place. This is his account of the incident:

"On March 14th the Recon Scout section was told that we would have to escort a group of Vietnamese wood cutters who were going out in the morning to cut logs for building the new NCO Club. Apparently, they had received word that the VC were going to ambush them for helping us build the club. The Vietnamese were often threatened by local VC for their cooperation with the Americans at Camp Rainier.

On the morning of March 15th, the Recon scout section met a group of woodcutters and their oxcarts and escorted them to a wooded area, which seemed to be North or Northwest of the base camp. It was close to the base dump.

I recall being in the lead jeep and we were going down a two-track road with thick bamboo on the sides. Visibility was terrible so we dismounted one man to walk ahead of the jeep as point. Nothing occurred until the logs were being cut and we heard a volley of automatic weapons being fired, which seemed fairly close. Later we received word on our radio that Clint and Tom had been killed. With the location being so close, I felt helpless that we were in the area but were not able to help. I always felt that the ambush that had been prepared for the woodcutters had been diverted to Clint and Tom when we showed up with our machine gun jeep escort. No proof, but I will always think that.

I took photos before we started with our patrol that morning, showing the security detail and the woodcutters that we were protecting.

It was necessary for someone from the company to travel to the ambush site to identify the victims. That assignment was passed to Jim Olafson, who was then serving as the A/2/12's Company Executive Officer.

When Jim arrived at the dump, the first thing he spotted was the windshield with the bullet hole through it. He looked inside the cab and saw both their weapons were still inside. He walked to the back of the truck and he saw both bodies. They were helpless when the ambush was sprung. Neither one had their rifles with them.

After Jim identified the pair, he returned the truck with the bodies first to Brigade Headquarters, where they took it from there. He had two more letters to write to family members of these men who had lost their lives while serving in A/2/12. It was his most unsettling responsibility. What can one say about what took place?

Apparently, Jim kept the details to a minimum. One family wrote back for more details and Jim wrote another unclear account back to them without giving them the complete picture of what took place. Of what use would that be? Better to keep the details primarily down to "they were ambushed and killed while completing a mission for the company."

After Alpha Association was formed in 2000, we visited the graves of both Tom Nickerson and Clint Smith. We met with the family of Tom first in Chatham, MA, located on Cape Cod. Before visiting the grave site, we met with Tom's dad and he told us how as Tom went through those 'breaking away' years, Tom and his dad's relationship became strained. Nonetheless, he said Tom came by the house to tell him that he was going to Vietnam and his dad became emotional. Tom's dad was a POW in a Japanese concentration camp for two years in World War II. He understood the dangers his son would be facing. After an hour or so he brought us to Tom's grave where we were finally able to say our goodbye to him, an opportunity not afforded to us in Vietnam.

Henry Osowiecki lived in Thomaston, CT, eight miles north of where Clint Smith lived when they were both drafted together in 1965. Henry was very close to Clint and he took the loss the worse. It would be five years after returning home from Vietnam when Henry summoned the courage to visit Clint's family in Oakville, CT.

He knocked on the door and a man answered. Henry introduced himself and his association with Clint in the service and said he would like to visit with Tom's family. The man paused for a moment and said, "If you didn't care enough to come sooner, we don't want to speak with you now." I guess neither side understood what the other side was feeling.

When Alpha Association had our first annual reunion in 2001, we traveled to the grave site of Clinton Smith in Watertown, CT. There were about fifty of us A/2/12 veterans visiting on that day, including the A/2/12 Commander, Executive Officer, and 1SG Sidney Springer. As we began to say a prayer for Clint, an army UH1 'Huey' helicopter flew about 200 feet above us over the site. We were startled but no one said anything at first. After the prayer was completed, Henry turned to us and said what we saw was quite remarkable. There were no military bases within thirty miles of the grave site.

The company returned to Dau Tieng on that afternoon and it wasn't long before everyone knew about the loss of their company mates. The company suffered no fatalities while on the Cambodian border experiencing sporadic skirmishes with the VC. Meanwhile, our friends had been killed in an ambush at the dump while disposing of company trash. It made no sense.

Before going back out into War Zone C on March 18, the company cleaned their equipment and pondered the next phase of our tour. We knew that we were going back to a very dangerous area, but the dump attack reminded us all that there were few safe places for troops in South Vietnam.

On the evening of March 17th, the night before we returned to War Zone C, I was sitting in my tent alone when Larry Barton walked in. Larry and I would often do that when in camp.

Larry's friendship with me went back to our early training days in the 3rd Platoon and our platoon trips into Seattle on weekends. We valued our friendship, even if we no longer served in the same part of the company.

Larry had been shot during the January 27th battle, returning to our company after recovering three weeks later. Larry brought that up during our time together. Larry was never one to take himself seriously and jokingly told me, "I got shot once. I sure hope I don't get shot again. That hurt like hell." Then he smiled. Then we talked about the tragic losses at the base dump. Larry said, "Can you imagine losing your life bringing out the garbage? Of all the different ways that one can get killed here in Vietnam, bringing out the garbage. Go figure."

That was the last time we had a chance to speak to one another. There were many ways one could lose one's life in a combat zone.

THE BATTLE OF SUOI TRE

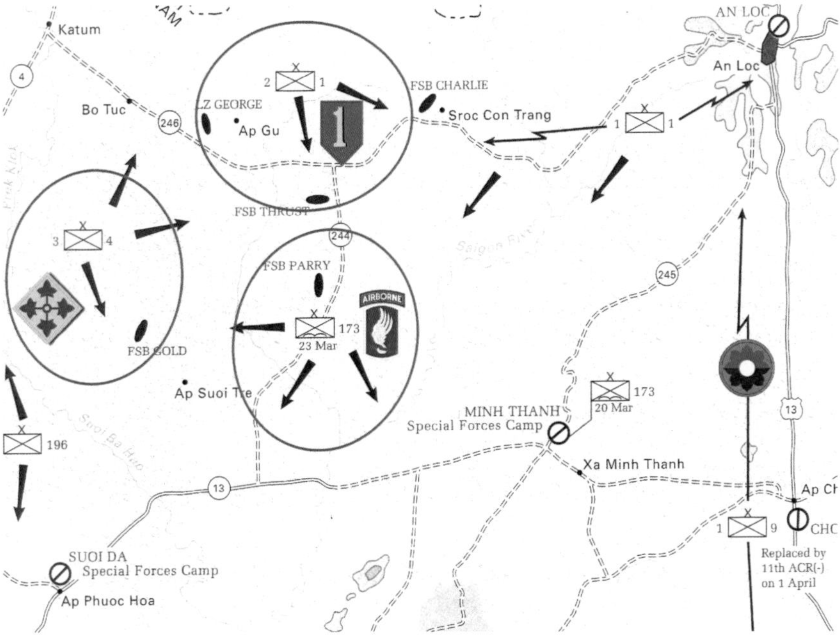

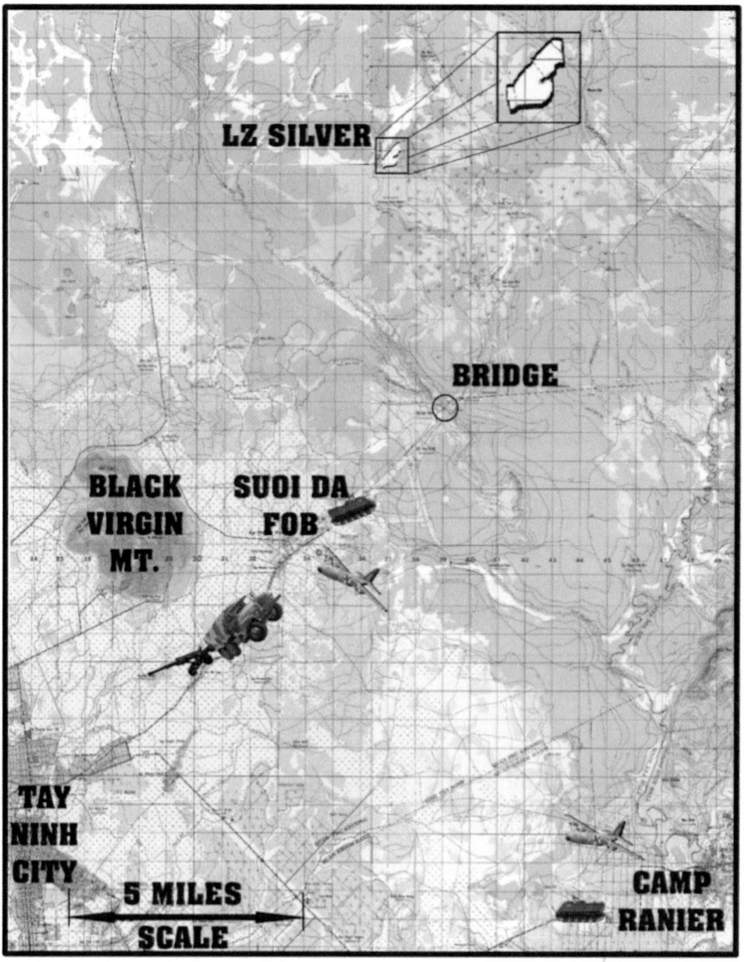
LZ SILVER
BRIDGE
BLACK VIRGIN MT.
SUOI DA FOB
TAY NINH CITY
5 MILES
SCALE
CAMP RANIER

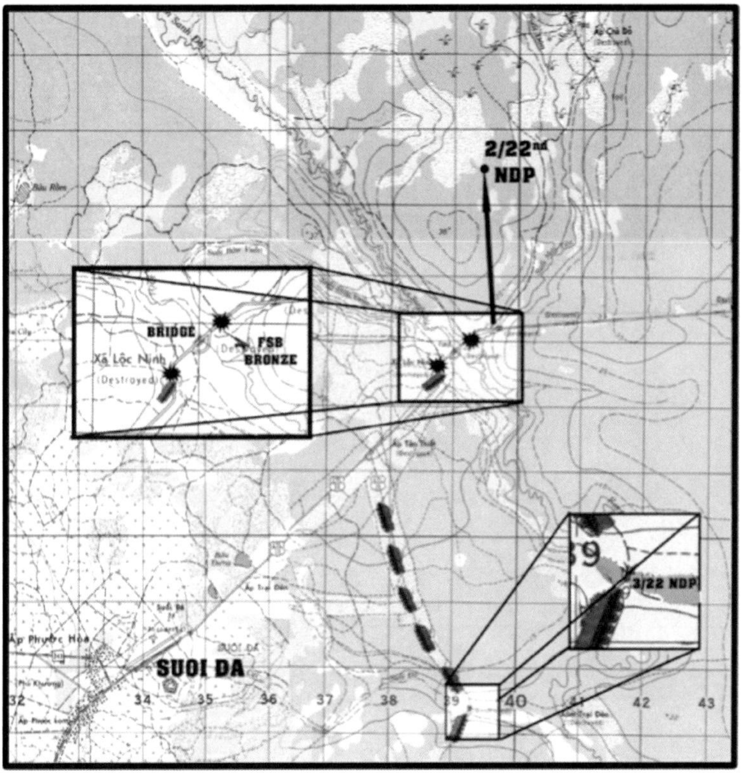

The Battle Book

OPERATION
JUNCTION CITY
VIETNAM 1967
BATTLE BOOK
Prepared For
Advance Battle Analysis
U.S. Army Command and General Staff College
1983

The Operation Junction City Battle Book was
Prepared for Course A660 Advance Battle
Analysis. Members of the CGSC Class of 1983
And Junction City Battle Study Group were:

MAJ G. C. Lorenz, Artillery
MAJ J. H. Willbanks, Infantry
CPT D. H. Petraeus, Infantry
CPT P. A. Stuart, Quartermaster
MAJ B. L. Crittenden Air Force
MAJ D. P. George, Infantry, Study Group Leader

Dr. Robert H. Berlin, Combat Studies Institute, was advisor to the study group.

The Combat Studies Institute provides a wide range of military, historical, and educational support to the Combined Arms Center, Training and Doctrine Command, and the United States Army. It is located at Fort Leavenworth, KS. This particular study, done in 1983, offered additional material that was not available in earlier documents I had examined.

I found it interesting that General D. H. Petraeus was one of the six contributors to the study and was only a captain at that time. Petraeus had graduated from West Point in 1974. He earned the General George C. Marshall Award as the top graduate of the U.S. Army Command and General Staff College Class of 1983 at Fort Leavenworth, Kansas, receiving a Master of Science degree in Military Science.

I discovered this study many years ago and after reading it, I put it away for future use in writing about Operation Junction City. As I will be exploring Operation Junction City II in the next few chapters, I thought that some of what they discovered by 1983 would be useful in understanding what took place in 1967.

One of the surprising revelations was the makeup and strength of our adversaries we would be facing in War Zone C during the operation. All contemporary sources from the period identified COSVN Headquarters and the 9th VC Division as the targets for extermination during the campaign. Knowing the history of the 9th VC Division, I assumed that included the 271st, 272nd and 273rd Regiments. Those were the original forces that made up the original division's units as we arrived in Vietnam in 1966. That all changed when additional troops arrived from North Vietnam to blunt the buildup of American forces in 1966-67.

By the time that we ventured into War Zone C to find and destroy COSVN and the 9th VC Division, we faced a considerably larger force than what I thought.

Here is the breakdown of the enemy units that we faced in what was later called the 'Iron Triangle of War Zone C'.

"The opposing forces were the 9th Viet Cong Division and elements of the Committee of South Vietnam (COSVN) Headquarters. The approximate strength of this force was 7,000 men. The task organization of this force is:

COSVN
9th VC Div
271st Regt
272d Regt
101st NVA
273d Regt
70th Guards Regt

Just before Junction City, VC forces had recently received a standard infantry rifle, the Soviet-made AK-47. They possessed some 7.62 machine guns in each battalion and very few .51 caliber machine guns, if any at all. They had a healthy respect for US armor as evidenced by a pronounced increase in the use of antitank weapons such as RPG-2s, recoilless rifles, and Chinese-manufactured antitank mines. Fire support was provided by 60-mm, 82-mm, and for the first time in the war, 120-mm mortars. The artillery supporting the 9th VC Division were 82-mm, 120-mmm, and 130-mm towed howitzers. Of these, it was the 130-mm howitzer which outranged the US 105-mm howitzer that proved most effective.

The corollary to firepower is mobility and again the US forces had the advantage of being able to move large numbers of men and supplies great distances rapidly—this being the result of the helicopter. JUNCTION CITY was initially supported by elements of three aviation groups which used 250 helicopters on D-day in displacing men, equipment, and supplies. This was the largest single day helicopter operation in the history of army aviation to date and also included a record number of Air Force sorties flown in a single day—575. Thirteen airmobile companies were used for the first four days of the operation. Thereafter the 12th Combat Aviation Group, with all its assets, was the sole source of support for JUNCTION CITY. By the end of the operation, army aviation had flown over 80,000 sorties and airlifted 19,000 tons of supplies. (source: Combat Operations After Action Report II Field Force Vietnam, 9 Aug 1967.)

The Viet Cong lacked the means to move great distances rapidly; however, they possessed intimate knowledge of the terrain which allowed them to "melt into the jungle" and thus escape decisive combat. Having occupied the area for over twenty years, they had been able to develop an extensive underground network of tunnels and facilities. More importantly, the Cambodian border played a vital role in mobility; for once the Viet Cong made it across the border, they had nothing to fear.

The Viet Cong main line forces fought according to the doctrine of avoiding decisive combat unless they were convinced that through the use of surprise and well-planned attacks, they could achieve the defeat of a US force. Otherwise, they were content to use their knowledge of the terrain and the cover of darkness to harass US forces with booby traps, ambushes, mines, and mortar fires. This tactic was well suited to an enemy who was outgunned and outmanned. It gave the VC forces an effectiveness out of proportion to their size. When the VC did conduct battalion or regimental size attacks, they showed evidence of careful planning and displayed professionalism in execution. (*You will read about that ability later in Junction City II when they decided to raise the stakes with major implications for both sides.*) They achieved tremendous volumes of small arms fire and advanced by leaps and bounds and normally conducted their attacks at night. The fortifications encountered were capable of sustaining very heavy artillery and air attacks and had well-planned defenses. Depending upon the circumstances, these fortifications could be stubbornly defended or simply abandoned.

Intelligence for both sides played an important role in Operation JUNCTION CITY. The VC depended upon an intricate and unusually reliable network of informants to predict US movements and to be able to either attack, ambush, or avoid US forces. Americans employed aerial observation and photography, SLAR, sensors, infrared devices, and patrol reports in their attempt to find the enemy and to forecast his actions. Pattern activity analysis came into vogue just prior to Operation JUNCTION CITY after its great success during Operation CEDAR FALLS. It was an intelligence system consisting of detailed plotting on maps of information on enemy activity obtained from a variety of sources over an extended period of time.

As more data was plotted, patterns of activity attention and locations emerged. It thereby became possible to focus on the areas of unusual activity. As a result of the additional information in January that indicated some movement of the 9th VC Division regiments, the thrust of the operation was changed from the eastern area of War Zone C to the west central portion.

A careful survey of US intelligence sources before the battle portrays an enemy Order of Battle that subsequent events were to prove fairly accurate. Therefore, it was decided to reflect the US perceptions of the enemy rather than speculate as to their location, strength, and intentions without adequate justification. The enemy Order of Battle was drawn specifically from the Periodic Intelligence Report (PERINTREPS) published by II Field Force in the two weeks preceding the initiation of Operation JUNCTION CITY. Until such time as VC/NVA sources become available, it is left to the reader to determine the accuracy of US intelligence. As stated previously, the intelligence of enemy dispositions was proved by later events to have been at least feasible.

Principal VC/NVA units involved in the battle were the Central Office in South Vietnam (COSVN), the 9th Viet Cong Division with three infantry regiments under its control (the 101st NVA, the 271st VC, and the 272d VC), the 273d VC Regiment under COSVN control, and the 70th Guards (or Security) Regiment, also under COSVN control. (*This is the swap I mentioned earlier. When the 101st NVA Regiment arrived in the south they were assigned to the 9th VC Division in replacement of the 273rd VC Regiment that was placed under direct control of COSVN with the 7th Guard Regiment providing security for COSVN. Supporting artillery units will be documented later)*

A brief description of each of these elements follows.

a. COSVN: boldThe COSVN was the major Viet Cong military/political headquarters in South Vietnam. As with most high-level headquarters, it was a diverse organization comprised of command and control, communications, logistics, civil affairs, transportation, and psychological operations elements. Its strength was believed to be 3,000 with its location in the northern portion of War Zone C. It was also believed that the actual command and control cell was a highly mobile group of 50-70 personnel, who when threatened would literally go to ground or infiltrate into Cambodia.

b. 9th VC Division: This division had operated in War Zone C since 1966 and was very familiar with the terrain. It encompassed three separate regiments.

(1) 101st NVA Regiment: This NVA Regiment was rated at C-3, with a strength of 1,250. The Regiment was believed to be located west of An Loc and north of Highway 246.5

(2) 271st VC Regiment: This regiment was rated at C-2, with a strength of 2,400. The Regiment was believed to be located east of Lo Go and west of Highway 22.6

(3) 272nd VC Regiment: This regiment was rated at C-3, with a strength of 1,850. Its location was plotted as south of Suoi Tre and west of the Saigon River.

c. 273rd VC Regiment: This regiment, under direct control of the COSVN, was rated at C-3, with a strength of 1,700. Its location was plotted in War Zone D, east of Lai Khe and Ben Cat.

d. 70th Security Regiment: This regiment was believed to have the mission of securing the COSVN and was directly controlled by that Headquarters. It was rated at C-4, with a strength of 1,000. It was believed to be located between Prek Lok and Katum.

This study did identify the unit that surrounded B/2/12 on February 26th. B Company was surrounded by a reinforced company of soldiers from the 3rd Battalion of the 271st Regiment, a security unit for COSVN. Was the headquarters' unit nearby? Maybe.

What I considered an oddity was the fact that the researchers noted how it hadn't been that many years since Operation Junction City had taken place when they began their report. Yet, with so many military personnel still serving the country, few could supply accurate details from their service during Operation Junction City. When asked, they would simply say that their recollections were blurry or forgotten entirely. Remember this is 13 years after the fall of Saigon. Yet in 2015, forty-five years after the Battle of Suoi Tre, Joe Engels (a 2/77th veteran of the battle) and I organized a reunion at Fort Carson. Every unit had a spokesperson who shared in great detail their unit's story in the battle. What do I attribute this to? Maybe too soon to look back? Or, just maybe, they were able to tell their story to men who understood the desperation of that battle in the jungle and understood its implication.

Finally, the final tally of Operation Junction City was documented in the study.

US Losses

Personnel: 218 KIA
1,576 WIA

Materiel Damaged

55APCs
27ACAV (Armored Cavalry Assault Vehicle)
1VTR (Vehicle, Tracked, Recovery)
50Tanks
42¾ Ton Trucks
32-1/2 Ton Trucks
4¼ Ton Trucks
75 Ton Trucks
3¾ Ton Trucks
4Tank Dozers
1M88 (Recovery vehicle)
1155 Howitzer, SP
5105 Howitzer
2Bulldozers
1AVLB (Armored Vehicle-Launched Bridge)

Material Destroyed

3Tanks
14APCs
1ACAV (Armored Cavalry Assault Vehicle)
2¾ Ton Trucks
22 ½ Ton Trucks
75 Ton Trucks
2155 Howitzers
3105 Howitzers
2Quad-50s (air defense gun turret with four .50-caliber machine guns)

1M577 (Command Post Vehicle or Armored Command Post Vehicle)

VC Losses

Personnel:
2,728 KIA (Body Count)
34 POW
137 Ralliers (Those that left the VC and rallied to the allied side)
65 Detainees

Equipment

491 individual weapons
100 crew-served weapons
754 artillery and mortar rounds
6,576 grenades
100,450 small arms rounds
508 assorted mines
811 tons rice
641 1bs medical supplies
17,361 assorted batteries
475,000 pages of assorted documents
4,313 bunkers
1,463 military structures
Now on to the history as it continues in Junction City II:

Junction City II, D-Day

The map above shows you the area of operations for the second phase of Operation Junction City. Note on this map FSB THRUST shown in the 1st Division's AO. Nine months later, that same clearing will be named FSB Burt. It was used by the 2/22nd Inf, 3/22nd Inf, and the 2/77th Artillery during Operation Yellowstone. The 2/12th was not there when a battle took place. They were on operation further south. It was there, at FSB BURT, that the Battle of Soui Cut took place. Their opponents were once more elements of the 271st and 272nd VC Regiments of the 9th VC Division. It was sometimes called the 'New Year's Battle of 1968', as it took place on Jan 1-2, 1968. Never heard of it? Maybe not, but you may have seen it depicted at the climax of the movie 'Platoon'. Oliver Stone was a member of B/3/22nd during that battle. Some of the movie depicted his time in B/3/22. Soon after this battle, Stone left the unit and volunteered for Long Range Recon Platoon assignment. It was during that LRRP period that the pervasive drug use set in the movie for dramatic affects came from. His former B/3/22nd comrades resented how they were portrayed in the movie. His former company commander, LTC Robert L Hemphill, wrote a book on the real B/3/22nd he led during that period to correct the record on the company that he led. He later served as the Historian for 'VIETNAM MAGAZINE' and researched the history of the Battle of Suoi Tre for the magazine in the December 1998 issue.

Saturday, March 18, 1967

Beginning early on this day, the 2/22nd left Camp Rainier and traveled to the designated LZ Silver. They were to secure the landing zone so that the two other maneuver units of the brigade could be safely inserted into the heart of War Zone C. It didn't go as planned, which led to a tragic airlift into an unsecured clearing further to the south; LZ GOLD.

As was documented in many AARs, the time that troops

needed to arrive at target destinations were frequently underestimated when setting goals for daily missions. They just did not comprehend the difficulty of traversing double and sometimes triple canopy jungle in these remote regions of the country. Many times, this eliminated the element of surprise and the Vietcong could easily adjust to changing conditions on the field of battle.

The following specific details of the action came from official AAR and Day Reports.

Suoi Da was the staging area for Operation Junction City II for the 3rd Brigade of the 4th ID. It was particularly important for its airfield, which could accommodate C-130 aircraft. The airfield and military base situated there were normally manned by Special Forces troops and local militia.

At 0630H A& B Company of the 2/22nd Mechanized Infantry left Dau Tieng. An hour later, C/2/22 and the Recon Platoon, then serving as part of the security at Soui Da, left Suoi Da and headed northeast along Route 13. The plan was for the battalion to assemble at the RP at the bridge.

The unit's mission was to head north from there and secure LZ SILVER, a clearing eight miles north and would be used to fly in the rest of the brigade's two other maneuver battalions by choppers. That bridge represented the midpoint for the expedition from Dau Tieng to LZ SILVER.

At 0750H, the lead APC of the C/2/22nd detonated a mine near the bridge, resulting in four US WIA with moderate damage to the APC. At 0800H, after crossing the bridge, an APC from the Recon Platoon detonated another mine, resulting in four US WIA with moderate damage to the APC. At 0855H, Recon platoon received two rounds from an RPG II, resulting in negative casualties. The platoon then continued to receive RPG II fire from the same location. Air strikes were immediately called in to attack the hostile area. All this happened even before the battalion formed and began their journey north.

Lt. Roger Frydrychowski, Recon Platoon Leader

"Actually, when we left Suoi Da on the 18th, we had no idea what our mission was on that day. All we knew was we were to travel to the RP near a bridge to meet up with the rest of the battalion.

"This day would be particularly damaging to my platoon. My unit was attacked three times within a two-hour period. This left us short-manned and not as effective as we could have been."

A& B Companies coming up from Dau Tieng continued their travel up Highway 19 to the RP and received no hostile fire during this period. They arrived soon after the Recon platoon attack and as soon as the air strikes ended, the whole battalion entered the jungle and began traveling towards LZ SILVER.

While This Was Occurring

The 2/77th Artillery Battalion, less Battery A, moved by convoy to Suoi Da. Battery C would pass through Suoi Da and established FSB BRONZE, near the bridge. Their mission was to support the 2/22nd Mechanized unit as they moved overland to LZ SILVER. The 2/77th technical headquarters and A and B Batteries prepared for an airlift to LZ SILVER at Suoi Da. That was set to take place on the following day. C/2/13 Arty.

Battalion, which was at Suoi Da, left for Dau Tieng to support Camp Rainier during this period.

Saturday, March 18, 1967, 0810H, Dau Tieng Base Camp

3/22nd Inf began Operation JUNCTION CITY, PHASE II, when their elements were airlifted by C-130 to Suoi Da. The airlift was completed at 0855H and a defensive perimeter was established vic coord XT394583. This was southeast of Suoi Da and near a bridge that the 2/22nd crossed heading to the meeting location. (See earlier map)

The 2/12th Inf was next on the C130s and flown into Suoi Da. We would spend the rest of this day at Suoi Da, waiting for the airlift into LZ SILVER.

At 0958H, Recon 2/22nd had another RPG II attack on one of their APCs, resulting in two US KIA and three US WIA. At 1100H, after six TAC air sorties, an artillery barrage took place, Co B swept the area in which Recon had contact. At 1109H, an APC from Co B detonated a mine, resulting in six US WIA and one APC destroyed.

It was turning into one heck of a morning for the mechanized unit. As they headed north, they were constantly harassed by enemy soldiers who were determined to undermine whatever the brigade was preparing to accomplish. Later we would learn that they clearly understood where we were heading and realized the dangers that the operation presented to their base areas.

At 1830H, 2/22nd Infantry laagered at XT394656. They had barely traveled a mile and a quarter after entering the jungle near the bridge. See last map.

Junction City II D-Day +1

Company Commander Palmer, on the phone, I am directly behind him, Johnny Martel, to far right with Walter Kelley rear

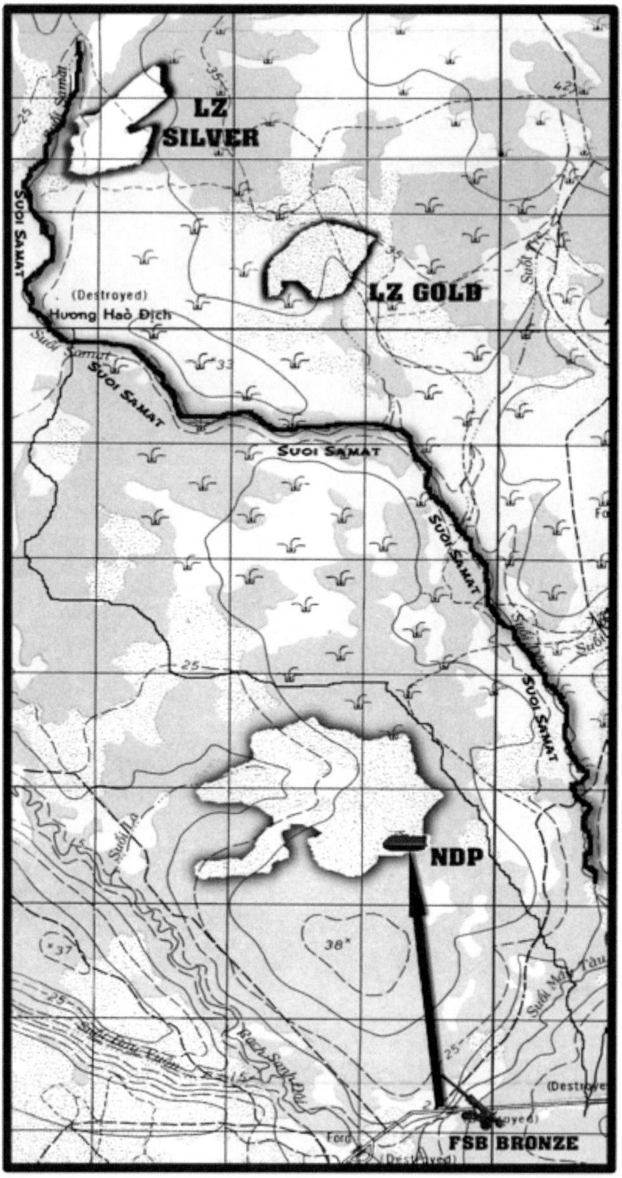
LZ
SILVER
LZ GOLD
(Destroyed)
Hương Haò Địch
Suoi Samat
Suoi Samat
Suoi Samat
Suoi Samat
Suoi Samat
NDP
38
25
FSB BRONZE
Ford

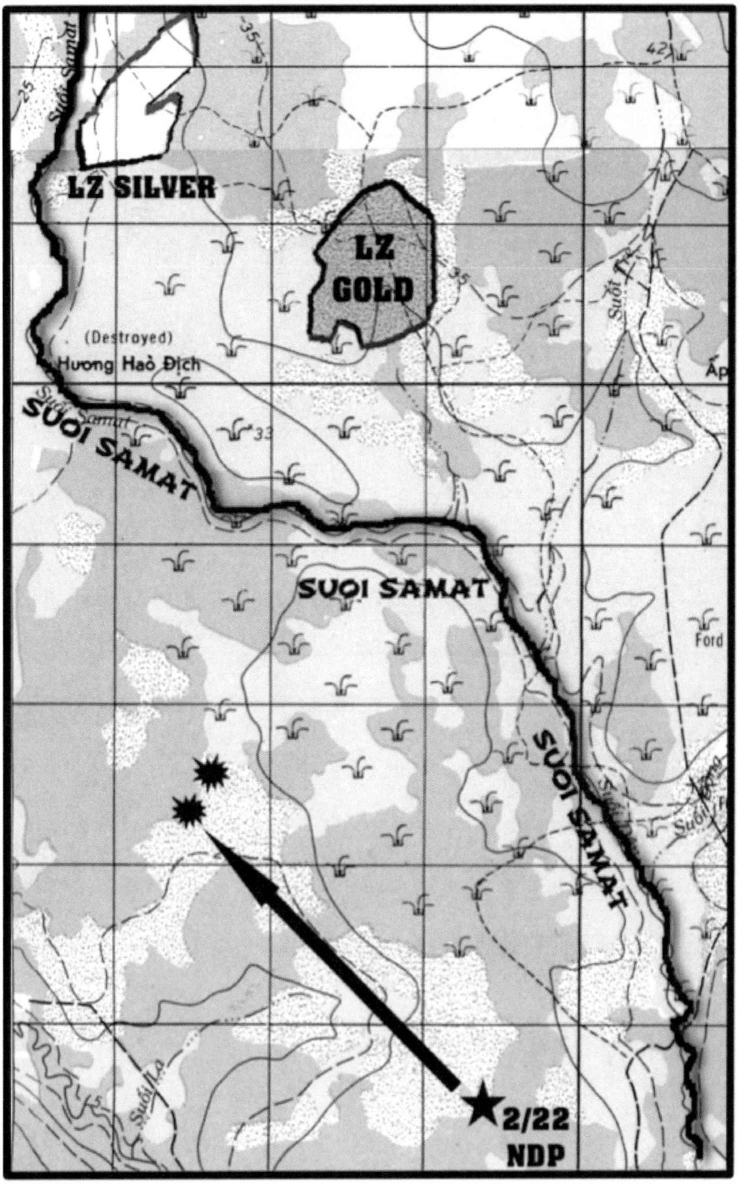
LZ SILVER
LZ GOLD
(Destroyed)
Huong Haò Địch
SUOI SAMAT
SUOI SAMAT
SUOI SAMAT
Suối Tre
Ford
2/22
NDP

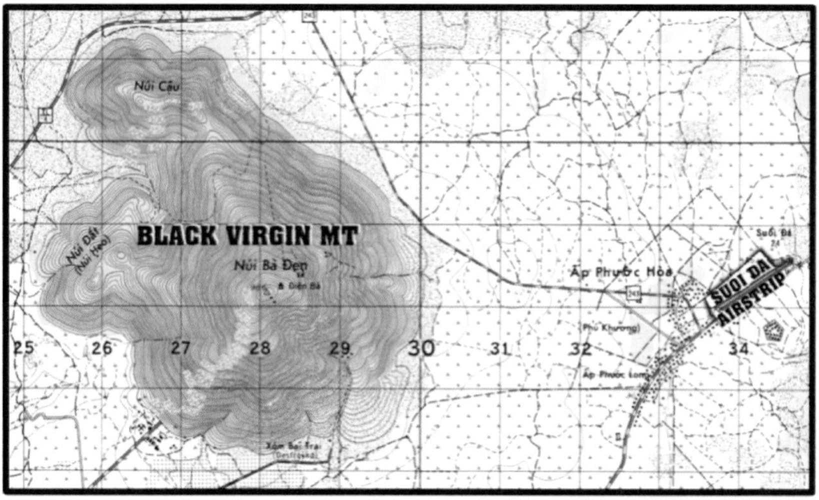

March 19, 1967

This title photo was taken at the airstrip at Suoi Da as we were awaiting our turn to board choppers taking us into LZ GOLD. There was a lot of hurry up and wait on that morning as operation planners were busy trying to adjust to the 2/22nd creep up War Zone C. It was taking much longer for the mechanized unit to arrive at LZ SILVER than planned. Colonel Garth, the 3rd Brigade Commander, was coming under heavy pressure to get his brigade into a blocking position. An alternate landing zone was further south and it was thought that the 2/22nd would be able to make it there to secure the clearing for the foot soldiers coming in by chopper. The new landing zone was called LZ GOLD and it was set a couple kilometers southeast of LZ SILVER. The cleared area was actually dry swampland (Official operation records indicated that there was less than an inch of rain during the operation). Many riverbeds were dry and actually appeared to be gullies with steep banks. In shallow areas, it was hard to distinguish the river from normal contour lines.

I exaggerated the features of the Suoi Samat River, snaking through the jungle, to show its importance in the scheme of things. It was a rather narrow river, and dry during this time of the dry season. The Vietcong were very well aware that it would become a significant obstacle for the Mech to overcome to reinforce GOLD in time to make a difference in their plans.

The 9th VC Division was now on the move to be in position for a major attack. The target was the 3rd Brigade of the 4th Infantry Division and all their attached units. As usual, the timing and location of the attack was in the VC's control.

Sunday, March 19, 1967 (Palm Sunday)

The 2/22nd Mechanized Battalion was awakened very early to try to get as far north as possible. They were still under orders to move to LZ Silver to secure the landing zone. They looked around and saw that they were in the southeastern corner of a very large dry marsh, ideal for the APCs to race through in their haste to make up time lost on the day before. The remaining journey had similar relatively light jungle terrain that could speed their progress. Charlie had other ideas.

0650H: The Recon Platoon of the 2/22nd assumed the mission of being the Bn advance guard and moved swiftly to a position two kilometers northwest of their NDP. It was then 0756H and the journey was progressing nicely. That was when the lead APC of the Recon Platoon was hit with an RPG II round and small arms fire at coordinates XT375673.

Co A was sent forward and swept the area and found a U.S. Claymore mine with blasting cap, one battery, and one Chicom grenade. The attackers melted away into the surrounding jungle. The battalion resumed its travel north.

0915H: Recon Platoon was once more attacked with an RPG II round and small arms fire at coordinates XT377677. This stopped the advance and 81mm mortar and artillery missions were called into the enemy position. After the supporting fire ended, the area was again swept, but nothing was found. The battalion had barely moved 200 meters in between attacks and it was apparent that the mech unit was going to have to slug it out for every inch of ground.

Lt. Roger Frydrychowski, Recon Platoon Leader:

"After this latest loss, my platoon was severely damaged. From this point on we were not placed at point for the battalion. This would change by a strange twist of fate on the morning of the battle, when we became the lead mechanized platoon entering the battle."

By this time, it became clear that even the LZ GOLD clearing was too far away for the mechanized unit to reach and secure. Colonel Garth finally had to make an ill-fated decision. The 1st Division and 173rd Light Infantry Brigade, were already very near their blocking position and would be setting up their base for operations. Operation planners, worried that the enemy would use the western escape route and slip into Cambodia again, leaned on Garth to make a decision that he would later regret. The two maneuver battalions would have to be flown into an unsecured clearing, not knowing what they would face. It turned into a disaster.

In fairness to Colonel Marshall Garth, he was between a rock and a hard place. Garth was no fool and realized the implications of each of his choices. As it turned out, his choice to fly into an unsecured landing zone may have been imprudent, but he had little choice than to throw caution to the wind.

After speaking to some chopper pilots who were a part of the effort to bring us into GOLD, I learned that they were not terribly surprised to see the enemy waiting for the Americans at LZ GOLD. They told me that for the entire weekend their helicopters were flying over the GOLD clearing, obviously reconnoitering the clearing in the jungle. The enemy had noticed and prepared the clearing with explosives.

The helicopter unit that would be flying us to LZ GOLD was the 145th Combat Aviation Battalion. The two aviation companies providing us those lifts were the 118th "Thunderbirds" and 68th "Top Tigers" Assault Helicopter Companies. Each of those were augmented by their own gunships and those of the 334th Armored Helicopter Company. Twenty helicopters in all were committed to that insertion into LZ GOLD.

As soon as we awoke and ate breakfast, we were marched to the airstrip. There we would await orders to board helicopters for another airborne assault. It wasn't a new experience for us grunts, but this looked different. Even for a novice soldier like myself, we couldn't help seeing so many high-ranking officers working out of a tent nearby. Interspersed with these general officers were short sleeved civilians carrying side arms. It wasn't long before whispers of "CIA?" made it through the ranks. They all appeared to be dead serious and spoke little, worrying a lot, and frequently rechecking their maps as reports were passed on to them from the 2/22nd vanguard into War Zone C. I think that even before we arrived at the strip, around 7AM, they knew that a dangerous gamble was about to take place. We didn't have a clue. Just as well, I guess.

At around 0930H, Alpha and Bravo Company of the 2/12th was told to stand by choppers that were sitting idly along the long runway of the strip. Alpha Company would be the first in the LZ.

When it became apparent that the 2/22nd didn't have a prayer to get into either LZ SILVER or GOLD the order was given to the chopper pilots to start their engines in preparation of the flight into LZ GOLD. The pilots of all 20 helicopters on the tarmac started their engines and checked their instruments in preparation for the flight. The grunts, lined up three on each side of the Bell UH-1 Iroquois (Huey) helicopters to expedite the loading. Each chopper carried a pilot, co-pilot, and two side door gunners.

Finally, at 0945H the order was given to man the helicopters. We jumped on board and shortly after, the lead helicopter began to lift his load. It reached maybe five feet in the air when it unexpectedly dropped back down to the airstrip and the engines were dropped to an idle state. Now THAT, was very unusual. It never happened before or since for us. A few minutes passed and we were ordered out of the choppers and were replaced with troops of the 3/22nd.

**After we formed into a veteran organization, we all wondered about that incident. Why didn't we go in first as planned? It wasn't until 2011,* *more than 45 years after the event that we possibly learned why the units were swapped at Suoi Da.*

Joe Elliott, our 2/12th Battalion Commander in 1967, had honored Alpha Association by attending our 2011 annual reunion. He knew the answer to that mystery and was eager to share it with us. Oh yes, the old Korean Veteran (*He was an Army officer during the Battle of Chosin Reservoir*) & Vietnam War Battalion Commander was still alive and kicking even as I am writing this book in 2022.

I won't repeat what I previously wrote about Joe other than to remind the reader that Joe Elliott was chosen by General William Westmoreland personally to lead the 2/12th Infantry in December, 1966. Previously he served on Westmoreland's General Staff.

After he was introduced to the group at our initial night together, he rose from his seat and proceeded to the podium.

I will paraphrase his speech on that Friday night assembly at the banquet hall.

"Since I arrived at the reunion, I have been approached by some men asking me if I knew why we were taken off choppers before they left for LZ GOLD.

"To explain what took place while you men were waiting on the tarmac, I need to explain what was playing out at the Command Center Tent.

"In preparation for the insertion into LZ GOLD, I met with the Command Group of the operation consisting of Colonel Garth, the 3rd Brigade Commander, General Weyand, the 25th Division Commander, his Assistant Commander, and two Major Generals from Washington. For me, as a Lt Colonel, that was an impressive group. All the battalion leaders and our staffs also were there.

"We were closely monitoring the radio reports coming through that morning. As the second RPG attack took place, not far from an earlier attack, it became obvious that the mechanized battalion had no shot at being able to secure a landing field where troops could safely land. Looking at maps, they realized that LZ SILVER was out of the question, but the 3rd Brigade needed to be placed into a blocking position for the operation, and soon. They chose a large marsh that was dry during the dry season and suitable to get our choppers into and the clearing was appropriate to set up a blocking fire base. It was closer to the advancing 2/22nd and the feeling was if unexpected difficulties arose, the mech would be close enough to reinforce the base. The clearing was called LZ GOLD.

"The tent went silent for a bit as commanders inspected the new landing site on a map. I was skeptical of the plan and I suspect that there were others who felt the same way, but hesitated to criticize the plan. I noticed a river that would have to be crossed for the mech to get into position should the enemy attack the men at GOLD in strength. It all seemed to me that for the sake of expediency and to relieve the pressure to get into position quickly, they were risking a very dangerous insertion into the LZ.

"The operation planners did not have the final say. That was Colonel Garth's call. After all, it would be his brigade who would pay the price if things went awry. Garth took a last look at the map and finally approved the change of plans. LZ GOLD would be used and the clearing was going to be unsecured. He thought he had little choice but to proceed with the riskier option.

"Finally, when no one else would question the new plan, I spoke up. I told them that this part of War Zone C was considered extraordinarily perilous and if my battalion was to be the first into the unsecured LZ GOLD, I wanted to be on the first airlift into the clearing. If the situation became dire for my battalion, I wanted to be on the ground where I could direct the response to an attack. Again silence.

"After a short moment, we ended the meeting and I returned to my battalion that was near the airstrip as LZ GOLD was prepped with artillery. When it was time to board the choppers, I climbed on board. The engines started and soon we prepared to leave the airstrip. In short order, the chopper engines were cut and we remained settled on them briefly before we were ordered to dismount and troops from the 3/22nd battalion took our spots on the Hueys. They would be the lead force into LZ GOLD.

I always felt that the organizers and leaders understood how dangerous a proposition they were approving. If something happened to one of Westmoreland's staff officers (and a friend), he may have not taken too kindly to the leaders who sent me into such a precarious situation. Politics prevailed and my unit was removed as the lead element into LZ GOLD.

"It was never my intent to have the 3/22nd draw the higher risk assignment of flying into GOLD first. I just wanted to draw attention to just how risky their decision would be to the men who had to fly into that unsecured clearing. If it was to be my men, I wanted to be with them.

"To the best of my knowledge, that explains why we were replaced as the lead battalion into LZ GOLD."

Walt Shugart was the Commander of B/3/22nd Company at GOLD. We met at the 2015 Suoi Tre event and became friends. I told him how his critical leadership at GOLD was a big part of the success of the battle. Although the entire perimeter of the base was fiercely attacked, the western perimeter, where his company was situated, suffered the main thrust of the attack.

We stayed in touch until his passing in October of 2017. Along the way he sent me a composition he had written describing in detail the days from March 19 through the battle itself.

The document was seventeen pages long and I will intersperse his thoughts on the period as they take place in the book.

Walt Shugart

"Now, even though battle ready, fully armed and equipped, the mission had been changed for my light infantry company of about 125 men. They were to provide half a perimeter of security for an artillery forward fire support base named Task Force 2/77.

"The task force was to occupy fire support base Gold at a landing zone northeast of Tay Ninh City in War Zone C. Its mission was to provide close and continuous artillery support for the maneuvering brigades of the 1st, 4th and 25th Infantry Divisions as well as the 173d Airborne Brigade.

"On the map, FSB Gold looked like marshland about a thousand meters long and five hundred meters wide, just west of a village, Ap Soui Tre. In the dry season, the dried marsh would be a good LZ. B Company would establish the eastern perimeter while A Company would secure the west.

"Now we were waiting on the tarmac in the heat and humidity of Vietnam at Suoi Da for the slicks. As we sweltered, I was reminded of Psalm 121 verse 6, "The sun shall not smite you by day…" The psalmist to the contrary, if not smitten, the men were fully abused by the sun with no shade to be found anywhere. The wait extended from thirty minutes to an hour to an hour and a half. Finally, I received word that while we waited, the LZ was being prepped with artillery and gunship fire to neutralize any enemy and make our landing unopposed.

"The company plan was simple. We'd land in four lifts with the four platoons going in sequence—first through weapons. The company command group of the Artillery FO, LT Pacheco, and his RTO and me and my two RTOs would accompany Lieutenant John Andrew's first platoon which would occupy the southern third of the company sector. 1SGT Jones and the company operations NCO, Richard Linneman, would join Lieutenant Hinton Whitehead's second platoon to site the Company Command Post behind it. The platoon would hold the middle sector with Lieutenant Jim Slinkard's third platoon on its left flank, tied into A Company. Lieutenant Mike Kaul's 81mm mortar platoon would be in the last lift to occupy a position about 100 meters behind 1st platoon and 50 meters south of the company command post.

"Task Force 2/77 was composed of the two rifle companies for perimeter security, three artillery batteries, two sections of air defense artillery quad fifties employed in a ground defense role, and the infantry and artillery battalion headquarters. The strength of the fire base would be about 500 soldiers. The 2/77th artillery battalion commander, Lieutenant Colonel Jack Vessey, would command the fire base. He had just assumed command days before when his predecessor was called home on emergency leave.

"While waiting, platoon sergeants and squad leaders, some no more than Spec 4's who had been draftees a little more than a year before, reviewed the plan with their men and checked and rechecked their equipment and gear. All were fully trained professional soldiers, and there was little else to do except wait.

"The brigade chaplain, Major Gene Adler, a Baptist, who in

close company referred to himself as "Rabbi Adler," announced a non-sectarian church call to mark the beginning of Holy Week for Christians at a point midway alongside the tarmac. The service would also serve to settle the nerves of those anticipating action with the combat assault. There were no palm fronds either to wave with shouts of Hosannah, provide shade for the bearers, or settle the dust on the tarmac. I attended the service to set an example and pray for personal strength, wisdom, courage, the accomplishment of the mission, and the welfare of the troops. I don't recall whether Palm Sunday was even included in the liturgy. The half hour service provided relief from the monotony of waiting, as well as the anticipated action.

"Around 10 AM, word finally came that the choppers were ten minutes out. The wait was over. The pick-up would go as planned with B company leading the way. The platoon sticks assembled on both sides of the runway since the Hueys landed in a column of twos. As the choppers flared for landing, the dust kicked up obscured the entire lift. In the resulting confusion, I ran head down toward what I thought was my assigned ship only to be jerked from behind by the Company RTO, SP4 Patrick Toyama, who yelled that the chopper I was about to board was for second platoon. Ours was the one just ahead to its left, the trail helicopter of 1st platoon…"

1015 H: C/3/22nd served as base security at Suoi Da, then the Brigade's FOB for this mission.

"The first airlift, composed of B Company men, lifted off from Suoi Da airstrip and on its way to LZ GOLD, seven and a half miles away. The first airlift was uneventful. However, as B Company's 2nd lift arrived at the LZ, Viet Cong hidden in the tree line triggered a command detonated mine constructed out of a 155mm howitzer round. This caused heavy damage to four helicopters and produced numerous casualties among the B Company troops and the aviators.

When the A/3/22nd soldiers on the first and second airlift arrived at GOLD, two additional mines were detonated, causing heavy damage to the choppers.

VC snipers in the tree line were causing chaos on the ground until they were flushed out by ground troops. It wasn't until 1300H, three hours later, before the landing zone was considered secure enough to fly 2/12th troops into the clearing…"

Walt Shugart

"…Once on board, we lifted off for the fifteen-minute flight at 1,500 feet to LZ Gold. The helicopter doors were open to provide free fire for the door gunners. but also brought a breeze to cool us from the sweltering heat endured on the tarmac just moments before. About two minutes out before landing, I found the LZ on my map and observed a pair of gunships making a final strafing run. The LZ was shrouded by smoke from the artillery and aerial bombardment. Nonetheless, 1st Platoon landed without incident, assembled, and headed in skirmishes toward the tree line to the east, which was a hundred meters short of its planned defensive position.

"Toyama and I jumped from a following chopper as it hovered a few feet above the ground, landed on our feet, struggled to maintain our balance, and headed toward the tree line. After moving about 50 meters, a massive explosion to our rear blew us face down to the ground. Turning around on my knees, I saw the lead helicopter in the 2nd Platoon lift and disintegrate in a fireball. It was the helicopter Toyama had stopped me from boarding. All ten members from the platoon's second squad as well as the four air crewmen were killed instantly.

"The following chopper, too, was damaged by the blast. The troops on board abandoned the aircraft from about twenty feet above the ground. The pilots attempted to fly the plane out of the kill zone, and as it passed over head, I could see the warning lights on the control panel flashing red before it crashed, bursting into flames after about 100 meters of flight, killing the crew on board. Simultaneously with the explosion, 1st platoon began receiving small arms fire from the wood line ahead. Return fire by 1st Platoon suppressed the enemy fire and allowed 3rd and weapons platoons to land with no further loss of men or equipment.

"The air assault continued bringing A Company, then both battalion command elements into the landing zone. These were followed by heavy lift helicopters moving the artillery pieces, ammunition, the quad 50 dusters, and communications equipment into place while the infantrymen provided security and policed the battlefield. Four additional five-hundred-pound bombs rigged for detonation placed strategically around the LZ were found.

"Had they been blown; the carnage could have been much worse. Within four hours of the initial landing, the LZ was secured and FSB Gold was fully operational, providing direct 105mm howitzer support to the maneuvering elements as the mission required.

"A Graves Registration team from II Field Force Vietnam removed the remains of the KIA from the LZ. Those who died in the helicopter crashes were scraped off the ground or what remained of the shell of the helicopters, scooped into metal canisters, and returned to parents, spouses, or next of kin. I had no time to mourn. That time would come with my personal letters of condolence to next of kin written as soon as time permitted ..."

Joe Boggs, Commander, 118th Assault Helicopter Company

I met him at the 50th anniversary of the battle reunion at Fort Carson. I asked him for an interview and he consented. This is how he recalled that day:

"After the initial losses from the mines, the remaining aircraft returned to Suoi Da. Two of our aircraft that were damaged staggered back as best they could. One went down in the jungle and the men had to be rescued. The other chopper just barely made it back to the pick-up point. That left me with five airplanes from my original seven that began the lift.

"When we made it back, the operation was held up a bit until it could be determined if we could continue with the airlift.

"We realized that it was going to become very difficult to secure the entire clearing, allowing us to get more choppers into the clearing for future hauls. The enemy activity at the site was still very fluid and the decision was made to secure a portion of the clearing and send in five choppers at a time to get the remaining soldiers of the 3/22nd into the clearing.

"We continued with the airlift using three groups of five aircraft. The "Top Tigers" of the 68th AHC, was the lead group, and they went in with two groups of five followed by my five choppers.

"While flying into LZ GOLD with my load, I was ordered to hover over the field and have the infantrymen jump down to the ground in case there were more mines in the clearing. I brought my aircraft into a hover a few feet over the clearing and as the infantrymen jumped to the ground, another explosion took place just forward of my ship. My plane was pressured backwards which blew out the chin bubble (these are plastic shields just under the windshield on each side which allows the operators to see forward and down as they land). My windshield was also fractured from the blast."

One of those infantrymen blown from the chopper was Mike Doolittle.

Mike Doolittle—FDC for A/3/22 mortar platoon

Mike was a five-year veteran of the Army at that time and arrived in the 3/22nd in early January 1967. He was then an E-5 just coming from Germany, where he served in the 8th Infantry from 1964-1966. He previously had served in Korea during late 1962 and part of 1963. He was next a drill instructor at Fort Ord early to mid-1964. He was no rookie when he went to jump from Joe Boggs' chopper on that day and was blown away.

Doolittle

"During the landing on the 19th, we lost some equipment on one of the downed choppers and were not functional, but told to dig in the tubes emplacements as they expected the equipment replacements. They never came." (You will be reading more about him later.)

Joe Boggs

"I soon got a call from Old Warrior Six, the battalion commander, who was in the command-and-control aircraft. He asked, 'Thunderbird Six, are you OK?' I answered that I was not sure. I pulled pitch, the aircraft responded, and I pulled out. The Huey was controllable and it could fly. I told him my aircraft was still airworthy and began the return trip to Suoi Da.

"I made it back to Suoi Da with the other four choppers from my company.

"We prepared to fly in another lift of soldiers but we held up until we could determine if my airplane could continue in the mission. The crew chief and I thoroughly inspected the aircraft and came to the conclusion that if it flew out of the LZ, it could fly back into the clearing.

"We made four or five more lifts into LZ GOLD. Each consecutive lift produced less enemy fire than the lift before.

"There were other combat aviation units involved on that day. I know that a medical helicopter unit provided lifts for the men who became casualties from the landing. Additional Iroquois ships were brought into the airlift to replace those that were lost in the clearing.

"After that day, the 145th Combat Aviation Battalion returned to our base and prepared to lick our wounds and get the battalion back up to strength.

"Typically, we did not do many battalion size operations. I was in Operation Attleboro as a pilot and was given command of the 118th in October, 1966, during that campaign. This particular mission was extraordinarily larger than most combat missions. They wanted to insert the troops into that clearing very quickly with a substantially large force of troops compared to most air assaults.

"I do remember that we arrived at Suoi Da on Saturday, the day before the actual lift day. We spent the entire day there. I don't know why. I can tell you that on that Saturday choppers were sent on reconnaissance missions flying over the LZ GOLD clearing. They must have suspected, even then, that they might have to change the landing zone to something closer to the advancing mech unit coming up to secure the clearing. We pilots were not too surprised that the Viet Cong were waiting for us the next day.

"As with most combat officers, I was reassigned out of the 118th not long after Suoi Tre. In those days, we were sent elsewhere after six months in front line combat. I was sent to Saigon where I served as a staff officer.

"Just before that transfer out of the 118th, a few of my men and I were called to Saigon.

"The commander of the 145th Combat Assault Battalion, had recommended us for medals. I was recognized for my leadership continuing with the mission even after my aircraft was damaged. I was awarded the Silver Star."

It was 1300H before the 2/12th was airlifted into LZ GOLD. That was over three hours after we first boarded choppers. For the life of me, I can't recall being told anything about the carnage that was taking place at GOLD. We didn't need to know; we were not told. To us, that three-hour wait was just another 'hurry up and wait' episode that was common for my time in the service.

When we landed, we viewed the carnage scattered around the clearing. It was a terrible site to behold. Some of the destroyed aircraft were still smoldering in the clearing and would be doing so for hours.

After the 2/12th arrived, we were directed to the northern end of the clearing to set in for the night. We didn't prepare for a long stay as we were scheduled to leave the base on the following day.

Gene 'Mac" McLemore, 2/77th Artillery, allowed me the use of the photo that he took on March 19, 1967 of the destroyed UH1 helicopters on LZ GOLD. You can see the smoke in the distance from the fires that were still smoldering in the clearing.

The 2/77th Artillery Battalion, minus C Battery which remained at FSB BRONZE, followed us into the clearing and began to set up their artillery pieces in the southern half of the clearing. The artillerymen soon began building bunkers and protective walls for their guns. One item missing from their guns were the shields. They were too heavy for the flight into GOLD by Chinooks so they remained behind. Many of the wounds suffered by the artillerymen on the day of the battle was to the upper torso. Those shields were very much missed once the battle began.

While preparing the fire base, the reaction force teams drilled in the event that they would be needed. These units would serve to reinforce the perimeter if it was in danger of being overrun. Some of the artillerymen were on those teams and were used to plug holes in the line. Every soldier is trained as an infantryman during basic for just such an eventuality.

The security of the base was provided by A and B Company of the 3/22nd battalion. As the day went on, these units would have a number of reinforcements brought in from other units to replace their losses from that day. The 196th Light Infantry Brigade sent a man named James Dale Brewer to GOLD. Dale was an original from the LIB who had arrived in country in June of 1966. He was plugged into the A/3/22nd Company where he knew no one. He would later step up and save many American lives when the outcome of the battle was in question.

Two Quad-Fifties were flown into the firebase. That weapon was normally used as an anti-aircraft gun. It was also a fearsome weapon that could be a formidable obstacle for mass ground assaults. It consisted of four .50-caliber machine guns mounted on a carrier with armor protection for the operator. These were placed at the north and south sector of the base.

The net consequences of that perilous airlift into GOLD

were staggering. Three UH-1 helicopters were completely destroyed. Three more were damaged. There were 10 U.S. KIAs, 18 WIAs, in addition to seven aviators KIA. A sweep of the clearing uncovered nineteen 18mm mortar rounds and two 175mm rounds rigged to command detonate. As you can see, the enemy had that clearing well prepared for our arrival and it could have been much worse.

It should be noted here that the army did learn a valuable lesson. Two months after the battle, a 1st Division unit was preparing to fly into a treacherous clearing three kilometers north of LZ GOLD. They experienced that same time issue as we did and considered dropping in a battalion into an unsecured clearing. With the experience of LZ GOLD fresh in their minds, they decided to send in a mech unit to determine if it was safe to come in. Sure enough, when they swept the area, they discovered the same type of mines and bombs that were rigged for command detonation in the clearing.

Final thoughts recorded by Walt Shugart about the March 19th insertion into GOLD and the preparations to continue on with the mission:

Walt Shugart

"The decimation of 2nd platoon during the assault disrupted the establishment of the B Company sector of the perimeter. Both 1SG Jones and Lieutenant Whitehead, who had jumped from the second damaged helicopter, were wounded. While wounded, they continued to provide leadership to establish the company CP and patch together with the 20 remaining second platoon soldiers to man its sector of the perimeter. After the chaos of the initial landing subsided, both were ordered from the fire base. They were evacuated along with 15 other wounded by Dust Off. B Company had sustained approximately 30% casualties during the assault.

"It was now about 1800 hours. The battalion headquarters was up and operational. The battalion commander, Lieutenant Colonel Jack Bender, summoned his commanders to the battalion Tactical Operations Center (TOC) for a staff briefing. Before departing for the TOC, I instructed the platoon leaders to dig in, coordinate fire plans, and allow the soldiers to eat a third at a time to maintain basic security. I also ordered 1st and 3rd platoons to begin security patrols two hundred meters into the woods to their front. The second platoon sergeant expressed concern that with his fifteen available troops he had little confidence in holding his portion of the perimeter in the event of an attack or even detecting infiltration by the enemy. The artillery forward observer had already begun to plan defensive concentrations of fire for the company sector before I left for battalion.

"The battalion S2 Intelligence Officer began the staff briefing with the weather report which every soldier in the brigade could have given based on his personal experience of Vietnam weather during the dry season. He followed with the intelligence briefing which included that security elements of the 272 VC regiment of the 9th VC division had mined the LZ and inflicted the casualties with no more than a platoon sized element. The enemy regimental headquarters and attached battalions had withdrawn to avoid contact. Whether the enemy had suffered any casualties was unknown.

"The S2 was followed by the battalion S3 Operations Officer who recounted the battalion losses suffered, noting that B Company had been the hardest hit. He went on to say that the battalion Scout Platoon would be attached to the company, absorb the remains of 2d platoon and in effect become B Company's second platoon. Further, the battalion operations NCO,

MSG Williams, would be assigned to B Company to replace 1SGT Jones. With those two actions, B Company was brought close to the strength it had at lift off from the tarmac seven hours before. The S3 continued saying that the 2/77 Artillery would provide a rifle platoon size reserve force for the fire base. A rehearsal of its commitment would be conducted prior to EENT that evening, reinforcing the B Company perimeter in the 1st Platoon sector.

The S4 Supply officer continued the briefing by review of the status of supplies and summarizing expected resupply throughout the expected two-week duration of Operation Junction City II. The troops could expect at least one hot meal a day plus a daily supply of sundry pacts which included candy, cigarettes, magazines, newspapers, cold beer, and soda.

"The staff briefing was concluded by the S1 Personnel Officer who began his portion on the prospects for replacements followed by his standard spiel regarding commanders' responsibilities for their troops' welfare. He could have read that portion verbatim from FM 101-5, Staff Organization and Procedures. Following the S1, LTC Bender concluded the meeting emphasizing the need to stay on the alert, that the enemy was obviously operating in the area in unknown strength, and that stand-to would be conducted 20 minutes prior to BMNT with all soldiers awake and alert, weapons loaded, and pointed down range. Stand-to would conclude ten minutes past sunrise. The meeting was over and I returned to the company command post with MSG Williams in tow. The Scout Platoon with Lieutenant Kaminsky, its leader, followed us. He quickly realigned the squad sized remains of 2nd platoon and filled the gaps in his sector. While I was absent, land line had been laid between the platoons, to the company CP, and to the battalion TOC providing a secure line of communication to every unit within the fire support base. Radio silence would be observed within the perimeter. Only hourly radio communications checks with the ambush patrols and listening posts would be breaking the silence.

There was scarcely time to brief the company chain of command before the rehearsal of the reaction force. From alert to deployment to withdrawal to its several gun positions, the rehearsal lasted about an hour. All the while, the platoons were digging in, reinforcing positions, and attempting to resolve the personal issues which had emerged following the contested combat assault.

At about 2100 hrs., I foraged a C ration meal from my pack. It was the first food I had consumed since breakfast that morning. I would have loved to heat it quickly with C4, adding a cup of hot coffee, but there was neither time nor space for that. LT Pacheco and I huddled under a poncho, used red lensed flashlights to preserve our night vision, and reviewed his fire plan for my approval. With the company radio watch being shared by the company and battalion RTOs, 1SGT Williams and I made a visit to each platoon to introduce him as first sergeant, check progress being made toward improving the platoon positions, assure their understanding and compliance with the mission, and exhibit a presence on the battlefield.

We were pleased to find both the readiness of the platoons and the state of morale at a satisfactory level. We did not disturb the rest of those who had already turned-in, anticipating their turn in two hours to be awake and alert. Returning to the CP, we exchanged thoughts about the mission and the status of the company. During that short time together, we forged a bond of mutual confidence, respect, and friendship which I had never achieved with 1SGT Jones.

After saying goodnight, I went to sleep around 2230 hrs. It

had been a long day to say the least. My sleep was interrupted by the hourly communications checks and outgoing H & I fires of the 2/77 Arty. But for me, day one at Soui Tre was over."

While this was playing out at LZ GOLD, the 2/22nd was still trying to get up to LZ GOLD. The VC were having none of it. After the 0915H attack on the 2/22nd, the movement was stopped and supporting artillery and 81mm mortar fire were called into the area. After sweeping the area to their front, the advance continued. This took a while and the journey began to move forward only to suffer another attack at 1205H. At that time, A/2/22 received small arms fire and a claymore was detonated one hundred meters north of the 0915H attack. This assault resulted in four U.S. KIAs and two WIAs. A Company was withdrawn and two sorties of TAC air were employed with medium and light artillery barrages. At 1545H Recon and C Company of the 2/22nd closed in to what would be their NDP for that night at XT377672. They had moved approximately two kilometers closer to GOLD but were still three kilometers away from the base at the end of the day.

At 1615H, A/2/22 swept the area of contact and came under small arms fire for a short while. No casualties were reported from this exchange of fire. B/2/22 destroyed two anti-tank mines found in their search area.

At 1700H, C/2/22 reported one of their men was wounded as a result of an artillery short round in their area sweep.

At 1710H, A/2/22 was sent all the way back to the bridge and FSB BRONZE. They were immediately placed under operational control of the 2/34th Armor Battalion, which was sent up to deal with the Viet Cong harassment. The 2/34th was attached to the 25th Division and would arrive at 0830H the next morning. At that time both units would head north to catch up with the 2/22nd Battalion.

Two Final Thoughts Relative To This Day

As the 3/22nd were duking it out with the remnants of the Viet Cong who greeted their battalion after arriving at LZ GOLD, overhead was an FAC aircraft directing air force sorties on the enemy. The plane was manned by two air force officers who were assigned to the 3rd Brigade of the 4th ID at Dau Tieng. Their names were Captain Walter Forbes and Captain Tonie Lee England Jr.

Walter Forbes was a three-year veteran of the air force and arrived in Vietnam on September 13, 1966. Tonie England was a 12-year veteran of the air force and had arrived in Vietnam on January 30, 1967.

At the end of this day, Tonie England wrote a letter home to his family that was shared with me by Bob Staib, another air force FAC.

Sunday
19 March 1967

"It is very gratifying to know that Uncle Berlyn and Aunt Mable now belong to your church. I feel certain that Aunt Mable has piece of mind because she publicly accepted Christ as her saviour. I have often wondered why people wait so late in life to fully realize how much God means to them.

Even though today was Sunday, I had a very busy day. I flew three times today as my brigade is out in the field again. They came in late Thursday afternoon and stayed until Saturday morning.

Saturday afternoon they moved out again and we are now engaged in a bloody battle with the Viet Cong in War Zone C. Today was a bad day as we lost quite a few men. I met two of the officers Friday night that were killed today. Some people might think this is a small war, but to the families of these two fine young officers it is big war. I am more depressed than I have been as I feel the closeness of death now. I was flying overhead when the battle started and I saw the company being hit. I put in an air strike, but I was too late to save them. If only I could have done more.

Actually, this past week has been as busy as can be. I fly at least twice a day and keep up with my additional duties. Maybe I am beginning to feel more the strain of flying combat missions. I am flying these times tomorrow and my commander said I could take a day off after that. I haven't had a day off since coming here at Dau Tieng. Who knows, maybe a day off would be good for me.

The weather is miserably hot. The temperature hasn't dropped below 86 in over a week. I am soaked from morning till night. I am grateful that I get to clean up every night. Of course, our poor men in the field don't get to clean up for days at a time.

I went to chapel Services this morning and I prayed as I haven't prayed in a long time. No, didn't pray so much for myself as I did for our men in the field. Even the Chaplain's face showed the strain. We have two Chaplains and one of them is always out in the field. The Chaplain just came in and he looked exhausted. I talked with him for several minutes tonight and I counseled him more than he did me. He knows most of the men who were killed today. Have your church continue to pray for the men over here. They are good men and they are fighting for your freedoms and mine. Because of these men and all who went before them, you and all Americans can go to church and worship as you please.

If my mood seems to have changed, it is because of what I am now seeing and doing. I don't believe I will ever become hard or callous when I see so many die, as I love life so much and I want to come home to my family.

I haven't had time to get any Easter cards as there are none here. In case I can't get one, I want to wish you a Happy Easter. Think of me on this special day.

Bye for now. I have the early flight and so too bad for me.
Your loving Son,
Tonie Lee

The above letter was mailed on March 21st and received on March 25th by Tony's parents. He would be killed during the March 21st, the day of the battle.

The 9th VC Division may have been done with the troops of the 3rd Brigade of the 4th Infantry Division on the 19th, but they were not done with the other participants of Operation Junction City II. The 273rd Regiment of the 9th VC Division was racing to the operational area of Junction City from War Zone D. They were in position by nighttime, March 19th to test the troops of the 9th Infantry Division who were providing security along Highway 13.

The 273rd VC Regiment attacked Firebase '14', 30 miles southeast of FSB GOLD. The Mechanized Unit, A/3/5th Cavalry of the 9th Infantry Division, 129 men strong, defended the firebase near Ap Bau Bang. The Artillery Battery manning the guns was B Battery of 7/9th Artillery. The attack, known as the Battle of Ap Bau Bang II, began at 2300H on the night of March 19th and ended in a rout of the enemy by dawn. The Vietcong had left 227 dead bodies on the field, but left only three wounded comrades there. Only 11 VC weapons were found at the site.

A high school friend who was drafted on the same day as

me fought in that battle as a machine gunner on an armored personnel carrier. His unit, stationed at a village north of the battle, was sent south at 0100H. Another mech unit from the south joined them. Memories of that night time battle affected him the rest of his life.

We met a few times after we got home and neither of us talked at all about our major battles fought within a day of each other. It was forty years before we made the connection. After realizing the coincidence, we rarely talked about it afterwards.

He died while I was writing this book. After more than 20 years of therapy, and 50 years after the combat that triggered his affliction, he succumbed to PTSD and took his own life with a pistol he had hidden in his garage.

FIRE SUPPORT BASE GOLD

The Base And Leadership

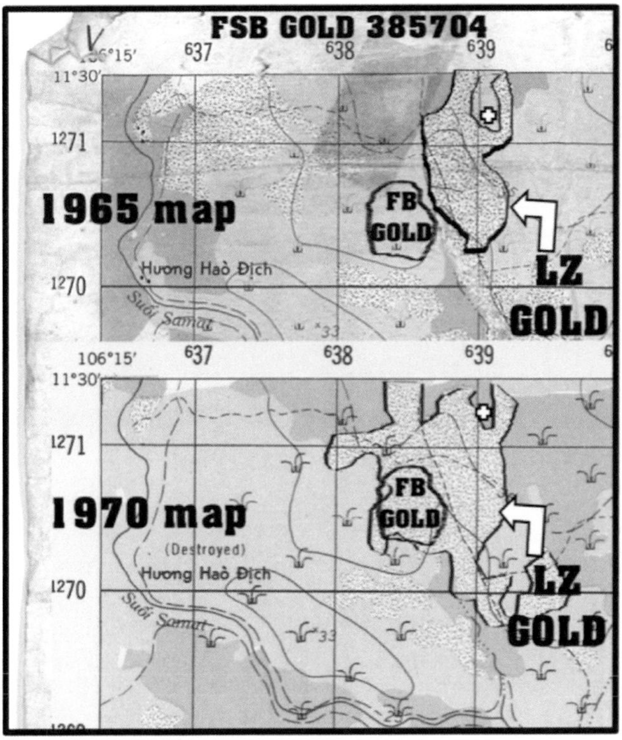

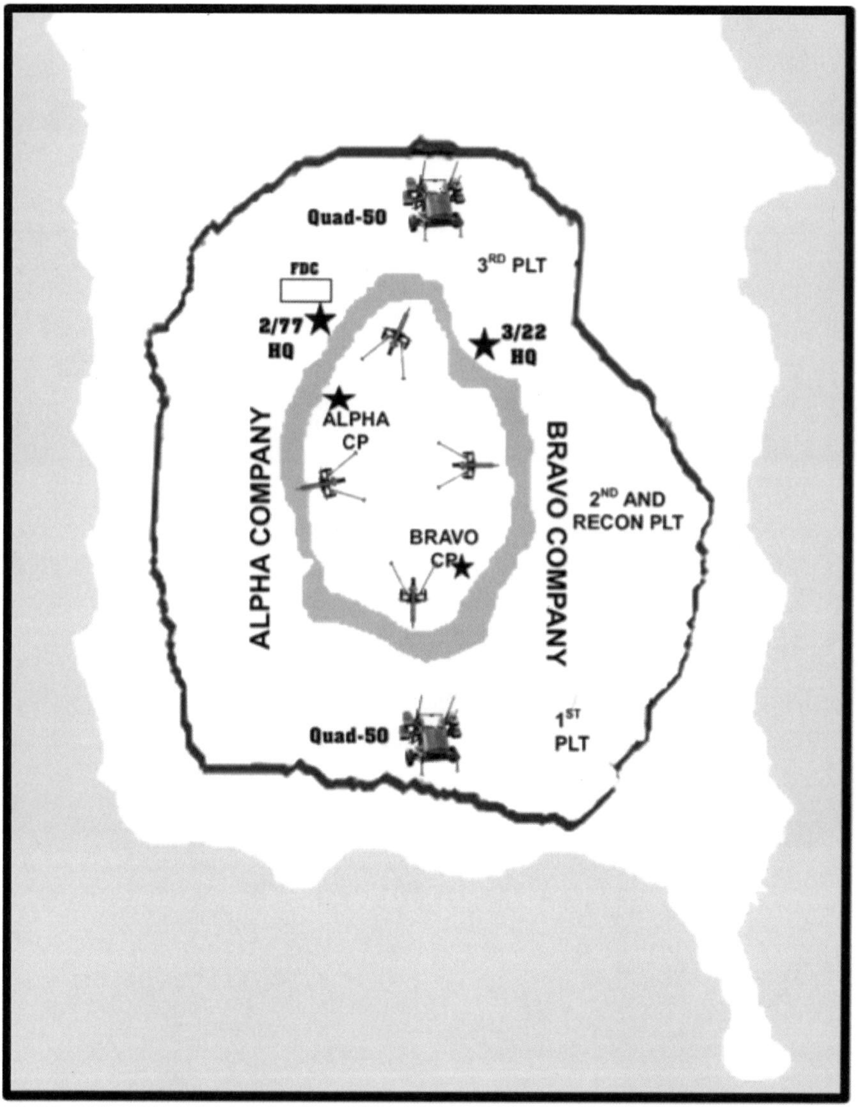

The men at FSB GOLD were blessed to have Lieutenant Colonel John W. Vessey commanding the fire base. He already had sterling credentials when he was sent there to command the 2/77th Artillery Battalion. He had been serving as the Deputy Commander of the 25th Division Artillery at Cu Chi when he was sent to the 2/77th to lead it on March 15th. The original commander was called home on emergency leave. Fate stepped in and Vessey would end up playing a prominent role in the defense of the base.

The infantry battalion which provided security for the base was the 3/22nd Inf Battalion and was commanded by John A. Bender, no rookie himself, as you will see. As I wrote earlier, C/3/22nd was not with the battalion at FSB GOLD as it was providing security for the Brigade Command Post at Suoi Da.

So, to say he was leading a battalion during the battle, is a bit of a stretch, especially after the losses suffered during the insertion of troops into LZ GOLD. The unit was partially reinforced on that day by others brought into the unit, but it would be fair to say that his unit was very much undermanned. Still, those who were at the firebase performed magnificently under tremendous pressure. Both the 2/77th Artillery Battalion and the 3/22nd Infantry Battalion, in my opinion, were the real heroes of the Battle of Suoi Tre. Everything depended on them holding on until relief arrived. Catastrophe waited for the entire brigade had they not so tenaciously defended that base on the morning of March 21, 1967.

John W. Vessey, Jr.

Vessey was born in Minneapolis, Minnesota, on June 29th, 1922. In high school, he joined the Minnesota National Guard; in 1940, his unit was called up for active duty. He was 18 years-old at the time. He served as an artillery sergeant in the 34th Infantry Division (Red Bull Division) and fought with distinction in North Africa, and received a battlefield commission at Anzio. The Battle of Anzio was an effort in Italy to swiftly capture Rome through a beachhead south of Rome. The battle turned into a stalemate that wasn't successful until allied forces from the south reached their positions and forced the Germans back. The Anzio battle costs were amazing. Allied losses were 7,000 dead, 36,000 wounded, missing, or captured of 150,000 troops; German, 5,000 dead, 4,500 captured, 30,000 wounded or missing of 135,000 troops. During this campaign, John Vessey received a battlefield commission.

Vessey next served in the Korean War.

After Korea, Vessey attended the U.S. Army Command and General Staff College, the Armed Forces Staff College, and the Industrial College of the Armed Forces. He also graduated from the University of Maryland at age 41. (In 1970, he graduated from the Army Helicopter School, where he was 15 years older than the next-oldest student). He never stopped learning and no challenge intimidated him.

I had seven more pages of accomplishments after his time in Vietnam. More on those achievements later.

John A. Bender, the Commander of the 3/22nd had a different military background, but that experience also began during World War II.

"John 'Jack' Bender

"Bender was born and raised in Bremerton, Washington, 40 miles north of Fort Lewis, Washington. His father, a military medical officer, died when Jack was six, and he was raised by his mother and grandparents. Jack was an avid outdoorsman, spending most of his time in the Olympic Mountains, fishing, hunting, hiking, and skiing.

After graduating from Bremerton High School in 1943, he enlisted in the U.S. Army and served in the 10th Mountain

Division and in Patton's 3rd Army in Europe until 1945, when he entered West Point. He was an outstanding cadet, excelling in boxing and wrestling.

At graduation in 1949, Jack was commissioned in the Infantry. He met and courted Charlotte Gilliam while taking advanced courses at Ft. Riley, Kansas. He won a bet with a classmate to be the first to date Charlotte and was so successful that, in March 1950, they married in the Infantry chapel at Ft. Benning.

Next, Jack was assigned as a platoon leader in the 1st Cavalry Division in South Korea. During the Korean War, his courage and actions under fire inspired everyone with whom he associated. He even made national news when he became the first American officer to lead his platoon across the 38th parallel. In addition to the Silver Star and two Bronze Stars, Jack came home with fragments in his leg from a mortar shell, which landed six feet from him and killed three men under his command.

In August 1954, Jack received a master's degree in business from Syracuse University. While at Syracuse, son John, Jr., was born. As a new captain, Jack was assigned to Heidelberg, Germany, in a staff position to the Comptroller of the U.S. Army in Europe. A second son, Mark Eric, was born in Heidelberg. After two years, Jack became the company commander of a heavy mortar unit in Berlin.

He then spent a year at Ft. Benning, before being assigned to the Pentagon in the Office of the Comptroller of the Army. While at the Pentagon, daughter Wendy Lee was born.

The next assignment for the Benders was Naples, Italy, where Jack was aide to GEN Ward, Chief of Staff, Allied Forces Southern Europe. This was a very broadening experience requiring interaction with diplomatic and military leaders from all NATO countries. Son Matthew Ward was born in 1962. In

1963, MAJ Bender and family were at Ft. Leavenworth, KS, while John attended CGSC (U. S. Army Command & General Staff College), after which he was assigned as training officer for the 4th Infantry Division at Fort Lewis.

In 1965, Jack was promoted to lieutenant colonel and given command of 3rd Battalion, 22d Infantry, at Fort Lewis. The battalion, then at 20% strength, grew to 100% with the influx of draftees in late 1965. He trained his men and deployed with them when the unit was sent to Vietnam in 1966. His greatest challenge came a year later at the Battle of Soui Tre, which occurred in connection with Operation Junction City. On 21 Mar 1967, the 3/22nd was defending Fire Base Gold, established and held by COL Jack Vessey's artillery battalion, playing the risky role of 'tethered goat.'

The largest, most frightening, and probably most violent of his bitter battles in combat took place at Fire Base Gold. American battalion commanders in Vietnam rarely woke to see their unit facing oblivion. LTC Jack Bender, the Infantry commander at Fire Base Gold, was an exception.

His wartime commanders and subordinates have said that John "Jack" Arthur Bender was the bravest man they ever knew. West Point's mission of producing leaders for the combat arms was well fulfilled with Jack's achievements. During his 30 years of active duty, he served with distinction and heroism. He had a 13-month combat tour in Korea and another 13-month combat tour in Vietnam, both times earning a Silver Star for gallantry in action.

Captain Walt Shugart

Captain Shugart, the B / 3/22nd Commander, was born on July 28, 1939. "Walt" Shugart was raised in Texas. He enrolled in the R.O.T.C program while he was attending Washington and Lee University, in Lexington, Virginia. Shugart received a regular army commission upon graduation as an infantry second lieutenant in 1961. After completing the Infantry Officer Basic Course and Ranger School at Fort Benning, Georgia, he was first assigned as a training officer at Fort Knox, Kentucky. Following that tour he was assigned to an infantry unit with the 8th Infantry Division in Germany. Shugart returned from Germany and was next sent to the 3rd Brigade of the 4th Infantry Division, then forming at Fort Lewis, Washington. Initially he served in the 3/22nd Infantry Battalion on staff as the S1 Officer. He oversaw the personnel administrative tasks associated with the battalion's training and deployment to Vietnam for combat operations. Once deployed to Vietnam, the Brigade Commander, Colonel Marshall B. Garth, reassigned him as the Commander of Company B, 2/22 Infantry. His company held the ground at the eastern half of the fire base.

Colonel Marsall B. Garth

Colonel Marshall B. Garth, the 3rd Brigade Commander at Suoi Tre, graduated from OCS in 1942 and was assigned to the 104th Infantry Division where he served from Platoon Leader to Battalion Commander until the Division's deactivation in 1945. He then graduated from the Command and General Staff College in 1946 and in September of that year he was sent to Nanking, China, where he served as an advisor with the Chinese Army until the Chinese Communists took over in 1949 and his departure became necessary. In 1949-1950, he attended the Armored Advanced Course at Fort Knox, Kentucky. Upon graduation he was assigned as an Infantry instructor at the Artillery School at Fort Sill, Oklahoma. In the summer of

1953, he joined the 45th Infantry Division in Korea as a battalion commander, and later the 24th Infantry Division where he served as a battalion commander, Assistant Chief of Staff, G-1 and Assistant Chief of Staff, G-3.

Upon his return to the States in late 1954, he was assigned to the Office of the Assistant Chief of Staff, G-3 (later DCSOPS) and at the completion of a three-year tour with the Army Staff, he attended the Armed Forces Staff College. In the summer of 1958, he was assigned to the Office of the Special Assistant to the Chief of Staff, SHAPE. Here he not only had the opportunity to work and associate with the three services of the U.S. Defense Department but with the military services of all the NATO nations. Following his SHAPE assignment, Colonel Garth attended the United States Army War College, Carlisle Barracks, Pennsylvania, and graduated in the class of 1962. Upon graduation, he was selected to help organize and man the new institute of Advanced Studies, part of the then recently organized U.S. Army Combat Developments Command. In the fall of 1964, Colonel Garth was assigned to the Army Concept Team in Vietnam. There he traveled widely throughout Vietnam working on ways and means to improve the conduct of the war against the Viet Cong. After Vietnam, in late 1965, he joined the 4th Infantry Division at Fort Lewis, Washington. Initially, he served as a special assistant to the Commanding General and in January 1966 he became the Commander of the 3d Brigade, 4th Division. He returned to Vietnam in September of 1966 and his brigade was detached from the Division and sent to the area north of Saigon near Tay Ninh in War Zone C, a traditional sanctuary for the Viet Cong.

Fire Support Base Gold was set up in the southern half of LZ GOLD.

Even comparing the clearing as it was shown in the 1965

dated topographic map, compared to the latest 1970 map, the area looks different.

If you look at the map that we used in 1967, you will see that the official coordinates were incorrect or the clearing was not situated as it was shown in contemporary maps at the time of the attack.

The 1970 map shows the layout of the fire base in a clearing, which is probably correct. In the 1965 map the base would have been in the jungle. It only proves that swamplands are difficult to precisely detail. The crosses on the maps shows where the enemy set up an aid station to treat their wounded during the battle.

The 2/77th battalion commander and his staff were spoken glowingly of by the artillerymen who I met over the course of the years. That was not the case with all their leaders.

FSB GOLD layout on the evening of March 19, 1967 before the 2/12th left on a search and destroy mission

Early in my research at the National Archives, I was able to attain a copy of an overlay transparency with reference coordinates clearly marked of Fire base GOLD. With the use of photographic software, I was able to lay the transparency over the maps, increase the transparency and line up the matching + of the two reference points and draw the outline on top of the transparency. With the removal of the transparency all together, I was able to see the base in the clearing clearly.

That was how I was able to determine the shape and actual location of the firebase so precisely. As you can see, the base was oval shaped and I was able to determine that it was 500 X 700 meters in size.

**Special thanks to 2/34th Armor Platoon Leader, Chuck March, who sent* *me a less detailed layout of the base. When he did that, it reminded of the transparency that I had forgotten about and I was able to dig it out.*

On the west side of the perimeter was A/3/22nd led on the day of the battle by Captain George Shoemaker. George could not make the 50th anniversary of the Battle of Suoi Tre event at Fort Carson, but I was able to interview him by phone and I made a video slideshow of his thoughts about that period.

George Shoemaker, A/3/22 Commander at GOLD

"After the loss of the helicopters in the initial entry into LZ GOLD, the choppers were ordered to not land their aircraft, but only hover 8-10 feet above the clearing and have the troops jump down from that height.

"B Company was the first company into the clearing, followed by A Company. The new company commander of A/3/22nd, Captain Williams who took over the company after I left to serve at battalion staff, suffered a broken ankle when he hit the ground. Falcon Six, the battalion commander, immediately called back to Dau Tieng and told the Battalion S-2 and S-5 (me) to grab our gear and get out to the LZ GOLD as soon as possible. The helicopter that brought us to GOLD hovered and I jumped into the clearing, executing a perfect tuck and roll. They would have been proud of me at jump school.

"I reported to Colonel Bender at the base and received my orders. I got up on the company radio frequency and ordered the Platoon Leaders and Platoon Sergeants to the company CP. The CP hadn't been built yet but as I walked in the direction of the CP area, and word went out that I had resumed command of the company, I could hear the men say, "Hey! The old man is back."

"On March 20th, preparations on the base were the order of the day. I sent observation posts out in front of our positions so we would not be surprised by an attack. The strength of our company on that day was 129 men."

Junction City II Day D+2

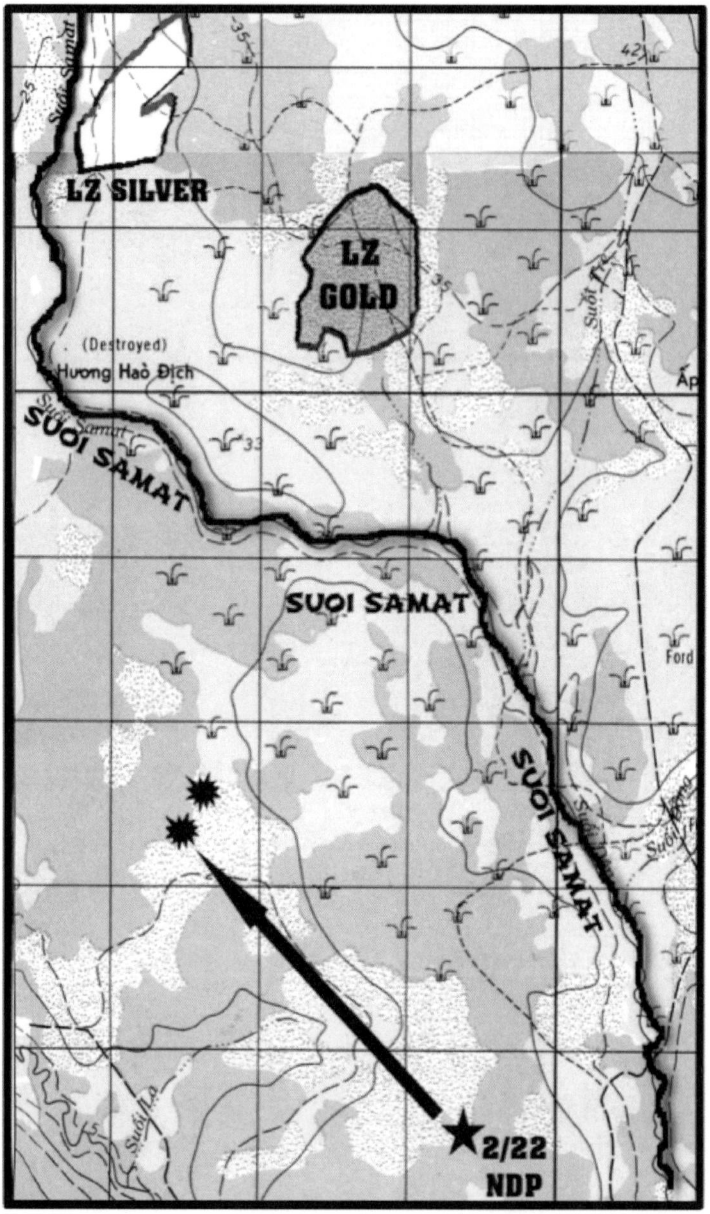

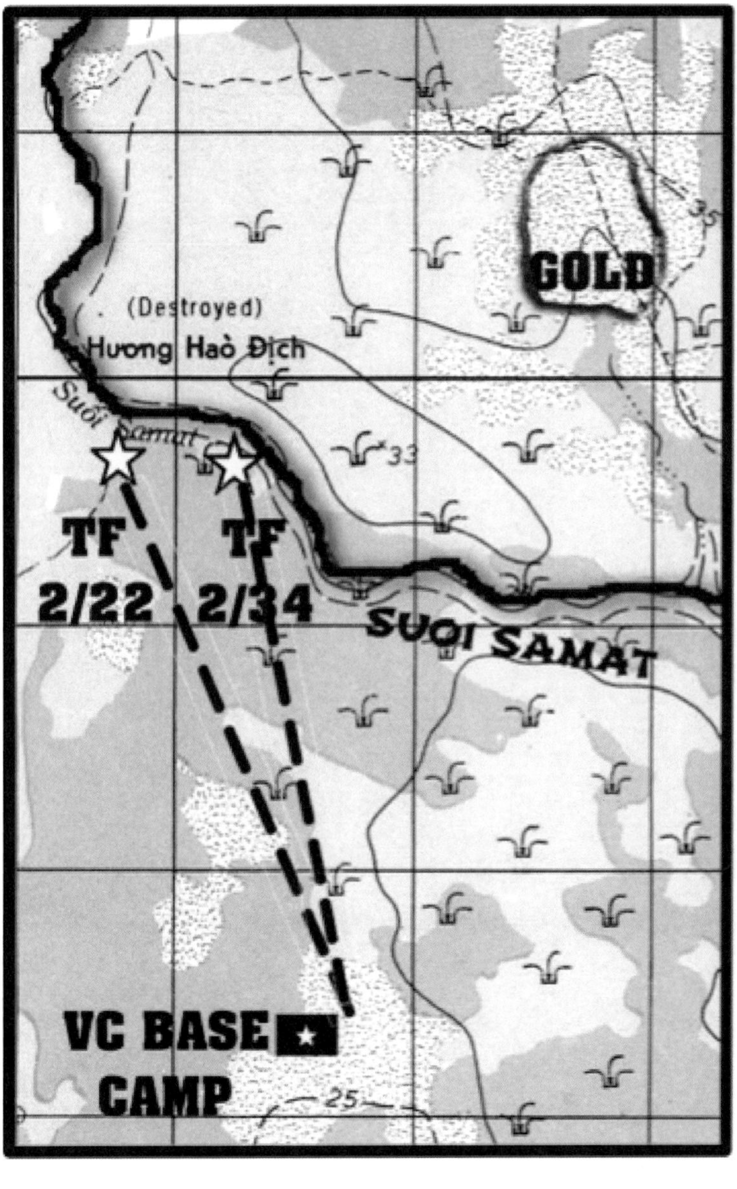

GOLD
(Destroyed)
Hương Hào Địch
Suối Samat
33
35
TF 2/22
TF 2/34
SUOI SAMAT
VC BASE CAMP
25

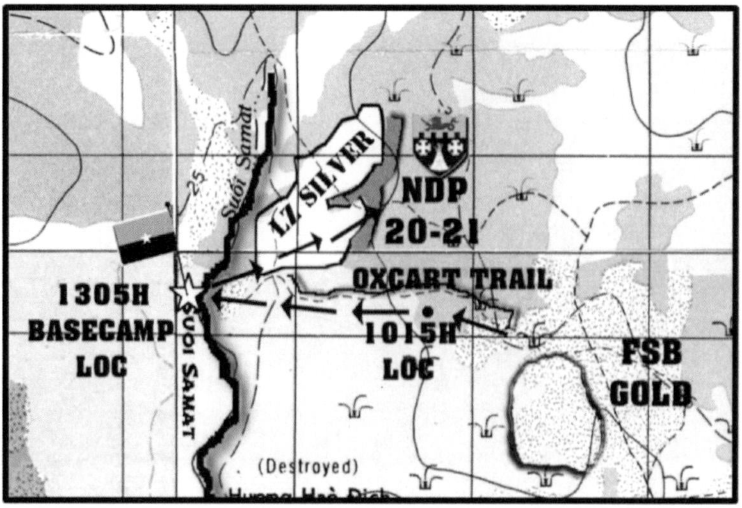

Monday, March 20, 1967

Walt Shugart At The Firebase

"The land line communication from battalion woke me from sleep at 0455H on Monday morning. It was 20 minutes prior to BMNT (Begin Morning Nautical Twilight).

"Toyama, who was on radio watch, answered it and passed the message for stand-to on to the platoons. He also reminded them that the ambush patrols and listening posts established as forward security would be returning to friendly lines. There had been no contact with the enemy overnight. None had been expected. Our experience with the enemy was hit and run. The VC recognized that attempting to stand and fight against the overwhelming fire power on call from direct support artillery and close air support was fruitless, if not suicidal. Mining an LZ to inflict maximum casualties by surprise was his best chance to stall our thrust into his area of operations.

“Once the patrols were secure behind friendly lines, the rifle platoons would, one at time, conduct squad sized security sweeps forward of the company front. The remainder of the company were to improve the positions initially occupied and prepare supplementary positions to their immediate rear for additional security. Even though an attack was considered unlikely, the hot LZ on Sunday provided motivation for the troops to ensure those positions were as well-prepared as were the primary ones. As time permitted, connecting trenches were dug between positions and adjoining platoons. Of course, two soldiers dug while one provided security.

“Following stand-to, I shaved, had breakfast, and visited the platoons to insure all understood and were following the stated priorities of work.

“I spent more time with the Recon Platoon since they had just been integrated into the company. Lieutenant Kaminsky proved to be a quick study. By the time of my visit, he had completed coordination with the adjacent platoons, established a forward listening post, and submitted to the first sergeant his supply needs. The recon platoon had become a fully integrated team member within Bravo company in less than 12 hours.”

The 2/12th Infantry Battalion broke camp at FSB GOLD at 0730 hours and headed northwest into the jungle to sweep the area called “AO ORANGE’, northwest of GOLD.

Meanwhile, the 2/22nd was a little slower out of the gate as they were waiting for their A Company to arrive with the units of the 2/34th DREADNAUGHTS” Armor Battalion who would be moving up to them from the RP (Rally Point) near the bridge.

The armor units arrived late in the day so both units spent

the evening near the bridge and waited for daylight before proceeding north to the rest of the 2/22nd Mechanized Infantry.

Chuck March, Platoon Leader, Charlie Company 2/34th Armor

"I had arrived in Charlie Company between the 13th and 17th of March along with Lt. Matt Giordono, an old friend from armor school. I was a 2LT and was supposed to replace 1LT Bob Brown as the 1st Platoon Leader, but I was temporarily assigned to 2nd Platoon, pending the return of their platoon leader from the hospital.

"On the 18th of March, our task force had moved from the vicinity of the French Fort to Suoi Da. The 2/34th task force was composed of 2/34 Battalion Headquarters, C-2/34, (A-2/34 would meet us at the bridge, coming from another location). B-2/34 was committed elsewhere and did not participate in this operation.

"HQ-2/34, A-2/34, C-2/34 met up with the A/2/22 Mech unit at the RP just as twilight was setting in. It was decided to wait until the morning before moving towards the 2/22nd NDP in the jungle.

"When we started, we had one of the battalion's AVLBs (Armored Vehicle Launched Bridge) with us. Presumably, this was to solve the problem of getting across the Suoi Samat river/stream. However, the AVLB soon had to turn back. The path cut by the lead tank was as wide as the tank chassis, but the AVLB bridge component, which was also on a tank chassis, overhung the side and kept getting caught up in the trees.

As soon as we arrived at the 2/22nd NDP, the group was broken up into two task forces. Both would be sent north, but taking different routes. Charlie Company, my company, was attached to the 2/22nd HQ Unit and their B and C Companies to form one task force. The other task force was composed of HQ Company-2/34 and A Company-2/34."

The jungle terrain took its toll on the armor unit's M48A3 Patton tanks as four of them had thrown their tracks during the journey. The rest of the tanks and A/2/22 arrived at 0955H.

Before the 2/34th units arrived, B/2/22 had been sent out to sweep to the northwest and traveled barely 2,000 meters when they came up on an enemy base camp.

They drew minimal enemy fire from rear guard VC defenders. Nothing came of the short exchange and when they moved up, they realized that they had discovered a large base camp. They asked for and received permission to pause their progress and thoroughly search the camp.

A detailed search of the base was made to determine its size and to seize what was left behind by the retreating Vietcong.

From that exploration, they discovered five large structures, bunkers, trenches, fifteen 60mm mortar rounds, two boxes of 30 caliber round ammunition, eight weapons, 30 rifle grenades, four directional mines, 40 lbs. of documents and approximately 3,000 lbs. of assorted items consisting of clothing material and gauze. Not a bad haul.

When they arrived on the scene, the 2/34th Armor battalion was placed under operational control of 2/22nd Battalion Commander, Ralph W. Julian.

It was decided that the newly enhanced fighting force would form two different task forces, each heading north but taking different routes. The first task force (Task Force 2/22) was composed of A Company- 2/34, HQ Company- 2/22, B Company- 2/22 and C Company- 2/22. The 2/22nd Recon Platoon was attached to this task force.

The second group (Task Force 2/34) was made up of HQ Company- 2/34, C Company- 2/34, and A Company- 2/22.

The new arrivals reached the Vietcong base not long after its discovery. While the enemy base camp was being searched by B/2/22, the task forces bypassed the camp, split up and continued to move north. Neither group ran into any enemy attacks or bases and eventually both units laagered at 1800H not too far from each other and near the Suoi Samat.

That NDP was still on the west side of the Suoi Samat obstacle. They were still two kilometers away from FSB GOLD at that time with no evidence of a crossing site during their patrolling on the west side of the river. I'm sure that enemy scouts were shadowing them and were well aware that they could not quickly arrive at GOLD to reinforce the defenders the next morning.

Lt. Colonel Ralph Julian, 2/22nd Commander recalls the day before the battle this way in an interview I had with him in 2017:

> "It was imperative that we had good terrain as the tanks could easily bog down where the 2/22nd vehicles could cross marginal terrain with caution. We had trouble finding a crossing site due to steep erosion of the Suoi Samat in its lower portions. The night before the battle we had reached an area southwest of the final site for FSB Gold. We had still not crossed an area we defined as a river bed, but pulled back into a laager for the night.
>
> "The 2/22nd Mech and 2/34th Armor began setting up their NDP near the Suoi Samat at 1632H and located it approximately two kilometers from FSB GOLD."

While the Mech and Armor barreled their way up towards GOLD, the 2/12th was traveling away from the GOLD fire base in a northwesterly direction. They were searching the jungle expanse in between FSB GOLD and LZ SILVER. Each company was sent on different routes but headed generally towards what was the clearing that was designated as LZ SILVER.

At 1015H, A/2/12 reached a jungle area near a trail and found evidence of enemy in the area. We found one poncho. Pretty insignificant, but that trail coming up would become noteworthy on the following morning.

This is how the day reports describe that trail after my company located it at 1138H: "A Company found an oxcart trail running N-S (*actually that trail ran mostly E-W*) 100 meters north of where we reported our last location. Trail appears to be unused."

It may have appeared unused, but the enemy was very aware of its location and how it could be used to travel from the LZ SILVER clearing to LZ GOLD quickly. Keep that in mind when reading the activity of the battalion on the next morning.

Note by the map that we crossed the Suoi Samat River to get to our next reported location. None of us who were there on that day recall seeing the river or anything resembling the remnants of a river, maybe a dry stream bed, but no river.

By 1200H C/2/12th reported their position at 1500 kilometers due west of where we last reported our location. They described finding suspicious items in the area and began a closer examination of the area.

At 1305H C/2/12th reported that they had located an enemy basecamp at the last location reported. "The basecamp consisted of six buildings, a trench line, one live cow, one live bull, fresh food, and a log crossing a small stream. (*That stream was obviously the Suoi Samat River, which was so narrow here that it was reported as a stream).*

No one recalls a cow, but the bull was fondly remembered.

Captain John Napper, Commander of C/2/12 recalls that bull as it was his unit that was given the responsibility of bringing the bull along with their company.

This is how John recalled the "Bull" story:

> "On the day before the battle, the companies headed north through separate routes. We had the westernmost route and we came up on a small stream (*Suoi Samat*) and on the other side of the steam there was one of those big round claymore mines about two feet in diameter. I would describe it as looking like a snare drum on a tripod. (*That was the suspicious item that was described in the day reports of that day.)* There were wires running from it, but there was no one there to detonate it. Still, there was evidence that someone was there not long before we arrived. There were pots filled with warm rice and a smoldering fire that had just been recently smothered. We found a "VC Bull", actually a male water buffalo, tied to a tree. I don't recall finding any cow in the camp. We brought it into the laager area off the clearing and we kept it with us because Joe Elliott wanted to have it taken out by Chinook at the earliest opportunity so it could be donated to villagers as part of the pacification effort.
>
> "I had a man who we knew only as "Cowboy" in our company. He was called up and was ordered to take care of the water buffalo and bring him along with us. The bull had a ring through his nose and Cowboy secured the bull by this ring and traveled with him through the jungle. He was not happy about that.
>
> "On the next morning, just as a barrage hit us, my men started to do a recon by fire into the surrounding jungle. Unfortunately, that bull was killed. He got in front of an M-60 machine gun and that was the end of him."

By mid-afternoon, the 2/12th had decided on a battalion size laager in the nearby jungle along the eastern border of LZ SILVER.

As we began preparing our NDP in late afternoon, the battalion began to get resupplied by choppers.

John Concannon, our earlier A/2/12 XO and later the Battalion S2 Intelligence Officer, was suffering terribly from kidney stones all this day. He was of little use to us in his condition so when the choppers arrived, he jumped on board to get transported to the rear.

He always regretted not being with the battalion when the big battle took place on the next day. However, he became a valuable source of information for Alpha Association when he was sent to Saigon to examine the documents that were captured during the battle. More on that later.

Ambush patrols and listening posts were established for the 2/12th before twilight appeared at 1900H.

Guam Summit

On March 20th at a location twenty-six hundred miles to our east on the island of Guam, a summit was taking place between President Lyndon Johnson and his advisors and South Vietnamese Chairman Thieu, Premier Nguyen Cao Ky, and other leaders of South Vietnam.

Johnson and his entourage landed at 1100H, local time. They were brought to Nimitz Hall for a working lunch. Following lunch, the participants were brought to Admiral Byrd's Headquarters Conference Room, Headquarters Naval Forces,

Guam via motorcade. The meeting began at 1500H and lasted just two hours.

Johnson, feeling that the allies gained the military upper hand since the successes of Operation Attleboro and Junction City, pushed Ky for a renewed focus on pacification and working to win the trust of the poor villagers who made up the bulk of the Vietcong insurgents. Ky was more interested in the cities and metropolitan areas, where his political support was the greatest. Ky argued for the mining and bombing of Haiphong harbor, where the Chinese and Russians were delivering aid to the North Vietnam. The premier criticized the allies for not following the enemy to their Cambodian sanctuaries. Neither group convinced the other to change tactics and the strategy remained the same for the foreseeable future. Johnson was not willing to expand the war into Cambodia or invite retaliation from Russia and, or, China by attacking Haiphong Harbor, whose ships always were offloading in the harbor.

Premier Ky, knowing where his power originated, refused to show deference to the many small villagers but was more inclined to beat them into submission. He did make one concession at that March 20th conference. He agreed to a popular election that would take place in September. It would be the first election since Diem was killed in 1963. Since 1963, the country had been ruled by Military Juntas. Ky was the sixth and final Chief of State under that rule. Nguyen Cao Ky won that election in a landslide and ruled until 1971.

Neither group convinced the other one to change tactics and the strategy remained the same for the foreseeable future.

As night time settled in for us, it was 2200H in Guam and the one day summit conference had concluded. Each participant had differing ideas on how to conduct the future strategy to win the war in South Vietnam. This was the third time that they had met since February 1966 for the same reason.

Two and a half hours later (2230H local time in Vietnam), Lou Urso, an Artillery surveyor in the 2/77th Artillery, was assigned to Fire Direction Control. He was also an RTO, a calculator, and he served as an Air Advisor (*he manned a radio to give data to any aircraft that would pass through our area*), his call sign at Soui Tre was "Square Lobster Gold".

Lou Urso recollection from the eve of the battle:

> "I was in my Air Advisory role on the night of the 20th. At about 10:30 PM a FAC called me on air advisory. He was directing an SLAR (Side Looking Air-Born Radar) mission through our area. He told me he had seen an entire grid square light up with camp fires three kilometers to our north. I alerted the officer on duty and he had me call one of our batteries for a fire mission. I asked the FAC to observe. When the first rounds landed, he called back to say all the fires went out, I wrote a report and gave it to the officer on duty (he failed to pass it on to our battalion commander). It might have been important information as to what was coming our way."

Senior Colonel Huang Cam, the Commander of the 9th VC Division, was having his own summit in the jungle north of FSB GOLD. Those fires probably were where the final battle details were being discussed.. He was under orders to take whatever units that he needed to destroy the defenders of FSB GOLD. The fire base was set up menacingly near the 272nd VC Regiment's main base to its north. (*I suspect that is where the 2/12th was headed after they rested at LZ SILVER on this night*). With the allied summit concluding on Guam the next day, a victory would be significant and would demoralize the American public who were, even then, beginning to have doubts about the war as the casualties were mounting in Vietnam.

Cam chose his most successful regiment, the 272nd VC Regiment, which never shied away from daylight attacks, to be the main force to achieve this victory. The regiment was composed of four attack battalions and they were supplemented by two additional battalions from the 101st NVA Regiment. Artillery support was provided by the U-80 artillery Battalion. All told, the attacking force was over 2,500 troops in strength. The defenders of GOLD numbered a mere 500 or so Americans. There were to be no survivors. The message to the American public needed to be overwhelming.

The pieces were in place. The units of the 3rd Brigade of the 4th Division, made up primarily of draftees, and only six months in country, would take on the best that the 9th VC Division could throw at them and would meet the challenge. The key to Cam's success would be a quick victory before the firebase could be reinforced. Cam counted on the 2/22nd and 2/34th getting held up by the Suoi Samat River. The 2/12th would be dealt with in a different fashion, but that also didn't pan out as they thought it would.

A pair of Vietcong passed in front of a listening post in front of A/2/12's position at 0255H. They were carrying flashlights and heading south. The observers fired off a burst of M-16 fire and the lights went out quick. Still, a few moments later, two more individuals passed by and were dealt with similarly. At 0400H, three more VC passed by A Company's position. This time they were told to use M-79 grenades to challenge them. Same result. At 0407H, A Company spotted four more VC passing by them. These VC must have been in a hurry. Flashlights were a dangerous target marker for anyone they passed by at night. Still, it happened on four different occasions that night. Usually once the first group was challenged, the VC rerouted their routes to get to where they were going without passing near our NDP. These soldiers were throwing caution to the wind. They were on an urgent mission.

The 4th Division brigade, for the most part, rested quietly this night. The enemy was busy moving their chess pieces into position for a winner take all gamble at daylight.

BATTLE DAY

The Attack Commences

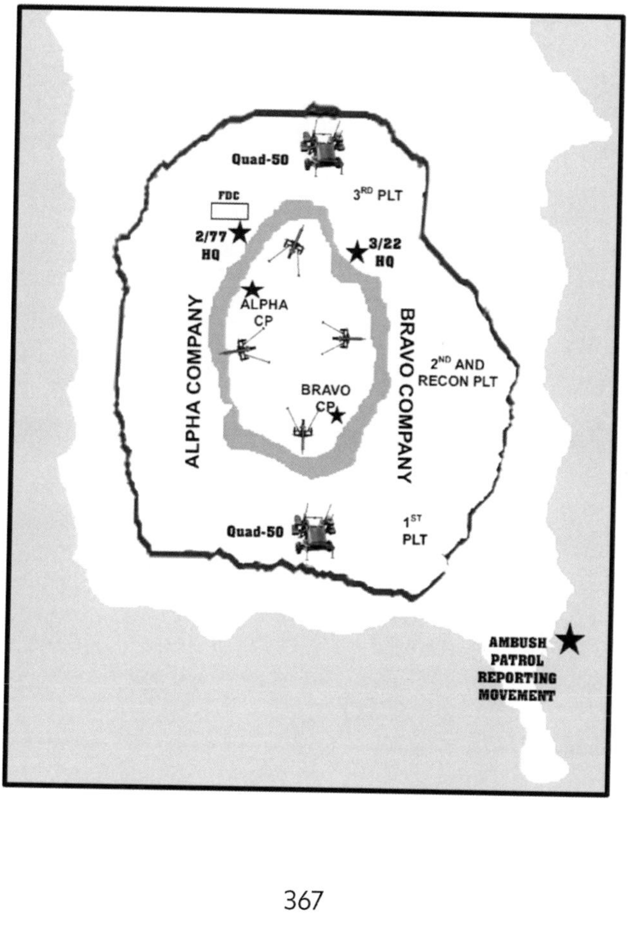

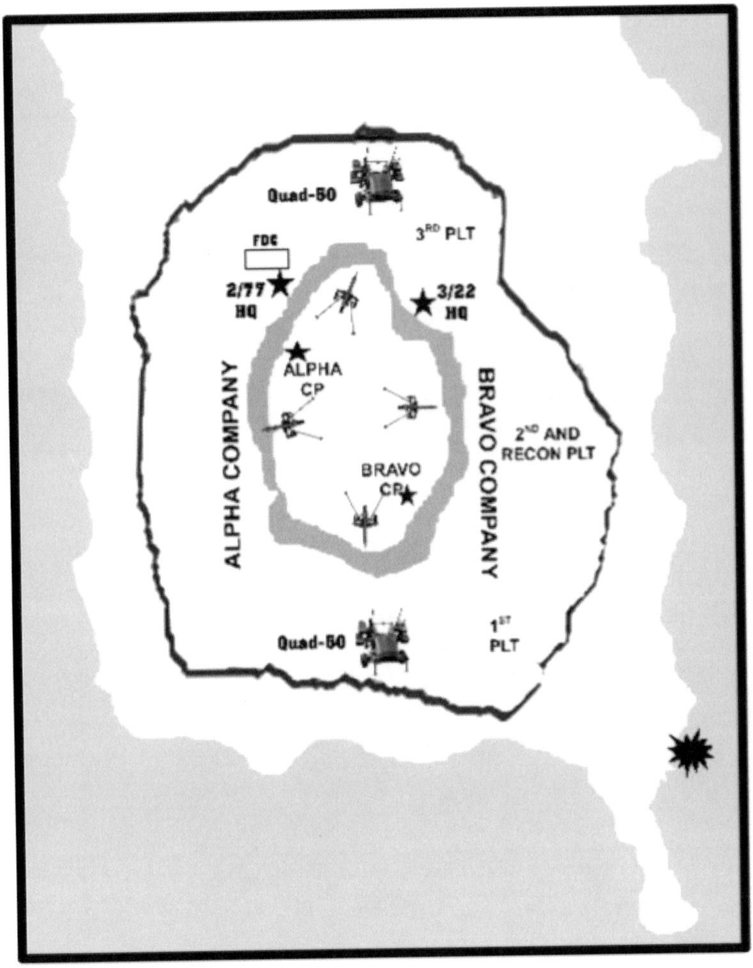
Quad-50
3RD PLT
FDC
2/77
HQ
3/22
HQ
ALPHA
CP
ALPHA COMPANY
BRAVO COMPANY
2ND AND
RECON PLT
BRAVO
CP
Quad-50
1ST
PLT

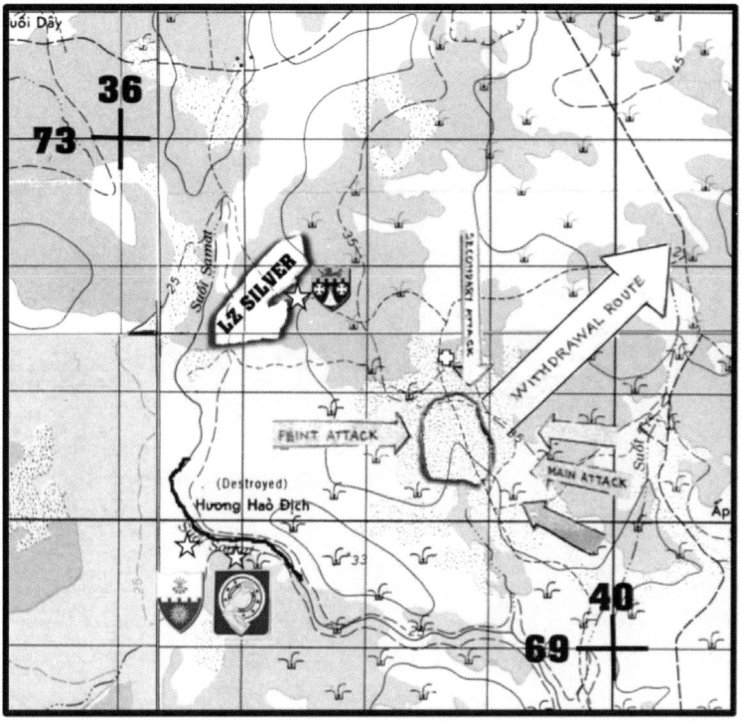
36
73
LZ SILVER
SECONDARY ATTACK
WITHDRAWAL ROUTE
FEINT ATTACK
MAIN ATTACK
(Destroyed)
Huong Haò Địch
40
69

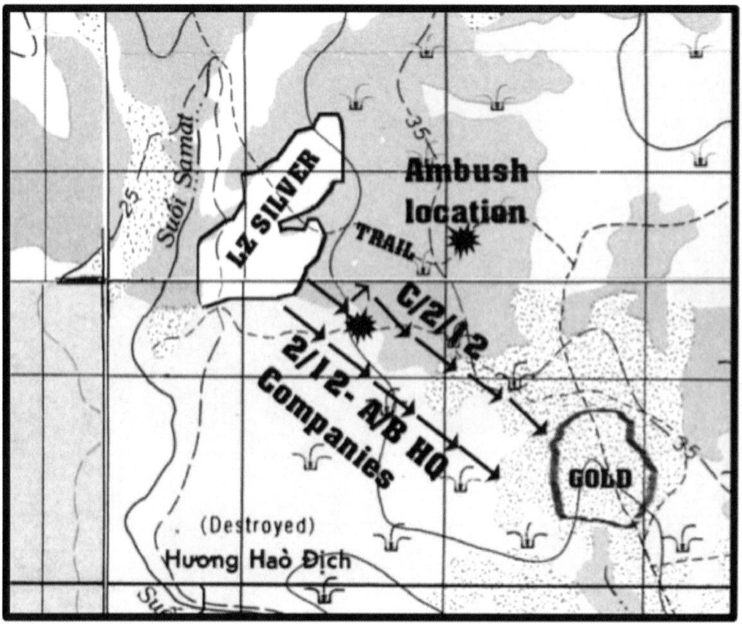

Tuesday, March 21, 1967

George Shoemaker, A/3/22 Company Commander:

"Master Sgt. John Senechek, A Company's first sergeant, both my RTOs, and I had been drinking coffee since 0330H. During this time, Top and I had checked our perimeter to make sure that all fields of fire were as good as possible. We also checked our reserve force to assure they were in a good position to react if need be. We also had three listening posts in front of our position to provide early warning of any pending attack.

"Our intel said that there was substantial enemy activity within War Zone C. We were as prepared as we could be for an attack. We just didn't know that such a large enemy force was on our doorstep."

There were a few night ambush positions established around the firebase. It was a quiet night for all of them until an ambush patrol sent 400 meters southeast of the base reported movement in front of their position at 4:29AM. The squad sent out by the 1st Platoon of B/3/22 signaled to each other silently to get on alert. Soon they heard movement not only to their front, but also all around their location. Soon the area went silent. As soon as it became quiet, they reported the situation back to the firebase. Spec 4 Terry Smith monitored the company net and received that call. He was told that if the enemy came back the patrol would blow their claymores and make their way back to the base as soon as they could. B Company commander Walt Shugart was informed on the situation and told his radio operator to keep him abreast of future activity outside the base.

It remained quiet the rest of their time on ambush patrol. The sun arose at 5:54 AM and the men began to draw a sigh of relief as daytime slowly replaced the darkness of night. They must have figured that what they heard was only enemy scouts sneaking up to see the layout of the fire base.

It took until 6:30 AM before the twilight of the morning dissipated into clear daylight and it would be safe for the ambush patrol to return to the firebase. As they began to return to the base, the sounds of numerous small arms and AK-47 fire filled the air with the echo of grenades exploding from the direction of the ambush patrol.

When B Company Commander Walt Shugart heard the noise, he ran out of his command bunker to determine what was going on. He was told that the ambush patrol was under fire southeast of the base. The ambush patrol returning from their night position came up on a couple of VC in the tall grass in that area and opened up on them. The enemy returned fire and a serious firefight broke out. Some members of the patrol worked their way back into camp, but five were pinned down outside.

Shugart called the 1st Platoon Leader Lt. John Andrews and asked for a report on the fire coming from his sector. He was told that the ambush patrol was under fire from an unknown number of enemy soldiers. He ordered Andrews to prepare a squad to reinforce the patrol. Almost instantaneously, the shooting stopped and they heard a short burst of AK-47 fire. He knew that the short, distinct burst of AK-47 fire meant that any survivors had been executed. Shugart located the company Artillery FO, 1st LT William Pacheco, and ordered him to call in some HE around the ambush patrol's last position. He wanted it near the position, not on it, in case somebody was still alive out there. Shugart told Pacheco to walk the artillery around the area in case any larger forces were nearby.

Pacheco never had the chance to call in that fire mission. At 6:31 AM, a massive mortar attack began by the recognizable thumps of mortar rounds leaving their tubes. The entire base was under attack by 61mm and 82mm mortar rounds, seemingly coming from everywhere.

The Battle of Suoi Tre had begun.

Brigade Tactical Operations 0631 AM

Ed Smith, who then serving as Assistant S3 at Brigade, described how Brigade responded to the attack on FSB GOLD:

"I went into the Brigade Forward TOC at Suoi Da earlier than normal. I was at the duty station of the Operations Watch Officer and called back to the Dau Tieng Base Camp and spoke to CPT Rich Craig, our Brigade Chemical Officer who was on duty in the Brigade Rear TOC. I called to see what activity they may have had overnight so I could brief MAJ Carrozza and COL Garth when they came into the TOC. At about 0630 hours, as I was on the phone with Rich, a radio call came in from MAJ Cliff Roberts, Operations Officer (S-3) for 3/22 Inf that was providing perimeter security for the firebase at LZ Gold. He reported in an unhurried, matter of fact voice, that the firebase was under attack. I asked Rich Craig if he monitored the radio transmission. He indicated he did and I asked him to dispatch our USAF Forward Air Controllers to the firebase. Rich said that the USAF FACs were in the TOC with him at the time and they were on their way.

At that point, I asked someone to notify MAJ Carrozza and COL Garth, and then picked up the phone and called the 25th Infantry Division TOC at Cu Chi and reported the attack. I also requested they dispatch COL Garth's Command & Control UH-1 Helicopter and helicopter gunships to support the firebase. The gunships were to contact 3/22 Inf on their radio frequency when they were on station.

As MAJ Carrozza and COL Garth came into the TOC, they were briefed on the developing situation. COL Garth departed for LZ Gold when his Command & Control Helicopter arrived and MAJ Carrozza remained in the TOC to coordinate our actions there."

Mayday, Mayday, Tay Ninh

It was just another morning for Captain Bob Staib, the former jet fighter pilot who on this day was serving as an Air Force Forward Air Controller and beginning a dawn reconnaissance mission over War Zone C.

It's important to learn the history of this man prior to his critical actions on the morning of the battle.

Bob was a graduate of the U.S. Air Force Academy. He was also a running back for the Falcons during his time at USAFA. In 1959, the USAF Falcons had a record of 5-4-1 under Coach Ben Martin who arrived at the academy in 1958 and stayed until 1977; a long period for college football coaches who normally were accorded little security.

Bob played in seven games. He was also a starter at defensive cornerback and 2nd team fullback when USAFA traveled to the very first Army/Air Force game. The Falcons were on a run going into the West Point game, winning four of the five games they played that Autumn. Three of those five teams were nationally rank in the top 25 at the end of the season.

Army had won three of their five games to date. The date was October 31, 1959, and the game was played in the original Yankee Stadium in the Bronx, NY. On a very rainy Halloween, Army and Air Force tied at 13-13.

When Bob graduated from USAF in 1962, he stood 19 of 298 graduates in the fourth graduating class of the United States Air Force Academy.

After graduating from the academy, Bob began his flying training in West Texas in supersonic T-38s and was assigned to the UK as an F-100 pilot. From England he volunteered for a consecutive overseas F-100 tours in Vietnam, as an F100 Fighter Pilot flying out of Bien Hoa airbase with the 90th Tac Fighter Squadron, call sign 'DICE'. That was in the summer of 1966, just a few weeks before the 3rd Brigade of the 4th ID arrived at Vung Tau.

After 120 fighter combat missions including CAS (Close Air Support) exercises during Operations Attleboro and Cedar Falls, Bob was assigned a different role and become a FAC (Forward Air Controller) attached to the 196th LIB, and flew out of Tay Ninh West, with the call sign 'Issue 44'.

On March 21, 1967 he took off from Tay Ninh at 6:30 AM in his Cessna L-19/O-1 Bird Dog observation plane for a quick recon mission over War Zone C. He was on his 210th combat mission and his 101st as a FAC. It was to be his most memorable.

Bob Staib, FAC

"Just as I left the dark runway at 0630 hrs., I received a" Mayday! Mayday!" call on Guard Channel. I acknowledged my proximity to the coordinates where a major battle had begun. I was redirected to the battle hidden to me by the Black Virgin Mountain (Nui Ba Dinh.).

"As I flew my Cessna 0-1 Birddog aircraft around the massive mountain, the battle site became clear to me. I raced to the battle site and immediately began coordinating air strike flights originating from the closest alert pad at my former airbase at Bien Hoa.

"Weather conditions on this morning were not ideal. Visibility was a couple miles. The ceiling was somewhere around 800 feet. It was not overcast, but cloudy with broken openings. If the ceiling is overcast, you can't see through it. If it was broken, you're able to see parts of the battlefield. That meant the jets and bombers could pass through those openings and see their target field. At times there were no openings and we had to wait until another opening came along before resuming our air attack.

"Regulations dictate that FACs fly at 1,500 minimum. That is to protect us from ground fire. However, with clouds down to 800 feet, I had to get below to see what was going on. That low ceiling was a problem for all the aircraft sent to the battle. The jets and bombers normally pulled out of their bombing run at 800 feet. They had to get below that to see what they were attacking. Once they dropped their payloads, they had to swiftly climb out of their dive, lest they crash into the ground. I can only imagine how many G Forces they had to experience when they climbed out of those dives.

"First of the sorties on the scene was a Tiger flight of F-5s off the alert pad at Bien Hoa. Tigers were mostly highly experienced fighter pilots selected carefully because the U.S. was trying to prove the viability of the F-5 as a primary Close Air Support (CAS) for the Vietnamese to acquire. The benefit of these exceptional F-5 pilots was that their bombs could be put in very close to the battle. The second unit to respond were B-57s. These planes were used like fighters but had the added advantage of carrying twice the load of the F-100s, which was the primary CAS aircraft in Nam. The B-57s were able to create havoc in the mass attacking waves of the enemy.

"Next, 'DICE 01' checked in, the first F-100 ("Hun") flights also off alert at Bien Hoa where ten more F-100s stood ready or had already been launched to support the Battle of Suoi Tre. The DICE pilots were likely some of my F-100 friends from my four years in Europe or Nam.

"By Air Force/Army agreement, USAF FACs were assigned to specific Army brigades to better coordinate critical CAS (close air support), much of what was needed at Suoi Tre. Adhering to that policy, as soon as 3rd/4th "CIDER" FACS were ready to take over, I had to turn over FAC duties to them.

The FAC O-1 airplane has three radios. Concurrently with coordinating with the ground commanders, I was briefing the 'CIDER 54' FACs, instructing the additional flights arriving and entering a holding pattern nearby, after marking the first target with a WP rocket delivered by diving through enemy fire at the chosen target, for example, the wave of VC massed in the northeast quadrant of the battle.

"As Tiger 01 acknowledged the WP burst of white smoke as his first target for 500 pound "Snakeyes," I cleared both Tigers in and I relinquished control to CIDER 54, the Sabre Jet fighters arrived, and then the bombers. After using up my eight marking rockets, I was replaced by Tonie England and Walter Forbes, the Air Force FACs for the 3rd Brigade of the 4th ID, stationed at Dau Tieng. Two FACs were flying together in one aircraft to enhance the CAS communications as the ground battle raged half a mile below their flight path. The CIDER FACs maintained the aerial attack on the marauding VC threatening the fire base.

"My job was done, or so I thought. I returned to my earlier recon mission."

The Relief Troops Alerted

Both units were up early, as soon as the sun rose, which was 0600H. Task Force 2/22 was sent early to sweep northwest of their NDP. Task Force 2/34 sent out a scouting position to find a crossing site across the Suoi Samat River.

The men of the 2/12th awoke with the sun and proceeded to brew up some instant coffee and maybe grab a cookie with jam spread on it for breakfast. (Breakfast of Champions). We all heard the racket that began at 0630H and it was obvious that something big was happening at GOLD, two klicks away.

Some of the men began filling in their foxholes, knowing that they would be heading back to the fire base. At 0640H, my Company Commander was contacted by our battalion commander and ordered to report to him immediately. Captain Palmer, Walter Kelley, his battalion net RTO, Johnny Martel, his body guard, and I hiked to the HQ Company's night position, just south of us. I looked down at the overly big foxhole that Martel and I dug for the night for the Company Command group and wondered if we would have time to fill it in. Walking past our guys I saw some had already filled their foxholes in and were eating with their backs to trees. Henry Osowiecki was cooking a cup of coffee and told us as we passed him, "Well if I'm going to die this morning, at least I won't be hungry." He was going to need that energy as he led one of the columns racing to GOLD.

We arrived at LTC Joe Elliott's location and we got the word that there was a massive battle taking place at GOLD and we needed to get everyone organized and ready for a forced march to relieve the defenders at the fire base. Captain Palmer started back to A Company's area, stopping at each platoon leader's location to inform them of the move. Knowing that we would need extra time to fill in the big foxhole, he allowed us to go ahead and take care of the foxhole that we dug.

Within three or four minutes after leaving the Battalion HQ area, we heard big guns firing south of us and then the whistling of rounds that suddenly stopped whistling for a second, followed by loud boom. We were under a barrage attack.

Johnny and I had just made it back to our foxholes when the salvo began. We both jumped in and hugged the southern end of the hole. To our back was about four feet of open foxhole. We swore to never dig such a large foxhole again. From where we were we could actually hear the unmistakable sound of artillery pieces firing in the distance. We were very aware of what that sounded like from when we called in artillery during past battles.

In the next foxhole, Lt. Willenbring, our Company FO, was

frantically calling for a 'check fire' (stop firing) to other artillery units in the area. Finally, after 16 rounds hit our position, the barrage abruptly ended.

We jumped out of our holes not able to see very far through the smoke enveloping the area. We counted our blessings and ran to Captain Palmer's position in a nearby hole.

When the dust cleared, we searched for wounded. Of the 16 rounds that dropped in on us, 10 of those landed in A and B Company's night position. A Company suffered seven casualties and one fatality from the barrage. B Company experienced one casualty. Six of the rounds fell near the Battalion CP group and they suffered five casualties. The Recon platoon had three men injured. C Company, set on the southern end of the NDP, had no casualties.

We quickly learned that my friend, Larry Barton, was the one killed in our company. Just a few days earlier he worried about being hurt again. This time he was killed instantly. Porter Harvey was near him when a round hit a tree over Larry, he said that an airburst of shrapnel rained down on Barton as he was trying to re-dig his refilled foxhole. He never felt a thing.

I would meet his sister, Pat, when I visited her while in Ohio for an Alpha Association reunion in 2006. She told me that the family was originally told that Larry was missing in action. Ten days later, they informed them that he had been killed. I have no idea how that could happen. I saw him get put on a chopper when he was sent to the rear on the day he was killed. Such a shame.

Larry was the man who replaced me as the 3rd Platoon Leader's RTO when I was sent to the Company CP group. I always felt that he took one for me. As time went by, I understood it just was meant to be.

Amongst the casualties in the CP Group was Battalion Commander Joe Elliott.

Jack Eldridge, Joe Elliott's Battalion Net RTO recalls that morning this way:

"HHC control group, including LTC Joe Elliot, Capt. Howard Paris, Capt. Winder, and all the other elements from HHC. Specifically, Bobby Guerra, who lost his arm shortly after ordering the days supplies (used my radio, the Battalion Net, more on him later). Tom Zarlenga was on the Brigade Net.

"When we received incoming artillery, Tom was trying to find out where it was coming from. I was catching hell from Lt. Allyn Palmer to 'shut the shit off'.

"That morning we were with A company and Recon Platoon and, of course, the HHC platoon.

"We got the artillery shut off (I was told that it was coming from an ARVN artillery base that was trying to fire at LZ Gold. If you look at the maps, the LZ's were very similar in appearances and they were firing at the wrong one. This may not agree with history, it is just a fact, in my opinion).

"During the early vollies, LTC Elliot was hit in the hand, more specifically, the thumb. He wrapped a handkerchief around it, refused extraction, and we headed one direction through the boonies, crawling through some of the underbrush, receiving directions from Col. Garth. A crew was left behind to medivac the wounded and the casualties. I think that Tom Gouty was one of the guys left behind. Tom Zarlenga was the guy who took Bobby Guerra's arm to him. It may be noted that LTC Elliot received a Silver Star for his efforts."

(It should be noted here that when Bobby Guerra's arm was blown off and it was buried in a hole. When it was time to get on the chopper, he said he wouldn't go on the chopper unless they dug up his arm and

retrieved his wedding ring. They retrieved his ring and gave it to him before they loaded him on the helicopter. True story, confirmed by three different witnesses who were present).

Decades later, there was some questions about the source of that barrage. All of us there had no doubt that it was friendly fire. Still the AAR documented that we were hit with 16 mortar rounds of unknow origin. We all knew what mortars leaving their tubes sounded like. We also knew the sound of artillery guns firing in the distance. Those were artillery rounds dropped on us. This entry in the Day Reports should have cleared up the question:

"Item 21 0715H- 2/12th requested a dustoff for two litters, six ambulatories. (An incorrect count). Note: casualties were result of counter mortar fire."

So at least on the day of the battle, the feeling was an artillery unit firing counter mortar fire to get the mortars slamming into GOLD hit us by mistake.

There was no time to waste so Joe Elliott sent C Company immediately to the fight as they were unscathed by the attack and they were the furthest south in our perimeter. At 0705H they entered the jungle and made a B-line to the battlefield.

The rest of the troops tended to our casualties and brought them to an assembly area in the clearing where Medivac choppers would take them for treatment.

Fate is a funny thing in war. Sometimes it works in your favor, sometimes not so much. Fate was on our side that morning as two attacks were avoided.

Captain John "Napper, Commander of C/2/12

"We were in the southern edge of the clearing when we were sent to GOLD.

"I remember the heavy bamboo specifically as when we arrived at a particularly heavy patch of bamboo, I ordered my point platoon to go around it 100 yards to the left. When we changed our approach to GOLD, some mortar rounds starting falling into where we would have been had we continued straight through the heavy bamboo. As a result, no one was hurt in my company.

"On that trip to GOLD there was a lot of confusion, with a lot of radio traffic indicating that it was imperative that we get to GOLD quickly. We violated probably every security protocol on the radios to get to the battle.

"In Vietnam there were days that we didn't cover two klicks through the jungle all day. On the battle date, we covered almost three kilometers in one hour and twenty minutes. That was quite a feat and required a lot of ignoring safety measures.

"We arrived at the northwest corner of the base at 0840H and were directed to fan out across the original northern perimeter and reinforce the defenders there."

A small group was left behind waiting for the extraction helicopters for our wounded. The rest of the battalion was put in high gear to get to the battlefield. We went through the jungle avoiding that trail to the north. Our point men never used a machete to cut through the jungle, instead we twisted our way through what turned out to be extraordinarily dense bamboo vines. Climbing over or under, we moved along non-stop. We never even put out flank security, a real safety hazard, but the jungle was so thick that we were afraid that we would lose contact with them. One more thing, we didn't use a compass. We just raced to the sound of the combat.

About a third of the way to GOLD, we finally made it through the bamboo that C Company avoided. Once out of that thicket we picked up the pace as the jungle was single canopy and easier to navigate. About halfway to GOLD, we were stopped abruptly by our Brigade Commander, flying overhead in his command chopper. Garth asked if we were anywhere near that trail leading from the SILVER clearing to the GOLD clearing. Capt. Palmer answered, "No, sir, we are deep in the jungle and nowhere near it." Garth replied, "Good, I spotted a platoon of VC on the side of that trail waiting to ambush you had you used that trail. I'll take care of them with gunships."

Walter Kelley was monitoring the battalion net when he heard Colonel Garth complaining how long it was taking the task forces to the south to find a crossing site across the Suoi Samat. Walter told me that Garth told them to sink an APC and ride over it to get past the river if need be. A short time later a crossing site was found.

The remaining companies of the 2/12th arrived at the battle at 0901H.

As I wrote earlier, the 2/22nd and 2/34th task groups were split up with the Task Force 22 with A Company-2/34, HQ Company-2/22, B Company-2/22, and C Company-2/22 that had the Recon Platoon attached to them in the rear. They were sent northwest, away from GOLD, before the battle broke out.

Roger Frydrychowski, 2/22 Recon Platoon Leader

"On the morning of the 21st the battalion was laagered southwest of Soui Tre. Because of my losses, Recon was attached to Charlie Company commanded by an outstanding officer and great friend, Capt. George C. White. After leaving the army, George was one of the men in charge of loading the last evacuees out of the embassy grounds as Saigon fell.

"Charlie company and Recon as part of Task Force 2/22 were given a mission taking us out to the northwest away from Suoi Tre, perhaps as bait? We moved out very early that morning and covered some distance with no opposition. This was undoubtedly because we were moving away from what would be the main attack against the fire base. Capt. White had recon trailing behind the Charlie company column as we moved out. Col Julian remained with Task Force 2/34th still some distance to the southwest of GOLD. Once hearing the gunfire at Suoi Tre, he ordered Charlie to turn back and move to the battle site. I do not recall hearing the guns but moving tracks make a lot of noise and earphones mask the rest.

"Col Julian at this point had received no orders from higher headquarters and only reacted in the best of military doctrine—to move to the sound of the guns. He directed the 2nd of the 34 armor units to intersperse the tanks with the tracks of the 2/22nd Mechanized Companies. Again—protocol dictated that tanks need to have infantry in a mutual supporting role. He then requested a chopper for his use.

"I was given no information and had no idea where everyone was or what they were all doing.

"I received a radio call from Capt. White saying that FSB GOLD was under attack and that we were to turn around to the east and move to GOLD. He made it clear to me that we had to move as quickly as possible. To save time, he had each track turn in place with it ending up with Recon Platoon now in the lead. Just after we had turned and started toward GOLD,

I saw an RPG round coming toward my lead element. It hit a tall termite mound about 15 yards away and failed to explode. I am sure the VC had an element trailing us with the orders to stall us if and when we turned toward Suoi Tre. An RPG round into the track would cause several casualties and disable the vehicle. Generally, this would mean a hold on the mission in order to find the enemy and secure and evacuate the wounded. It didn't happen.

"As we moved, George would update me on the intensity of the fight taking place at GOLD and our need to push on swiftly. I passed all this on to my men and my drivers in the lead tracks did a spectacular job of pushing down trees, bamboo, and everything in their paths. Each man in that platoon shared an extreme level of intensity to get through to the base. Soon Col. Julian arrived overhead in a small open chopper carrying him and an artillery forward observer and piloted by a 19-year-old warrant officer. I was detached from Charlie company and received direction guidance from Col. Julian who flew at tree top level and varying a flight path to avoid VC fire though taking small arms rounds into the undercarriage of the chopper.

"His directional guidance was critical in our movement through some very thickly covered ground. My lead drivers handled those tracks with an intensity that needed no prodding from me. As I recall, I was in the third track from the lead. At some point in the thickest part, we came upon two tanks stopped with one of the men saying that it was too thick to move the tanks or the tank had thrown a track after hitting too large a tree. I ordered my lead tracks out of a column and they worked separately to clear a path wide enough for the tanks.

"Frankly, I didn't give a damn if they moved and told the lead tracks that our movement was critical and when faced with doing a slow wide path or a faster narrow one, we did the narrow for speed. My objective was to get my platoon with ten fifty calibers (machine guns) and Charlie Company's twenty more and infantry men through to Soui Tre.

"Recon broke through the tree line southwest of the fire base. What I saw was an eerie image I will never forget. we were still a couple of hundred meters or so from the perimeter. I saw a thin strip of wooded area jutting out east into a vast open area. At the woods and out for what may have been a hundred yards a low hanging gray cloud was all that identified the fire base location. As we were about to break out into that opening, I asked Col Julian for directions and orders. He said that he wasn't sure of the tactical need and for me to check with the unit commander at the fire base."

Note that Roger never speaks of crossing the Suoi Samat. I can only assume that they lucked up on dry river bed that was not steep at all and they could cross there not even knowing that had done so. In conclusion, I never met anyone who thought they passed an actual river.

Some of the mechanized and armor veterans of the battle didn't know they crossed the river, others talked of tanks being pulled across the river with the help of chains and towing from APCs that made it across. Fog of War. Especially after 55 years.

No matter. They made it in time and that's all that matters because by the time the relief forces arrived, the defenders were running out of ammunition and the VC seemed to sense, "Just one more push."

The Combined Attack

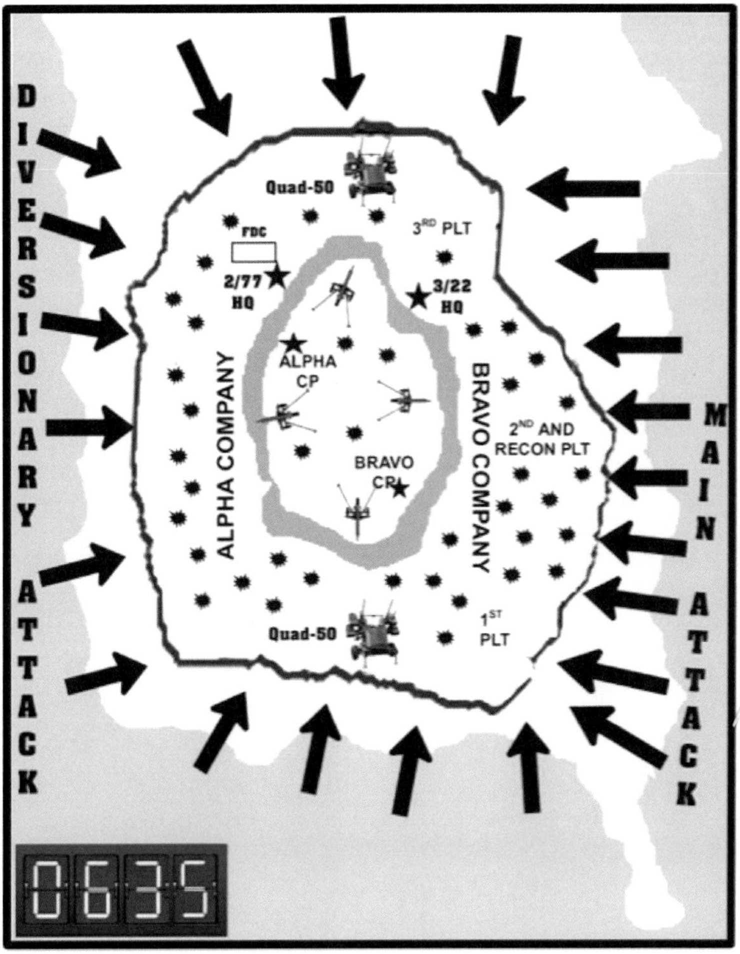

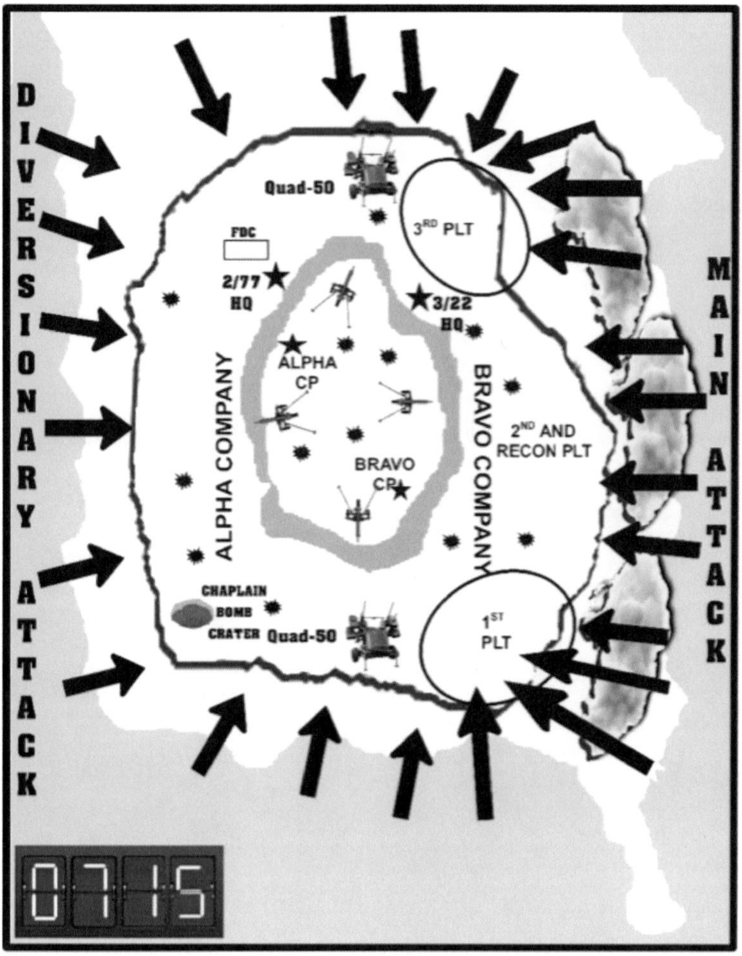
DIVERSIONARY ATTACK
MAIN ATTACK
Quad-50
FDC
3RD PLT
2/77 HQ
3/22 HQ
ALPHA CP
ALPHA COMPANY
BRAVO COMPANY
2ND AND RECON PLT
BRAVO CP
CHAPLAIN BOMB CRATER
Quad-50
1ST PLT
0715

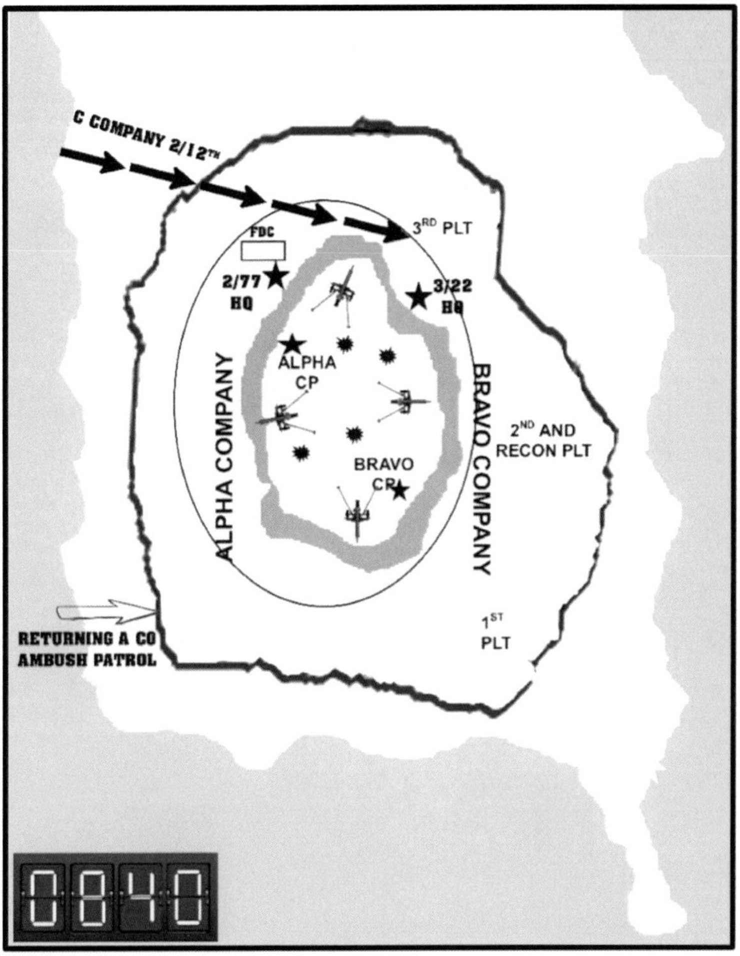
C COMPANY 2/12TH
FDC
3RD PLT
2/77 HQ
3/22 HQ
ALPHA CP
ALPHA COMPANY
BRAVO COMPANY
2ND AND RECON PLT
BRAVO CP
1ST PLT
RETURNING A CO AMBUSH PATROL
0840

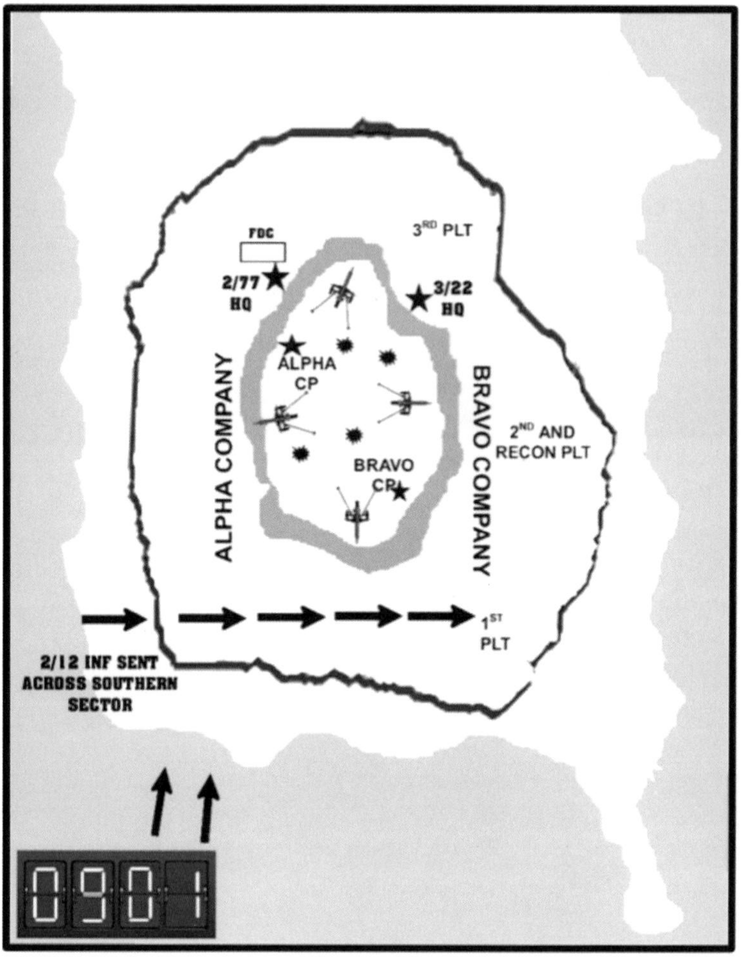
FDC
2/77 HQ
3/22 HQ
3RD PLT
ALPHA CP
ALPHA COMPANY
BRAVO COMPANY
BRAVO CP
2ND AND RECON PLT
1ST PLT
2/12 INF SENT ACROSS SOUTHERN SECTOR
0901

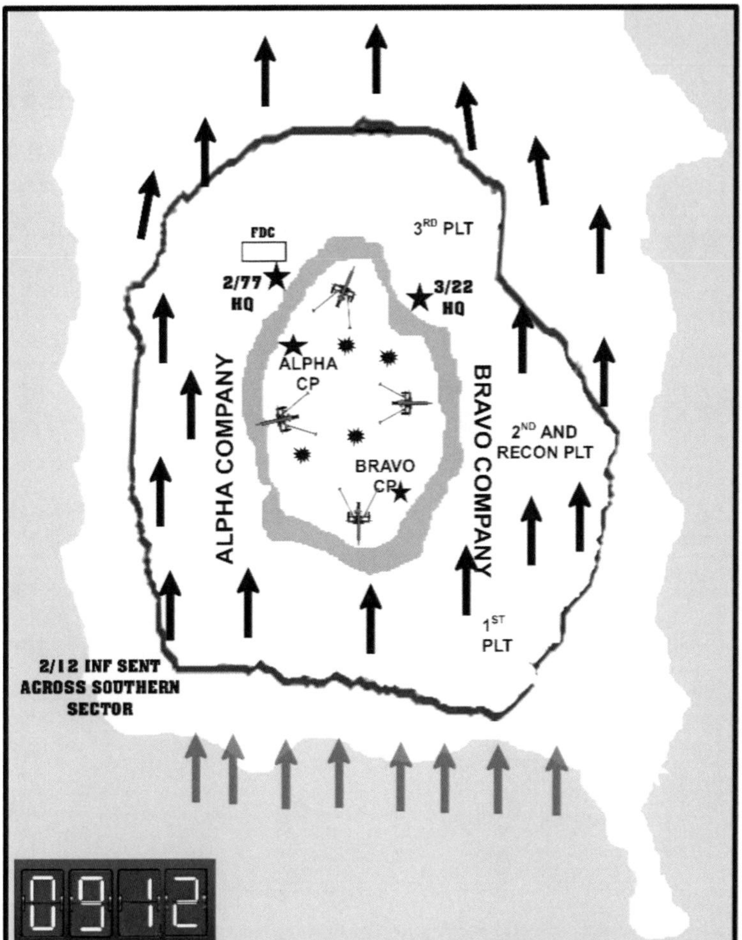
FDC
3RD PLT
2/77 HQ
3/22 HQ
ALPHA CP
ALPHA COMPANY
BRAVO COMPANY
2ND AND RECON PLT
BRAVO CP
1ST PLT
2/12 INF SENT ACROSS SOUTHERN SECTOR
0912

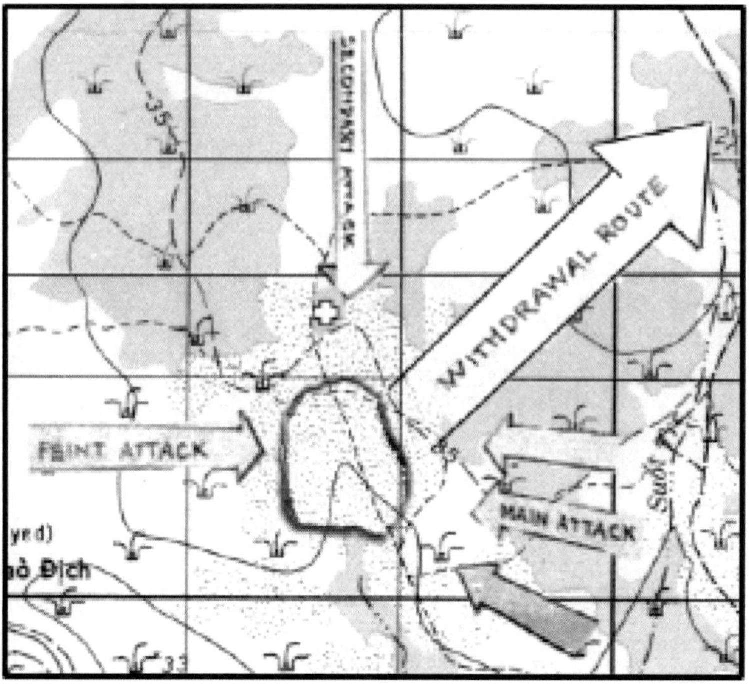

At 0631H FSB GOLD came under heavy enemy 60mm and 82mm mortar attack as the ambush patrol survivors in the southeastern sector had escaped back to the firebase.

At 0635H the Reconnaissance Patrol, 3/22 Inf engaged a large VC force which had approached to within thirty-five meters of FSB GOLD's Southeastern perimeter. An estimated 650 mortar rounds showered on the firebase. While the barrage was taking place, Vietcong attackers creeped up on the perimeter by crawling in the tall grass around the base. Soon the base came under attack by waves of VC firing recoilless rifles, RPG II rocket launchers, automatic weapons, and other small arms. The heaviest attacks were concentrated on the northeastern and southeastern portions of the battalion defensive perimeter around FSB GOLD.

As the attack continued, the three artillery batteries of the 2/77 Arty began firing counter mortar fire in an effort to neutralize the VC mortar concentrations which continued to rake the entire fire support base.

By 0638H the amount of mortar firing had diminished after some vigorous anti-mortar fire was directed nearby.

With the attack intensifying and his troops coming under increasing pressure, Capt. Shugart told his RTO to contact Major Cliff Roberts, (Falcon 3) the Battalion Operations Officer, that his company was fully engaged and was providing final protective fire. He asked for a FAC Overhead to supply air support. He was told that he was on the way and the fighters were already scrambling at Bien Hoi.

During the initial assault, Co B reported that 1st Plt positions (southeastern perimeter) had been penetrated and that the reaction force from 2/77 Arty was required to reinforce this sector.

At 0701H the reaction force began moving to the 1st Platoon's portion of the perimeter. At 0706H the first FAC arrived at GOLD (Capt. Bob Staib's birddog aircraft) and within five minutes airstrikes had begun. Additionally, close supporting fires from two batteries of 105mm had been brought in to within one hundred meters of the battalion's perimeter. At 0711H, Co B reported that its 1st Platoon had been overrun and surrounded by a human wave attack. Airstrikes were called in along the Eastern woodline to relieve the pressure on B Co's perimeter. Soon the first of the fighter jets had arrived, and as directed by Bob Staib began dropping bombs on the VC in the northeast corner. First on the scene was a Tiger flight of F-5s off the alert pad at Bien Hoa. As I wrote earlier, Tiger Pilots were highly experienced fighter pilots. The benefit of these exceptional F-5 pilots was that their bombs could be put in very close to the battle. This was imperative with the enemy not only within the perimeter but moving up to flood the zone. These attacks took out the massed soldiers who were attempting to join their friends who had already broken through the 1st Platoon perimeter. The massed VC were mowed down before they ever reached the perimeter. The VC inside the perimeter were on their own.

Walt Shugart on the First Half Hour of the Battle

"The attack began with a massive mortar barrage from 60mm and 82mm mortars of the U-80 artillery Battalion. The first mortars began on the eastern B Company side of the perimeter and drifted towards the west side of the perimeter and the artillery positions. By this time, the Vietcong had snuck up to within 30 meters on the center of the B Company's perimeter. When they were spotted, a massive firefight broke out all around the base.

"The din of battle and ferocity of the enemy assault made communication with either platoons or battalion nearly impossible. While Lt. Pacheco made calls for fire and close air support, I directed the battle solely from gut reaction. FM radio communications among the troops on the ground was impossible. All press to talk buttons were depressed at the same time by those seeking support, guidance, assistance, ammunition resupply, medical support, and a myriad of other calls relating to the combat. This completely blocked radio communication.

"My first inclination was to dive into the foxhole prepared and improved upon the last two days. Finding no way to direct the battle from the bottom of the hole and assisted by 1SGT Williams, I attempted to conduct the defense above ground where I could perceive how well the company defense was being conducted.

"Above ground, I could see enemy hoards advancing through 1st Platoon and also over running the quad 50 at the south end of the company perimeter. The attempt by the enemy to turn the weapon on us occupying the interior was stopped by its crew and supporting fire from weapons platoon."

On the other side of the fire base, that Quad Fifty emplacement was also being threatened by desperate VC attackers.

You may recall James Brewer who was transferred to the 3/22nd from the 196th LIB. He was in the northern sector of the fire base, not far from the northern Quad-Fifty. That Quad was a vital target for the enemy and they desperatley sought to put it out of commission. Within the first thirty minutes, an RPG round exploded near the quadmount and blasted the operator out of his seat, landing a few feet behind the weapon. This is when Jimmy Brewer raised to the challenge.

From his Bronze Star proclamation:

"Specialist Brewer's position was under heavy concentration of automatic weapons and mortar fire. Specialist Brewer, assessing the Quad Fifty machine gun position being overrun, maneuvered through intense enemy fire to the beleaguered position and placed highly effective fire on the oncoming enemy force until he was mortally wounded. His actions were directly responsible for keeping friendly casualties to a minimum."

Jimmy was awarded the Bronze Star Medal with "V" Device and Gallantry Cross Medal for heroic act on that morning. Soon both Quad-50s were destroyed.

Shugart Continuing His Story

"Still, nothing stopped the continuing human wave assault

across the company front. Fearing that the penetration of 1st platoon would cause the defense to collapse, I called for the reserve from the 2/77 Artillery to reinforce 1st platoon. Providentially, the action took place exactly where it had been rehearsed on Sunday evening.

"Even though the platoon position was reinforced, it was not stabilized. I ordered all platoons to withdraw to the supplementary positions. The weapons platoon provided covering fire for 1st Platoon which was most heavily engaged. LT Kaul coordinated the withdrawal of the remnants of LT Andrew's platoon. About that time, I shouted into the din, "Fix bayonets!" I have no idea whether anyone either heard me, obeyed the command, or took the time to disengage from hand-to-hand combat, draw bayonets, and attach them to their M16's. Quite frankly, it didn't matter.

"All of B Company was engaged in a life-or-death struggle."

Meanwhile George's Shoemaker's Company was also under increasing pressure during most of the battle even though the focus of the attack was on the eastern side of the perimeter. The enemy concentrated on pinning down his company, looking for weak spots to penetrate through the fire base. The attacking force was large, but the attackers pinned their hopes on the collapse of the eastern sector and just rolling over the rest of the base.

George Shoemaker, A/3/22 Commander's Thoughts on the Battle

"Top and I yelled at the same time. 'Incoming' and he, my two RTOs, and I dove into our command bunkers. The first mortar round fell in the jungle followed by the second which landed around 15 meters in front of our command bunker, destroying our supply of soda pop and beer from the last evening's resupply. (*They'll pay for that!*)

"One thing for sure was Top and I checking out the company's defenses and the fields of fire contributed to the company's overall defense. Everyone was focused and luckily, we were able to get our LPs back to the base before the ground attack began.

"The explosion of the second mortar round triggered a human wave attack coming from all directions around the fire base. The attack was challenged, but no matter how many enemy were killed, the attack proceeded unabated.

"No sooner was an attack repulsed than the enemy dragged their wounded back to the treeline, given doses of opium, and prepared for another attack. This sequence went on almost unabated for over the next two and a half hours."

The Infantrymen had their hands full that morning, but remember…the real threat to the VC was from the artillery Howitzers. As a result, the VC were targeting the gunners with mortar fire from nearby clearings and RPGs and 57mm recoilless rifles fired from the wood line. The artillerymen responded by lowering their guns and firing point blank HE rounds into the attackers. They may have been trying to crush the defenders, but the artillery position was their ultimate goal.

When we arrived in Vietnam, there were not enough M16s for all the soldiers. Artillerymen were equipped with M14s and usually carried a load of 200 rounds per person. Not much ammunition if they got involved with a battle of long duration.

Bob Deshaies, worked out of the artillery CDC. He was quite talented and was able to accomplish many tasks within the battalion. On this day, his job was to maintain the communications equipment for the 2/77 battalion. That included all the antenna and generator hookups. That job got him exposed to quite a bit of dangers as he moved about the artillery base during the attack.

Bob Deshaies

"I was in the battalion FDC when the battle began. Colonel Vessey was with us when the attack began. He immediately ran out of the tent and headed for the 2/77th Command Center. From where I was located, I saw Vessey continually exposing himself to danger as he moved from gun to gun. He was encouraging the artillerymen and at times helping the troops repair damaged guns with parts from the nearby damaged Howitzers. He was fearless.

"As the casualties mounted, wounded soldiers began to arrive at the FDC. The officer in charge told them to get the wounded out of the Operations Center Bunker, telling them this was not an Aid Station. I never forgot that. These wounded men were carefully carried to a large bomb crater in the southeast corner of the base where Major Gene Adler, the 3rd Bde Chaplain, was receiving them and doing what he could for the wounded, both spiritually and physically. Somehow the Vietcong bypassed them in their attacks and he survived the morning.

"It was imperative that communications were functional for the artillery units. I had strung out all the commo wire between the units and it was my job to assure that they were operative. On more than one occasion during the battle we lost communications due to wire being damaged by the attacks. I had to scramble to find the breaks and repair them as best as I could, not even having electrical tape with me to cover the repair. Somehow, they held.

"During the battle, I was told that one of our main antennas had gotten knocked down. Off I went into the chaos to remount the antenna. Communications were essential and the land lines using commo wire were most effective as the radios were jammed with multiple callers.

"At another point in the battle, I noticed a gunner from Joe Engle's gun who was down. When I went to aid him, I realized that he was beyond hope. He was wearing a flak jacket, like all the artillerymen, but it was open in the front. That was where a sniper shot him, killing him instantly.

"For me, the scariest part of the battle was the first fifteen minutes. That is when the mortar and ground assault was at its peak. It was then that I made my peace with God. From then on, I put my life in His hands and just did my job."

Bob would later be awarded a Bronze Star for Valor for his actions at GOLD.

Mike Doolittle, FDC for A/3/22's Mortar Platoon

"During the battle the enemy was everywhere and inside the perimeter, there were maybe a half dozen killed around our gun pits, I was out of ammunition, two grenades left, and my bayonet fixed, scared to death, and knowing I was about to die… Then a very strange thing happened in all that chaos, tracers zinging overhead, explosions everywhere, but I heard this voice very clearly, it was coming from the105 artillery gun nearest our mortar pits. I had no idea who it was, but when I turned to look, I could see these two men at that gun, and one was looking right at me, motioning me to get over there. I was scared to get out of my hole, but dying alone was not a good feeling (in hindsight), and I crawled over maybe 20 yards to the gun…

"It was obvious to me that he was an officer, older like my

father, and he ordered me to unpackage these rounds sitting behind the gun as they both worked to prepare it to fire. I remember one moment as I heard the pinging of rounds off of the gun, explosions of incoming and rounds from the ammo piles, fires flared from the excess powder bags. When I looked down the side of that 105 out towards where it was pointed, a mass of enemy headed our way (hundreds in my mind), all shooting at us and everywhere else. I did not want to see my own death coming so I buried myself in the task of fumbling my way through opening these canister containers and handing them to the younger of these two officers. I did not know who they were by rank or name, and as a young soldier, obviously this being my first experience with intense combat, scared shitless.

"These memories are not like a movie, but still moments in time, as there are also many blank spots of that day . . . It seemed like forever, yet it was but very few minutes I would guess, it was hard to recall how many rounds were fired, four maybe five . . . But then the entire attitude of these two men changed, elated—excited—relieved hard to put just one word to that moment. "They left the gun and I was still sitting behind it, when I did stand up and looked out at the area that the enemy had been coming from, it was full of smoke and haze, but no enemy anywhere close, then the sound of high revving engines filled the air. I remember going back to my mortar pits, and found an aid kit, and another one of my guys, we began to make our way out towards the perimeter to help with our guys who may have been wounded, we found maybe six to eight dead enemy, and two severely wounded, I remember picking up their rifle as I had no ammunition, as we continued going out."

It would not be until the year 2015 at the Suoi Tre reunion in Colorado Springs that he learned that the 'old' soldier who was at the gun that day was Colonel John Vessey, the base commander. Another revelation to him on that day was the fact that all the units which participated in the Battle of Suoi Tre had been awarded the Presidential Unit Citation. It was awarded on October 21, 1968 almost a year after the battle. None of us got the memo at home.

Shugart was busy trying to shore up his southeast sector, but after that he radioed his center group leaders made up of his 2nd Platoon and Recon Platoon. The Recon Platoon Leader told him that so far, they were holding their own against a tremendous push by these massive waves of enemy attacking. He then spoke to his Third Platoon Leader in the northeast sector who told Shugart that in the center of his position VC had made it through the defense and were fighting his men foxhole to foxhole. He told him he didn't know how much longer they could hold. Both units indicated that they had expended a lot of ammunition and would need resupply soon.

Shugart

"I was angry. No, I was damned mad! Those guys were killing my troops, my guys, my soldiers. I was determined to inflict as much damage on the enemy as possible to preserve as many of my soldiers as possible with all means available. But those means, especially small arms ammunition and grenades, were being quickly depleted. 1SGT Williams and I moved together to redistribute what ammunition we had to 2nd & 3rd platoons. While low crawling out of the CP with enemy bullets snapping over our heads, a Chicom grenade fell about five feet in front of us. We both ducked with our faces in the dirt and the grenade exploded harmlessly."

It was just around this time that the reaction force had arrived and were attacking across his front. He warned his men to be alert for friendlies entering their zones and told them not to shoot them up. Within a minute, the Air Force fighters arrived.

The Air Force F4 Phantom jets arrived and dropped their loads on the massed enemy troops in front of B Company's position. Bob Staib's FAC plane, situated to the southeast directed a devastating attack on the VC positions just outside the eastern perimeter and within the immediate treeline.

Shugart watched the explosions and directed the FAC to drop ordinance closer to the 1st Platoon's position in the southeast corner. Staib acknowledged and the next run began right in front of the struggling platoon's position. After that attack, at least VC reinforcements would not be arriving soon. They were lying in the field and treeline.

Capt. Staib remained at the site directing air sorties until the 3rd Bde FAC came on the scene. It was now Capt. Toni England and Capt. Walter Forbes turn to direct the air attack for the Air Force.

Shugart

"At 0745H after four sorties, the FAC was hit by enemy fire. As it spiraled to earth the pilot calmly told LT Pacheco, "Well, I guess you'll have to get a new FAC." Both the pilot and co-pilot were killed in the crash. A new FAC was on station almost immediately. (*That was when Capt. Bob Staib was recalled from his reconnaissance mission near the Cambodian border.*) The sorties continued with strikes progressing ever closer to the 1st platoon area. In desperation, I contemplated calling in the next strike on top of the platoon position. I began to think that I might have to write them off as an unavoidable combat loss.

"Recon platoon and third platoon were being pressured even though they had withdrawn to their supplementary positions.

The battalion CP, positioned behind 3rd platoon, saw its line faltering and called A Company for reinforcement. A platoon sized element from A company joined the fight linking Recon and 3d platoons. Although aware of its arrival, I relied on their soldier skills to conduct that portion of the battle with whatever was available.

"With the battle raging at its peak, I saw Sp4 Linnemann, the company operations NCO, walking around in front of the company CP firing well aimed shoulder shots at the enemy. I was astounded, but Linnemann went unscathed despite wave after wave of attacking enemy advancing without letup.

"At some point, I was visited by LTC Vessey and his sergeant major. They crawled to the CP bringing small arms ammunition with them. They were the only friendly visitors we had during the battle."

When the FAC was shot down, Shugart directed his Artillery Officer Pacheco to tell the artillery unit to load beehive rounds in their guns. Shugart then contacted his 1st Platoon Leader for a situation report. He was told that a reaction force had just arrived in his sector, however the unit was barely holding its own and ammunition was getting low. He then checked on the beehive round readiness.

Shuster had to gain control over his platoon's position in the southeast corner of the battle field. In some cases, the action was down to hand-to-hand combat. The Vietcong were running freely inside the original perimeter and were moving towards the secondary positions, closer to the Howitzers. When the beehive rounds were ready, he informed his men in the overrun positions to duck down into their bunkers and stay down. Once his men acknowledged that message, he ordered the artillerymen to rake his positions with beehive rounds in the 1st Platoon sector.

Shugart

"The sustained attack found the enemy throughout the company. The fight was desperate. We were down to our last bullets with nothing left but our bare hands to continue the fight. With rank upon rank of enemy infantry still advancing, the 2/77th lowered its tubes for direct fire and unleashed successive volleys of beehive rounds. Each shot sounded like a massive swarm of bees had been unleashed. The advancing troops were mowed down like a reaper harvesting wheat. Nonetheless, succeeding enemy files continued to advance, disregarding the fate of their predecessors."

The Vietcong were suffering devastating losses, yet when the charging enemy were mowed down, a new set of attackers emerged from the woodline to take their place.

At 0813H, the northeastern section of the perimeter was once more overrun by another human-wave attack. Two minutes later, elements of Company A which had established an ambush just outside the perimeter the previous night charged into the camp's perimeter and assumed defensive positions. Somehow all the men had managed to elude the surrounding Vietcong.

At 0820H, Captain Napper and his C/2/12 troops reached the rear security of the attackers in the northeast. The Vietcong realizing that it was imperative to prevent the reinforcement of the base put up a stiff resistance that kept the relief forces from reaching the clearing.

The Vietcong realizing that relief forces would be arriving soon made another push. They were able to takeout three guns of B Battery and one A Battery gun. The artillerymen, including Colonel Vessey and his staff, scampered behind the sandbagged positions immediately to cannibalize what they could to get more of the Howitzers back into the fight.

At 0840H, Captain Napper and his C/2/12 troops were able to battle their way into the clearing. Coordinating with the A Company Commander, he was told to pass through the northern positions and shore up the defense on the northeast corner of the perimeter. They worked their way up through the troops on the northern sector and teamed up with B Company's 3rd platoon to move the enemy out of the zone and reestablish the original defensive perimeter.

Talking to a few defenders of GOLD over the years, most didn't have any idea that a relief force was headed their way. That is why the arrival of C Company men in that northeast sector was such an unexpected surprise... and much needed. A similar situation took place when the rest of the 2/12th arrived in the southern zone 20 minutes later.

This victory was made possible through an array of military men whose specific talents all contributed on that day. Here is a story on some of the chopper pilots who were fearless and contributed in saving many lives.

George J. Stenehjem, Centaurs, D Troop (Air), 3rd Squadron, 4th Cavalry, 25th Infantry Division.

Excerpts from a speech written by Bain Cowell using the Centaurs website stories to tell their part of the aerial battle. The speech was delivered by George Stenehjem at the 50th Anniversary of the Battle of Suoi Tre Event in Colorado Springs:

"In March 1967, I had just assumed command of the Air Cavalry troop. During Operation Junction City, Centaur helicopters inserted Long Range Reconnaissance Patrols (LRRPs) in the jungle to detect enemy movements along the eastern flank of War Zone C as U.S. armored and mechanized forces drove north. Our helicopters were standing by at Tay Ninh to insert, support, and extract the LRRPs if they made contact. One of the patrols warned of a possible attack on Fire Support Base Gold.

"At dawn on 21 March, the Centaurs became multitaskers. We had to continue supporting recon patrols while responding to the crisis at LZ Gold. I got a request from the Air Force to rescue a forward air controller L-19 spotter plane that had been shot down. I radioed the Service Platoon Leader, Captain Tom Fleming (now, like me, a retired colonel), to crank up the maintenance and recovery helicopter, a Huey nicknamed "Stable Boy," and four Huey gunships to search for the lost aircraft. Tom, looking for the L-19, spotted tanks crashing through the jungle; a kilometer away he saw "an immense conflagration" of air strikes and exploding artillery rounds on LZ Gold. Soon I took off in my command-and-control Huey with my crew of four. Looking north past Nui Ba Den I was startled to see a mass of gray smoke pouring out of the jungle.

"Meanwhile, another Centaur "slick" commanded by the Aerorifle Platoon Leader, Major Harold Fisher, was already in the air to get a situation report from a recon patrol. But he saw the smoke and explosions at LZ Gold, and D Troop Operations told him that scores of wounded GIs needed medical evacuation. He and his crew immediately volunteered, no matter what the risk. As they circled in a holding pattern and listened to frantic radio transmissions from ground troops shouting that they were being overrun and their howitzers destroyed, the Centaurs wondered if they could survive a descent into that inferno. On their final approach under low clouds, they ran an obstacle course of incoming mortars, outgoing artillery, close-in airstrikes, and the risk of being targeted by the enemy. Upon landing, they encountered a "thousand-yard stare" of exhausted troops who had fought for hours, unflinching as they calmly loaded their wounded onto the chopper despite incoming fire all around.

"Service Platoon Leader Fleming, unable to find the downed Air Force plane, repositioned his chopper for medevacs. He headed in at treetop level amid mortar, RPG, and machinegun fire, almost collided with then-Lieutenant Colonel John Vessey's small C&C helicopter, *(which was being used by Brigade Commander Colonel Garth to direct the defense of the base)* and landed near the exploding ammo dump and smoldering wreckage of two Hueys. On the ground 'everyone was in holes or crawling around.' They began dragging the wounded to the medevac helicopter.

"As each severely wounded man was lifted into one side of the ship, a less-wounded man would get out the other side. Then Tom saw a nearby quad .50 machinegun, 'its barrels jammed and overrun with VC, blown away by direct fire from a howitzer,' and U.S. infantrymen firing from behind VC corpses. As he pulled his loaded Huey up, U.S. armored personnel carriers burst from a tree line and drove into the ranks of attacking VC.

"Stable Boy delivered the casualties to the Army hospital at Tay Ninh West Basecamp and returned for a second medevac run. On a third trip, they learned the location of the Air Force crash site south of LZ Gold. Eight Centaur gunships made daisy-chain gun runs around the site to protect Stable Boy as it hovered over the wreck and lowered medic SSG Kelly 75 feet on a rescue hoist. Kelly found two decapitated bodies and the Centaurs informed the Air Force. U.S. ground forces later retrieved the remains.

"During the medevacs, the Centaur crews worked to keep the wounded alive enroute to the hospital. Lieutenant Colonel John Bender, commanding the 3/22nd Infantry, directed the medevac crews to casualties on the ground. On Fisher's "slick," a soldier with an abdominal wound cried out for water, but crew chief Specialist Mike Vaughn, trained in first aid, correctly said no. The man was in traumatic shock, so aircraft commander Fisher took off his own shirt to be used as a blanket. Another man with a gaping thigh wound asked to die because his girlfriend wouldn't like him anymore; the crew chief and gunner encouraged him to hang on and pull through.

"All told, Fisher and his crew made three trips and carried a dozen WIAs to hospitals at Tay Ninh West and Bien Hoa. With mortar and small-arms fire still coming into the LZ, the Stable Boy crew was asked to medevac a wounded VC prisoner as a priority—requiring other casualties to be off-loaded—and also medevac a badly wounded U.S. soldier from a listening post. Fleming hovered his ship across the LZ beyond the perimeter to locate and pick up the wounded GI, who had lost both legs and an arm and had a sucking chest wound. After takeoff this soldier tried to push the VC prisoner out the door, but was restrained by our medic. Both patients later died.

"The Centaurs' gunships came back repeatedly to support the ground units and protect medevac choppers. Some of the gunships were diverted from other missions. Pilot Rick Arthur, 20 years old with only three weeks in country, and a light fire team of two Hueys armed with rockets, flex machineguns, and a grenade launcher had just finished an all-night standby on counter-mortar duty at the Dau Tieng airstrip. As they prepared to return to Cu Chi, a crewman from a Chinook helicopter at the refueling point asked the Centaurs to escort his big bird loaded with howitzer ammo into a hot LZ not far away. The Centaur crews agreed and followed the Chinook, which made it in, unloaded, and departed.

"All hell was on the radio," Rick recalled, so "there was no way to coordinate and we held our fire." Watching the airstrikes and artillery, "we simply picked the common target and fired our load of rockets on it."

"The Centaurs' artillery observer, First Lieutenant Bain Cowell, was riding with me, but didn't fire a shot at LZ Gold; many artillerymen on the ground were doing that. Instead, he helped me keep the Centaur helicopters away from friendly fire. Hunched over his maps and PRC-25 radio, he kept track of the azimuths and max ordinates of incoming artillery rounds, which he passed to me whenever there was a brief lull. The helicopter whipped from side to side as we dodged rising tracers uncomfortably close to our rotor. Suddenly I heard crew chief Specialist "Willi" Williams on the intercom: 'Sir, high performance at nine o'clock.' We looked left to see an Air Force jet heading straight at us. I remember an F-4 screaming in so fast and close I could almost read his nametag.' Our co-pilot at the controls plunged the Huey out of harm's way. Back at our Cu Chi base, "Willi" found a hole in the floor under Bain's seat in the back of the helicopter where his radio had been, and a smashed bullet. The PRC-25 had served as the artillery observer's 'chicken plate' armor. The radio was dented but intact and still worked. We saw "Willi" three years ago at a Centaurs reunion and thanked him for saving us; sadly, a few months later he died."

A/2/12th, leading the other battalion's units not yet at the battle site, reached the southwestern corner of the battlefield at 0901H. Unlike our brothers from C/2/12, who had to fight their way into the clearing, we met no resistance. The enemy seeing their chances slimmed by the quick moving relievers threw all their attackers into the field. They probably knew that the APCs and Armor would follow close behind and they needed to send the last attackers (rear defense troops) into the field then.

Lieutenant General Bernard William Rogers wrote this about this period of the battle in his study,

> "VIETNAM STUDIES, CEDAR FALLS, JUNCTION CITY, A TURNING POINT."

> "At 0901 another relief column from the 2nd of the 12th (A, B, HQ Companies) broke through and linked up with the battered Company B. With the added forces and firepower, the units were able to counterattack to the east and reestablish the original perimeter. But the Viet Cong were still attacking. As they advanced, many of the soldiers could be seen wearing bandages from earlier wounds. Some were so badly wounded that they could not walk, and were carried piggyback into the assaults by their comrades."

Shugart

"Just as all seemed lost, a stream of 3rd Brigade tanks, APCs, and infantry troops swept through the firebase from the southeast to the northwest, driving the enemy out of the firebase. Like the cavalry in a western movie, the friendly forces arrived just in time to save our beleaguered force. Using shouted commands as well as hand and arm signals, I directed B Company to follow the counter attack to occupy its original positions. The battle was over. It had concluded three hours of close continuous combat.

Lt. Roger Frydrychowski, Recon Platoon Leader, 2/22

"Colonel Julian was flying overhead in a bubble helicopter and directed us to the southwestern edge of the LZ GOLD clearing.

"We broke through the treeline in the southwest of the clearing. Appraising the situation at the battle site, the first thing I did was try to determine where the base was through the smoke.

"Contacting Julian, I awaited orders. He told me he was not aware of the technical needs of the defenders and I should check with the commanders at the fire base. Upon receiving word to proceed forward and was told that they would give us further orders when we were at the base. We then rode into the clearing in a column formation. We began receiving small arms fire almost immediately. As we moved forward, we realized that the enemy to our front had remained as we would at times run over them and the dead VC that littered the field.

"We stopped at a clump of trees and by then we were taking fire from inside the fire base and from outside the base. I got out of the track and looked for a base commander to find out what was going on and how I should proceed. I was unaware of the situation at that time. As I began to do that, I began to realize that bullets were whizzing by me at a heavy clip. I began to second guess that decision and paused to reconsider. Just about that time two officers from the base with oak leaves on their collars ran to me. I shouted over the noise, "Which way should I go?" I was told anyway that I wanted to go. So, I began to circle around to what I thought was the eastern edge of the base perimeter with my platoon.

"C /2/22nd, to my rear, had formed a skirmish line before they attacked. I can't confirm that but that was what I was told.

"I recall seeing American infantrymen who were maneuvering out along the field who today I assume were men of C/2/12 who were on the north side of the base camp.

"During our race up the field, there were some foolhardy Vietcong who tried to jump on our APCs to get to us. They were shot off with 45s and M16s. At one point they surrounded us, but we dealt with that with small arms fire.

"I know that the armor unit did sweep up the field attached to the mechanized units. Most of the attack came from the eastern woodline and the woodline up north.

"When I came up the field, I could see the dead VC still in skirmish formation where they laid in a neat row. The discipline that they displayed was remarkable."

At 1405H, Recon Platoon and C/2/22 were sent to retrieve the bodies of FAC Forbes and England, east of the firebase. They received one RPG round that missed its target and the attacker fled. They moved on and arrived at the crash site at 1437H and found both men still in their aircraft. They were pretty mangled and crowbars were needed to remove them from the wreckage.

Shugart:

"Now began accounting for losses. To my surprise and gratitude, the air strike called on 1st Platoon fell short, not killing any of our troops.

"Surviving 1st Platoon soldiers told me that they wanted the strike to land closer than it did.

"Four of the eight-man 1st Platoon ambush patrol members were killed as they attempted to withdraw to the perimeter. The remaining, though wounded, played dead as the enemy advanced through their position. Of the thirty-three friendlies killed, ten were B Company's. An additional number from the Recon Platoon members were also killed; however, Lt Kaminsky reported these losses directly to HHC, so I never knew how many were his.

"At the evening briefing, I learned that enemy losses were far greater—596 KIA by body count with an additional 200 estimated killed, drawn from battlefield evidence. The enemy mission was to annihilate the firebase, as it had done to the 7th ARVN Regiment in the Michelin Plantation in December of 1965, taking no captives and leaving no survivors. Although the enemy killed by conventional artillery and air strikes will never be known, the 40 beehive rounds proved the great equalizer. Enemy dead were found arms pinned to their chests by a multitude of flechettes. VTR's (Tank Retrievers) with bulldozer blades dug trenches which served as mass graves for the vanquished."

Walt Shugart reflected the true strategy of the attack. Lt. John Concannon, the 2/12th Intelligence Officer was sent to Saigon to investigate the vast number of documents that were collected from the enemy during the battle. At an Alpha Association reunion many years after the battle, he described what he learned during the two-week investigation of the documents.

John Concannon

"To understand why Fire Support Base GOLD was chosen by the 9th VC Division Commanders as their target, you need to go back over a year and a half to the Battle of Ia Drang (*depicted in the Mel Gibson movie "WE WERE SOLDIERS"*). That battle took place in November of 1965 when an American unit was first targeted for destruction in the Vietnam War. The enemy soldiers were unsuccessful, but they learned some valuable lessons that they applied at the Battle of Suoi Tre. One thing that the NLF learned from that battle was the fact that American Artillery units were not only powerful but very accurate. Most of their losses in the Battle of Ia Drang were the result of very effective artillery fire. They decided to target an artillery unit to take them out of the picture. They learned different lessons from the battle at Fire Support Base GOLD.

"They were aware that the unit that they would attack was a recently deployed unit that had been in country for only six months. In addition, they knew that the bulk of the unit was made up of draftees. 'How tough could they be,' they asked. Actually, those soldiers were remarkable in their steadfastness. With the proper leadership and outside support, they became an unsurmountable obstacle to the 9th VC Division's plans to annihilate an American unit and achieve a significant victory that may have tipped the balance of the war in the NLF's favor.

"Another lesson they learned was you may have tied down the artillery batteries at GOLD, but that did not make them defenseless. The batteries just lowered their guns and fired point blank at the onrushing attackers using their smallest charge, which exploded into the onrushing human waves of VC. When the attackers had reached a point where they were close to overrunning the 3/22nd infantrymen, the artillerymen loaded their howitzers with 'Beehive' rounds. These rounds were very effective in destroying the human wave formations.

"Finally, the plan was to overrun the firebase very quickly and then roll over the remaining units in short order. Effectively, the rescuers were not only fighting for the defenders, but also for their own survival. They were coming after us next."

As shown on this map, the enemy was driven from the battle site and left the clearing and entered the jungle northeast of the base. They purposely did so to divert any attention away from their aid station that was set up in the jungle to the north of GOLD. It was very difficult to see everything that was going on through the smoke and carnage, but I personally recall vividly the sight I saw as the APCs and armor unit chased the VC survivors into the jungle. It was easy to follow the action as the

Air Force was pounding the enemy's escape route with bombs and napalm. That scene did not last long before Brigade Commander Colonel Garth called them back from their pursuit. The pursuit barely went 200 meters into the jungle. Garth could not account for one 9th VC Division unit that was not engaged in major battles over the last two days; the 271st Regiment. He was worried that our forces were being led into an ambush and pulled them back.

Within an hour of the arrival of the Mech and Armor units, the battle was over. Garth had landed on the field and was directing the activity, which included the extraction of the wounded and KIAs.

On a very personal note, I need to tell you about a friend that I made from the western defense perimeter. When the pursuit was taking place, I was standing at the southwest corner of the field. Out of the nearby A/2/12 bunkered positions popped a soldier of small proportions. To me he looked like a kid beginning high school. He trotted up to me and threw his arms around me. He kept repeating, "Thank you, thank you." Remember these men didn't even know that we were racing to save them from annihilation. They were overwhelmed. I told the 'kid' not to worry, we were here and this battle was over. I never forgot that man.

At the 2017 Battle of Suoi Tre Event in Colorado Springs, I mentioned that story to Richard Hazel, an A/3/22 veteran of GOLD, who had attended Alpha Association reunions in the past. He told me excitedly that he knew who that man was and he was, in fact, attending the reunion that weekend. Richard brought me over to him and I identified myself to him and how we had met 55 years earlier at the end of the Battle of Suoi Tre. I asked him to replicate that hug and he agreed to do that. His name is Dennis Baldauf and since that day, we have been Facebook friends and keep in contact.

Dennis, upon learning that I would be at the dedication of the 12th Infantry Regiment Monument, as I served as the Historian for the project, traveled 85 miles from Newman, Georgia to be with me at Fort Benning for the dedication. I was quite moved by his thoughtfulness.

The only thing left for us relief forces to accomplish after the battle was policing up the dead VC bodies which littered the field. However, the men of my company had one more surprise awaiting us before the day ended; the despised "Infusion Program" would visit the battle site less than two hours after the end of the battle.

The Battle Aftermath

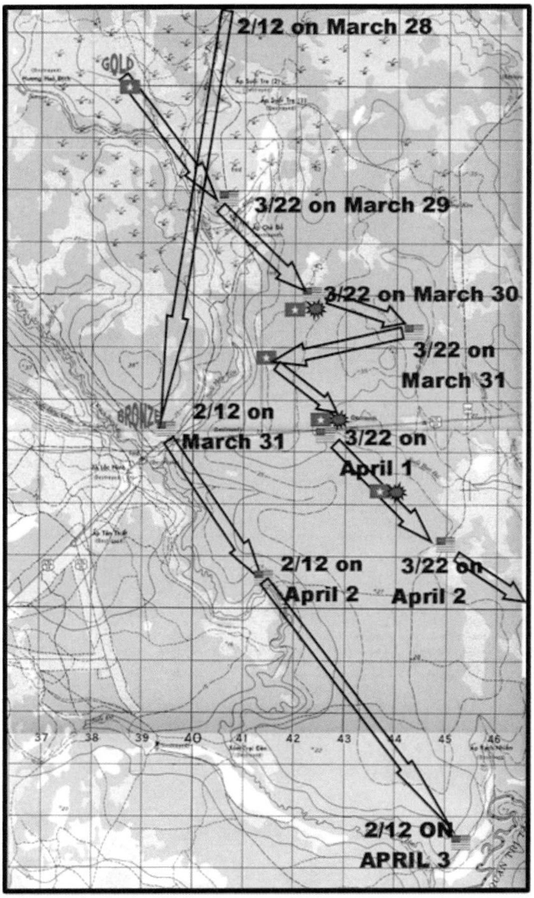

Burnie Quick would later become A/2/12's Commander, in July 1967, but was on Brigade Staff when the battle took place. He shared with me the official brigade collection that was taken on the day of the battle. This photo is one of them.

Here is the story about that big surprise that we only learned about two hours after the battle ended:

I don't think any of us 3rd Brigade 4th ID troops had given any thought to the significance of the day of the battle. That March 21st date was exactly six months to the day that the brigade had boarded the Nelson M. Walker troop ship and set the clock for our return to America exactly one year after that day. That condition presented a real problem for the Army. How to prevent a unit from losing all their experienced soldiers being replaced at the same time, a year after they started their Vietnam tour when they needed to be sent home. The intelligentsia of the Army came up with an answer to the problem.

The solution was to break up the units and send some soldiers to other units where they could filter out of those units with different DEROS dates, without creating major manning problems. The program was called the "Infusion Program'. It may have solved the problem for the war planners, but was bitterly resented by the common grunt who had to leave their buddies and move onto another unit where they knew no one and had to prove themselves all over again after six months in combat.

Around two hours after the battle ended, First Sgt. Springer arrived by helicopter to the battle site. He had a list with the names of maybe 15 of the A/2/12 soldiers who were gathered together. They were told to gather their equipment and assemble near where the choppers were evacuating the American wounded and the dead from the battle. Porter Harvey was one of those unfortunate victims of the Infusion Program.

Porter Harvey

"I had become injured from the barrage of that morning when I was slammed into a tree by one of the explosions. My knee was wrenched, but I could limp with it, so I joined the rest of my company and marched into the battle. My leg was throbbing, but there were those worse off than me so I figured, I will take care of my leg after the battle. I can do this.

"After the battle I was told to get ready to be taken back to Dau Tieng with some of the others from A/2/12. There were maybe 15 of us who assembled by the chopper areas where the dead and wounded were being taken out of the field. Needless to say, we were not placed on the choppers together and they fit us in wherever there was an open spot on the aircraft. I shared a chopper with wounded and dead veterans of the battle.

"When I arrived at Dau Tieng, I was sent to the aid station where they braced up my leg and sent me on my way back to the A Company area at Dau Tieng.

"I spent a week or so recovering (partially only) before I was sent to a 196th LIB company in Chu Lai where I reported to the First Sergeant. He looked at me and shook his head, "You ask for replacements and all they send you is the limp or lazy." He wasn't pleased.

"I was sent to a platoon and tried to intermingle with them, but they were not impressed by my arrival. To them I was just another FNG.

"After participating in one of the most chaotic battles of the war, I was alone to contemplate what just took place. I was not near any of my A/2/12 brothers. When I tried to explain how violent that battle was, all they said was, "Yeah, we've been in plenty of big battles."

"It was not until Alpha Association was formed in the year 2000 that I was able to unwind from that battle 33 years earlier. A/2/12 veterans understood and appreciated our common history."

It took decades before the troops that witnessed the 'Infusion Program' came into power in the army and vowed "never again". Today troops move into combat as a unit and remain together until they return stateside. That program only destroyed unit morale and cohesion.

There would be a few more "Infusion" cuts over the next few weeks. By the time that the April A/2/12 roster (a month later) was constructed, 55 of the original men who landed together at Vung Tau were gone. By that summer, so many new men had come into my company that I recalled very few.

For those of us who survived that cut, it was off to work collecting the enemy bodies scattered around the battle site.

My company was sent to the northern end of LZ GOLD to clean up the battlefield of dead Vietcong soldiers. This was about the time that General Westmoreland arrived with his entourage of MACV VIPs who arrived to witness the results of the carnage that took place. They took along representatives from the press with them. Peter Arnette, who was with AP at the time was one of the reporters invited to the battle site. He would later gather acclaim for his reporting from Vietnam through the Gulf War. One of his noteworthy accomplishments was being granted an interview with Osama bin Laden in March of 1997 while working for CNN. He asked Bin Laden, "What are your future plans?" Bin Laden said, "You'll see them and hear about them in the media, God willing."

The defenders of the base were assembled and I could see where they were meeting along the southern LZ GOLD vicinity.

In the distance I could see Westmoreland standing on the hood of a jeep addressing the base defender survivors.

After Westmoreland addressed the defenders of Fire Support Base GOLD, 2/12th Infantry Battalion Commander LTC

Joe Elliott moved towards General Westmoreland to present himself to the General. As he approached, Westmoreland recognized his old Staff Officer and smiled. Joe went right up to him and once in front of him, he raised his hand to salute. Joe had been bleeding quite a bit since he lost part of his thumb in the morning barrage and never completed his salute. He lost consciousness from the loss of blood and dropped at Westmoreland's feet.

Joe was given first aid and carried to a dust-off chopper. He would spend the next two weeks in a Saigon hospital recovering from his wounds.

He returned from the hospital and continued to lead the 2/12th Battalion until mid-June. He was replaced by LTC James F. Greer, the 3rd Bde Executive Officer.

At the Change of Command Ceremony, Joe was awarded a Silver Star for Gallantry, and a Purple heart for his leadership and wounds during the Battle of Suoi Tre. He was also awarded a Bronze Star for Meritorious Service for his time in the battalion.

Joe eventually would be sent to the United States to serve at Fort Leavenworth. Kan., in the Department of Divisional Operations. He served as an instructor at the Command and General Staff College until he retired three years later.

A little while later a chaplain held services for this group. Mike Doolittle allowed me the use the original copy of the picture taken during that religious service for this book. He had friends who had connections with AP and acquired a copy of that photo that was plastered across newspaper front pages across America. Mike told me that when that photo was taken, he was looking at the chaplain's shoes. That chaplain arrived from Saigon and his shoes were spit shined. He couldn't remember the last time that he saw spit shined shoes. Seemed out of place on that day. The bodies were being collected and placed on the plate on the front of the APCs that is used to break water for amphibious operation. I took three souvenirs that I found during this exercise. One was an empty magazine for a carbine rifle. Another was a pair of Ho Chi Minh sandals. Finally, I found an unusual memento, a home-made VC gas mask.

That gas mask was a rare find in Vietnam. Gas masks were not standard equipment for NVA soldiers. It appeared to be made out of a brown shower cap, with a clear shield for the wearer to see through and a gauze strip at the bottom which somehow was supposed to filter out any gas that we might use in the battle.

I asked John Concannon, our battalion Intelligence Officer, if VC were normally given gas masks. He said no, this item was probably made by a mother, wife, or sweetheart of Vietcong. It was a very personal item made with love.

The photos at the rear of this chapter were pretty much typical for what we saw while collecting the dead VC bodies around the battlefield. These images are from the official 3rd Brigade collection.

At the end of this long day, a summary was organized by Brigade describing the results of this battle.

DELAYED ENTRIES FIRE SUPPORT BASE GOLD

UNIT TYPE# ROUNDS

2/77 Arty105 mm2200 HE- 40 Beehive Rounds

C1/18 Arty105 mm1008 HE

B/3/13 Arty155 mm357 HE

B/2/35 Arty155 mm357 HE

B/2/32 Arty175 mm20 HE

8 inch22 HE

3964 HE artillery rounds fired during the battle
Also conducted 14 air sorties with assorted ordinance

ENEMY TOTAL BODY COUNT AS OF 2143H
VC
596 KIA (100 credited to air strikes)
Another 200 KIA probable due to air and artillery

ATTACKING UNITS
5 VC Bns. 4th, 5th, 6th Bns. of 272nd VC Regiment, plus two unidentified regiments (*later identified as units from the 101st NVA Regiment.*).
Total strength of attackers: Reinforced regiment.

FRIENDLY LOSSES AS OF MARCH 21, 1967 at 1615H
2/12—1 KIA, 23 WIA; 2/22—2 WIA; 3/22—21 KIA, 44 WIA, 3 MIA; 2/77—8 KIA, 44 WIA; USAF—2 KIA
Total: 32 KIA, 113 WIA, 3 MIA—(79 WIA evacuated)

Many of the defenders received valor awards from their action on this day. Among those receiving Silver Stars or Bronze Stars for Valor were all unit battalion commanders (including Colonel John Vessey) and company commanders and Brigade Commander Colonel Garth.

Colonel John Vessey spoke to the defenders after the battle and told them: "I was in action in World War II and the Korean War and I have never led such courageous troops during a battle. I would take you men with me anywhere."

Vessey had the distinction of being the only man to enter the army as a Private and outside of a Sgt. Major position (*he received a battlefield promotion as a First Sergeant after the Anzio Campaign*) held every rank leading up to Four Star General. He said through it all, being a sergeant was his toughest job in the military.

Vessey's leadership during this battle was rewarded by his being chosen by President Ronald Reagan to serve as the tenth Chairman of the Joint Chiefs of Staff in 1982. He retired from the army near the end of his second tour as Chairman in 1985. By then he had served forty-six years in the service, the longest of anyone at that time had ever served in the Army.

As a civilian, he was chosen by both Presidents George H. W. Bush and Bill Clinton to serve as a special emissary to Vietnam on the question of American service personnel missing from the Vietnam War.

In a speech given in Colorado Springs to the veterans of the Battle of Suoi Tre he addressed that assignment.

"I never expected that job to last as long as it did (*over ten years*) but it was rewarding and an interesting assignment. On occasion I was asked to give a speech to former NLF soldiers in Hanoi. During one memorable occasion, I finished my speech and after the gathering was over a frail elderly gent approached me. He said, "I know you." I didn't think much of it at first as my name and picture had been in Vietnamese newspapers on several occasions.

"I said, 'Oh you do?' He then said that he was the leader of the 9th VC Division which attacked me at Suoi Tre. We both felt a bit uneased by the revelation until Cam finally said, 'You know, you killed a lot of my men on that day.' I came back with a quick, 'Yes, and you killed a lot of MY men.' We both smiled and left it like that. We killed 638 of his men and he killed 33 of my soldiers. He realized pretty quickly that it was no contest. I never saw him again."

Two days after the battle, the armor and mech units headed north looking for remnants of the VC who attacked on the 21st. They found some of them but most of the attackers headed south to recover from the battle.

The 2/12th and 3/22nd troops remained patrolling at FSB Gold until Saturday, the 25th when both battalions were sent south to search the jungle for the VC survivors. \

We marched to Dau Tieng base camp through the jungle and arrived there on April 8th. During that two-week period, we discovered a few VC dead bodies lying near their weapons. These were soldiers who could no longer walk and were left to slow us down. They didn't make it. These were the VC that got our total up to 638 from the original 596 counted on the day of the battle.

We arrived at Camp Rainier 21 days after we left camp. We were exhausted and looking forward to well-earned rest. It turned out to be only a three-day respite as we were once again sent out into jungle. This was typical for our tours in Vietnam. Twenty-one days out and three days in for maintenance and a few beers. It was no wonder that our infantrymen lost vast amounts of weight while serving in Vietnam. I went to Vietnam weighing 158 lbs. and returned weighing 127 lbs.

We did not know it then but all the units at the battle were recommended for the Presidential Unit Citation, which was finally awarded the year after the battle. That award to the soldiers was equal to each man earning the Distinguished Service Cross. The DSC is the United States Army's second highest military decoration for soldiers (*just below the MOH*) who display extraordinary heroism in combat with an armed enemy force. Most of those who qualified were not even aware of the honors until decades later.

Ed Smith, who served as the Brigade Assistant S3 Staff Officer and later A/2/12 Commander, wrote this about the days after the battle:

> "The next day after the battle, I was directed to begin preparing the paperwork to recommend 3rd Brigade and its assigned and supporting units for the Presidential Unit Citation. I flew back out to LZ Gold and walked the firebase and talked to a number of survivors of the battle. I recall a conversation with LT John H. Andrews, the B/3/22 First Platoon Leader, the company that took the brunt of the enemy's main attack. He told me he was overrun three times during the battle, and that he thought his time had come.
>
> "Some survivors thought the attacking VC were 'juiced up' on drugs or something. They reported that some VC were wounded, were patched up, and came running back into the fight. Other VC came running into the fight with Chinese Claymore type mines strapped to their chest, with another VC running behind them with the detonator.
>
> "After gathering information at LZ Gold, and going to 25th Infantry Division Headquarters in Cu Chi to get information about the format and procedures for the recommendation, I was sent back to Dau Tieng ahead of the 3rd Brigade to begin writing the recommendation for the Presidential Unit Citation. As I walked into the quarters area where the officers from Headquarters, 3rd Brigade lived, a couple of the Vietnamese house girls came up to me asking about some of the people within 3rd Brigade that they knew, wanting to know if they survived the battle. The house girls told me that some VC came through Dau Tieng after the battle and told the civilians there that the VC had killed over 2,000 Americans. I was able to assure them that the VC grossly misrepresented what had happened, and that the VC were the ones who suffered the horrendous casualties.
>
> "Later, our Brigade S-2, MAJ Wright, told us that a copy of

the VC Attack Order had been found and translated. According to the captured document, the intent was for the VC to attack during darkness, in the wee hours of the morning. For some reason, the attack came at first light when our troops were up and alert. I hate to think what the result might have been had the attack taken place at night when it was more difficult for us to maneuver our reinforcing units and to coordinate our supporting fires. The result may have been entirely different."

I am not sure that the commanders knew it at the time, but for the rest of us we were unaware that most of the VC survivors headed straight to the Michelin Plantation to recover from their losses. The plantation was so huge that many soldiers could disappear in the vastness of the plantation. In addition, the plantation villagers could support them adequately as they recovered.

Ten days later, on March 31st, Lt. Colonel Alexander Haig, commanded his 1st ID battalion, the 1/26th when it was inserted into a clearing named LZ GEORGE, a few short klicks north of LZ GOLD. Actually, close to where Oliver Stone fought in his New Year Day Battle of 1968 which was depicted in the movie '*PLATOON*'.

Remember that unit Garth could not account for at LZ GOLD? That unit, the 271st VC Regiment, supplemented by some survivors of the Battle of Suoi Tre, attacked Haig's troops in force. The VC originally outnumbered Haig's force by 3-1. The 9th VC Division once more attempted that elusive historic victory that would score major political points.

The Battle of Ap Gu, as it was named, was a resounding victory for the Americans. After the scope of the battle was determined, the 1/16th Infantry was sent to LZ GEORGE to reinforce the 1/26th Battalion. The battle began at 1300H on the 31st and when the battle ended at sunrise on April 1st, the VC reportedly suffered 609 KIAs left in the field and surrounding jungle. Once more American firepower proved decisive.

Between March 19 and April 1, the 'Dragon' 9th VC Division suffered 227 killed at the Battle of Ap Bau Bang II (273rd VC Regiment), 638 killed at the Battle of Suoi Tre (272nd VC Regiment), and another 608 killed at LZ George (273rd VC Regiment). That's a grand total of 1,473 of their fighters lost during a period of 12 days.

Those losses must have boosted draft quotas in Hanoi as the 9th VC Division was effectively taken off the playing field in War Zone C for weeks.

By November, 1967, the 9th VC Division was strong enough to attack the provincial capital of Loc Ninh. The 1st Division fought them off for three days before the 2/12th was sent north to reinforce the units there. When the battle was nearly at an end, the 2/12th was sent north of Loc Ninh to serve as a blocking force for the enemy who were trying to escape to Cambodia. They barely arrived in position near a plantation road when darkness fell on them

Shortly after dark, LPs were reporting small groups of VC moving in front of their positions and some even accidentally entering the perimeter. Soon it was apparent to the invaders that they had come up on an American blocking position and prepared to do battle. By 0340H mortar rounds were raining down on the 2/12th Battalion laager site. It was an unnerving night for the men who were made up of our replacements. Nobody dared to move that night, not knowing if it was friendlies or enemy moving about in the darkness. A/2/12, led by Captain Burnie Quick at that time, had his Weapons Platoon bring along one mortar tube. It was manned by a mortarman named Charlie Page. That one mortar tube made a big difference in the outcome of the battle as it convinced the VC to by-pass the 2/12th and find a different route into Cambodia. In December, Charlie Page was sent to Cam Ranh Bay where LBJ was visiting the troops. While decorating soldiers with medals, Charlie Page had his Silver Star pinned on him by President. That was quite an honor for him and all the Vietnam veterans of A/2/12.

Jim Bisson, who arrived in A/2/12 in August of 1967 as a PFC, would later attain the rank of Brigadier General. He was there for that battle in the jungle north of Loc Ninh. He is a much-respected member of Alpha Association and one of the leaders of the men who replaced the original boat people in Vietnam. Jim still remains in close contact with the platoon members that he served with in Vietnam. Jim told us that the originals had our Battle of Suoi Tre. For the replacement A/2/12 soldiers, that night in the jungle north of Loc Ninh was their iconic battle while serving in our company.

Oh yes, the unit that snuck up on the 2/12th north of Loc Ninh was the 272nd VC Regiment. Amazing stuff, huh?

Two more big battles for the Dragon Division took place in the next three months, the New Year's Battle and of course the Tet Offensive of 1968. At the end of those campaigns, the 9th VC Division was effectually destroyed as an NVA entity. The local VC were replaced by North Vietnamese troops who continued to carry on the war until its ending. Essentially the "Dragon Division" died when the American firepower overwhelmed it in War Zone C and the suburban villages of Saigon.

Back To The "World"

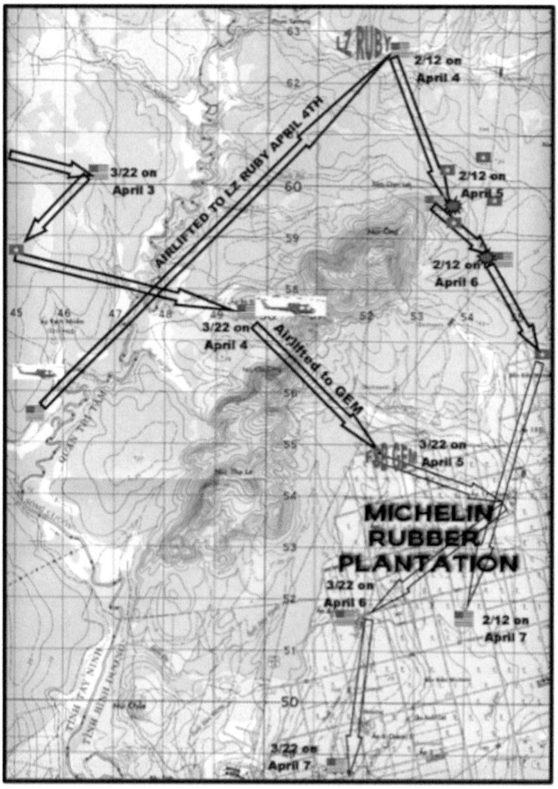

I learned in mid-May that my DEROS would be coming up approximately on August 21, 1967. That was an early out, but was necessitated by the fact that the entire original brigade soldiers had to leave Vietnam by September 21, 1967. Some left as early as August 1, 1967. Others, like valuable NCOs, remained in Vietnam until the last day, September 21, 1967.

That being the case, my "Short Time" calendar began counting down on May 21st. I had a pocket calendar where I marked off the last 90 days in country. It was the longest summer that I spent in my life.

By late July, I was then RTO for Captain Burnie Quick. Our battalion was sent into the badlands west of Dau Tieng where Operation Attleboro took place and we found that bomb factory. Just north of the factory, on the night of July 31—August 1, the battalion set in a laager position just north of an enemy base we found in January. That night our battalion received a 120 round mortar barrage. Both Captain Quick and I jumped into a small foxhole together. As the mortar rounds dropped closer and closer to us, Captain Quick asked me if I was scared in his southern Alabama drawl. I said, "No sir." He replied, "The reason I asked is I can feel you shaking." I came back with, "I thought that it was you that was shaking." We laughed about that decades later. Fact is we were both shaking. After the barrage ended, we checked to see how close to our hole the last round that dropped in was. It was approximately 15 meters from our little foxhole and right in line with it.

We walked back to Dau Tieng on August 3, and thankfully Burnie allowed me to stay in camp for the remainder of my tour.

Burnie Quick later served at many levels, climaxing with the position of Chief of Staff of United States Army Infantry Center, Fort Benning, Georgia. He was also inducted into the Alabama Military Hall Fame in October 2014. That was quite an honor. Jim Bisson became a personal friend of Burnie and attended the ceremony.

I thought I had it made in the shade when on August 20, I was kicked out of my cot, while sleeping, by a new NCO in the company. He told me to get my gear together and report to the battalion motor pool. I was riding shotgun on a resupply convoy going to Saigon on a deuce and a half. I told him I was due to leave that day. He didn't care and sent me on my way.

I was riding two vehicles behind a Patton tank near Cu Chi, when the tank was lifted in the air by a powerful landmine. I checked around for an attack out of the jungle, but none came. Great! One day to go and I get it riding shotgun in a truck!

When I arrived at Saigon, I contacted Skip Barnhart, our Company Clerk, to ask if my orders had arrived. He told me yes, they had. I needed to get back to Dau Tieng by chopper to get my orders as soon as possible. I needed to return to Saigon's Camp Alpha to spend the night there for an early morning flight out of Tan Son Nhut Air Base to America.

I caught a chopper heading to Dau Tieng and ran to the clerk's office where Barnhart was waiting with my orders. He gave them to me and we bid each other a fond goodbye. We never spoke again for four decades and reunited in Alpha Association. We are still close friends today.

After receiving my orders, I went to my tent to pack my bags for the trip home. There waiting for me was Ron Del Orefice, "Wolfie", who shot this photo as I was packing my bag to go home. He gave it to me at a 2005 reunion. He also gave me the last few months of my pocket calendar that I gave to him upon leaving Dau Tieng. I told him that he could count down his days in Vietnam on that same calendar. He promised to return it to me on that day in August 1967 and kept his promise in 2005.

I took a last photo of the Black Virgin Mountain in the distance as I walked to the chopper pad. That is how I want to remember Camp Rainier.

I arrived at the chopper pad and when a Huey arrived, I asked if they were going to Tan Son Nhut Airbase. They said yes, and told me to jump onboard.

I was alone on that flight, sitting on the floor of the helicopter. It gave me time to think and digest what was happening. I had been dreaming of that flight for many months and now it was here.

A feeling of melancholy overtook me as I began to think about what had happened to me and my brothers of A/2/12 since we arrived in October 1966. I saw my company turn into a unit that I hardly recognized after being filled with replacements. They seemed even younger than we were when we arrived in country.

I recalled 1SG Springer's speech at Fort Lewis telling us how it is not too difficult to get a man to go to war, but very hard to get him to return a second time. Now I understood.

I had one last thought as I was nearing my destination. I was always told that I would be leaving for war as a kid, but coming back as a man. I had just turned 21 years old three months earlier. I still felt like a kid and what's more, I felt that I had gone to war with a lot of other kids. We were kids fighting a grownup's war. We did the best we could, understanding little of what took place over the last year. That is what convinced me to become an Historian for our unit. I needed to learn more about our mission other than "Go in the jungle and if it moves, kill it". This is also what motivated me to write this book.

After a restless night at Camp Alpha, I joined a group of other men heading home and boarded a Braniff Airline 707 jet. It seemed a bit odd because it was painted in purple. What was even more surprising was that the airline stewardesses wore miniskirts, which we had never seen before. Some of the guys would keep dropping things to see the girls bend down to retrieve it for them. It was all in good fun and the young girls didn't seem to mind.

When our jet ran down the runway for takeoff, we all held our breath and no one said a word. Once we arrived at a safe altitude, a cheer went up. We looked down and saw the South Vietnam coastline pass by and we were soon over the South China sea, homeward bound.

Our first stop was at Yokota Air Base near Tokyo, Japan. Ironically, my son would serve three years there while serving in the Air Force. After refueling, we set course for Travis Air Force Base in California. We arrived there early in the morning on the day we left. We regained a day after passing by the International Date Line that we crossed on the troop ship heading to Vietnam. At one point in the flight, we had to change course because of a typhoon in the area, but we just flew around it.

In the middle of the night, I plugged my earphones into the music port on the arm rest of my seat. The first song that came up was Nancy Sinatra's "You Only Live Twice". How appropriate. I really did feel as though I was beginning a new life, completely apart from how I had lived earlier.

As the sun arose on the Pacific coast, we could see the coastline of California. We all went nuts, stretching to get a better view of America. There it is, just as we left it the year before, we thought. However, that was not the case, as we would learn shortly.

The plane landed at Travis Air Base and we stepped down the portable staircase to the tarmac. There were military buses there which we boarded to take us to a nearby terminal. Once there we unloaded the buses and walked along a nearby walkway that paralleled a fence about 30 feet to our right. We were surprised to see a group of maybe one hundred war protestors who yelled epithets at us. It was hard to distinguish what they were trying to tell us as each tried to outdo the other with their protests. All we knew was they were not pleased to see us. The Air Force NCO who accompanied us suggested that we just ignore them as he walked us to the brick terminal.

Once inside, we took seats awaiting some paperwork to be completed by our guides. Soon OG colored Army buses pulled up and we boarded them for our final trip to Oakland Army base, about an hour away.

To this day I still do not understand how war protestors were allowed to enter a U.S. Air Force Base to meet us at the gate. I don't know if it changed after August of 1967, but it was traumatic to us soldiers who were inexplicably jeered at after serving in Vietnam. Some of us just brushed it off as just a bunch of lunatics who have no idea what we had just left in Southeast Asia.

I'm sure any of my Vietnam War Army brothers will recognize the next step in our processing. When we arrived at the army base across the river from San Francisco, we were guided into what I can only guess looked somewhat like an athletic hall of some kind. It was a large building with bleacher seats along the side where we were instructed to sit.

Soon an Army Officer arrived and he instructed us on the process we would need to complete before we were released and were alowed to go home. He assured us that the army only had our interest at heart and if we felt we had any physical or mental issues that needed to be addressed, we would need to stay at the base for a month or two before we could be sent home. There were few who took him up on his offer to stay at Oakland AB for treatment.

Then the paperwork began which we had to acknowledge that we were fine and did not need treatment. I think many of us we would have signed a contract with the devil to make our way back home. We signed. We're good. Release us.

Before we were dismissed, that same officer gave us a piece of advice. "Once you leave here, I suggest that you remove those summer khaki uniforms and put on civilian clothes if you want to head into the nearby cities." We had heard about war demonstrators, but we didn't understand the depth of hatred that some of these people felt towards us. After seeing the protestors at Travis, it seemed like wise advice.

Still, I was anxious to go home. I had thought about it continuously while in the jungle during those long hours of radio watch in the dark. How wonderful would it be to feel safe, really safe in a dry bed indoors? Sleeping without the constant noise of artillery firing indiscriminately at any crossroad or other clearing where enemy may or may not be lurking. How magnificent would it be to be able to finally get to share what it was like to be with these great heroes and how they changed my concept of what it means to live, really live? How we all clung tenaciously to life when all the evidence pointed to the fact that we could be killed at any time. We learned how dangerous a mission we were sent to where even friendly fire was not always friendly.

At one point in the process, I saw an Army doctor stand in front of each of us and asked us if we could hear alright. We all answered, "No sir. Send me home. I am fine." There was another soldier on the side of him with a clipboard who checked our names off of a roster. Funny looking back at it now. The Army was processing a half a million men out of Vietnam a year. There was not a lot of time for long research to discover hidden ailments. "No, sir. Send me home, I'm fine."

When I was 60 years old, I was sent to a hearing specialist to check my hearing. Before we began the tests, she told me that when I first returned from Vietnam, I was given a hearing test and no problems were discovered at that time. I thought back to that officer standing in front of us to self-evaluate any loss of hearing and smiled.

The lady technician giving the test asked if I thought I had a problem with my hearing. I thought to myself, if I didn't have a problem in 1967 why would I have one now in 2009. I clammed up and answered that I suspected so, as my first wife always said I did. She called that "selective hearing".

As it turned out I was diagnosed with tinnitus at that time. Ringing in my ear was something I lived with continuously, but heck, isn't that what should be expected when we were exposed to all those large explosions and firefight sounds? Of course. Man up and live with it, I thought. After a while I learned to live with it and ignored the ringing.

I was released from the Army with orders to report to Fort Dix to complete my two-year conscription period in 30 days. That would be a cakewalk for me. It would give me time to get free dental care for my neglected teeth in Vietnam. Maybe too many cookies with jam on it?

We were directed to the base transportation area where we boarded buses to our destination. My destination was the nearest airport in San Francisco. I took the bus there and passed over a huge bridge which I naively thought was the Golden Gate. It was the equally magnificent San Francisco–Oakland Bay Bridge. I didn't figure out that mistake for decades. Life is an endless series of lessons.

I arrived at SF International Airport and bought my ticket to Boston on a redeye flight that would require me to change planes in New York, but would get me into Boston at 8AM.

I called my mother to tell her of my arrival in Boston, but she got confused when I explained to her that I had to change to another airline from the trip to Boston. She never wrote anything down and thought I would be arriving at a different gate. She told the rest of the family and they all made plans to greet me at the airport. The only problem with my 8AM arrival is my youngest sister was passing papers to buy a house that morning at 9:30. The ride home would take an hour so there was little time to waste. It would be no problem as long as I arrived as scheduled.

I hopped on the plane at 10PM local time and set back to enjoy the ride home. I arrived in New York on time, soon after daybreak, and ran to the next gate that would take me to the plane that would be taking me to Boston. Everything worked like clockwork.

I had been planning for a year how I would be seen coming off the plane and yelling, "The kid is back" with my arms in the air. Sort of like the scene from the movie "Stripes" when Bill Murray made a similar entrance out of a plane. When we landed, I purposely waited until everyone left the cabin then I made my way to the plane exit door. I stood at the doorway with my arms wide apart waiting for my cue once I saw my family there. All I saw was a large group of flyers with their greeters walking away to the terminal. This wasn't what I planned for at all.

I entered the terminal and sat at the empty gate waiting for my family to arrive. An hour went by and I began to wonder what might have happened. For the entire hour a young lady manning the gate check-in station eyed me sympathetically. How sad she must have thought. This soldier probably returning home from Vietnam had no one to greet him. Fifteen minutes later, she walked over to ask if I had anyone who would be picking me up. I said that I had expected that my family would be here. She asked me for my name and told me to stay seated at the gate area. She then used the airport terminal intercom to inform my family where I was located. Ten minutes later, I saw that receptionist break out in a smile. She saw my family walking quickly to the gate. I think she was as happy as I was. I turned and saw the family smiling and quickly walking to my seat.

Once there and we greeted each other, I was told that my sister Cookie could wait no longer and headed home to pass papers on the house that they were buying that morning. Soon we headed to the parking lot and I jumped in the back of my brother-in-law Ernie's car. He thought he had the hard job of telling me some bad news.

As soon as we left the airport entrance, he told me that my high school girlfriend Carol had gotten married while I was in Vietnam. She was the one who saw me off from the draft board. I had broken off with her during the summer before, not wanting to tie her down when there was a possibility that I may not make it back. She was terribly broken hearted. After all, we had been a couple for almost four years. I told her that if it was meant to be, we would get back together when I got home. Now I learned from Ernie that "It wasn't meant to be." Ernie said they didn't want to tell while I was serving in combat. I heard many worse homecoming experiences and I took it well. I was alive and was about to embark on my "Second Life" as described in Nancy Sinatra's song.

We arrived at my mother's apartment and were joined by Cookie and her husband Danny who had passed papers earlier in the morning. It was a joyful reunion and all those months of worrying and lighting candles in our church were ending. Indeed, the "kid" was home. Yet, somehow a piece of the kid remained at the Draft Board on December 13, 1965.

After an hour or so of small talk, (*no mention of Vietnam or what transpired since we last saw each other)* everyone left. After a long night on a plane, I decided to take a nap.

I awoke at 5 PM or so and ate dinner alone with my mother.

After cleaning the dishes, my mother went to watch TV and I told her I wanted to go for a walk. My car was sold while I was gone. She said OK and went back to her TV.

I walked a mile to the local bowling alley where the nightly team bowling was taking place. I walked in and sat in the stands to watch people bowl. Nothing changed here, but what did I expect? Life goes on.

I don't know why I chose a bowling alley to go to on my first night. I was a teetotaler in those days so it would not have occurred to me to go to a nearby bar. I guess I wanted to go to a building with a lot of noise and see if I could meet up with anyone I knew. Unfortunately, I did.

At the end of my second hour, Richard Martin, a grade school friend of mine, came down the aisle to join his team that would be bowling soon. He spotted me and with a smile, came over. I stood up to greet him. He said, "Hi Bill, where have you been? We haven't seen you in a while." I was about to tell him I just got back from Vietnam, but I never got a word in before he came back with, "How about those Red Sox?" (This was the Year of the Impossible Dream).

I smiled and just said, "Yeah, how about those Red Sox," and left.

The two decades of research, I hope, answers that question, "Where have you been?" Additionally, it answers who was with me, which truly heroic men crossed my path in Vietnam, and the amazing accomplishments that we recorded in U.S. military history. It was a long time coming, but I hope the reader gets a better perspective of what it was like for the men who searched the jungles day after day.

In conclusion, they now know all about the adversity we faced, the infamous "Dragon Division".

CAST OF CHARACTERS

Babcock, Sam: Third Platoon member and Association Board member

Barney, Gary: Third Platoon infantryman

Barton, Gene: Third Platoon Sergeant

Barton, Larry: Third Platoon RTO

Barnhart, 'Skip': A/2/12 Company Clerk

Bender, John A. LTC: Commander 3/22nd at GOLD

Berard, Lou: A/2/12 cook and New Bedford resident

Bergeron, Ron: Fourth Platoon vet and first man to go searching out the men of the company

Berardi, Fred: A/2/12 rifleman who also served as flamethrower operator

Bisson, Jim: A/2/12 enlisted man and later Brig General, Texas National Guard

Brablec, Doug: A/2/12 Supply Section veteran, 1967-68

Bradbury, Bill: 2/12th HQ Company Recon Platoon Leader

Bremner, Tom: Weapons Platoon infantryman, p 15

Comeau, Bill: CO's (Commanding Officer) RTO (radio/telephone operator) on the company network

Concannon, John: Original A/2/12 XO later serving as Battalion Intelligence Officer

Davis, Homer: 1st Platoon NCO and Drill Instructor

Del Orefice: Known as "Wolfie", he was a replacement that arrived in May, 1967.He got his name because he arrived from the 25th Division's "Wolfhound" Battalion

Deluco, Jim: Weapons Platoon veteran and Alpha Association Photographer

Deshaies Bob: 2/77 artilleryman in charge of generators and antennas at GOLD

Doolittle Mike: FDC for A3/22 Company mortar platoon

Eising, Ken: A/2/12 basic graduate who was sent to HQ Recon Platoon after basic

Eldredge. Jack: RTO for 2/12th Cmdr. Elliott

Elliott Joe, LTC: 2/12th Commander assumed command Dec 1966-June 1967

Evans, Donald: A/2/12 Medic, Medal of Honor recipient

Farris, Andy: First Platoon Leader and later Company Commander of A/2/12

Filous, Peter: First Platoon Veteran and Association Treasurer

Garth, Marshall B., Colonel: Commander, 3rd Brigade 4th ID, January1966–April, 1967

Gonzalez, Jose: 2nd Platoon Leader and veteran of the Cuban Bay of Pigs invasion

Hanna, George: third platoon sergeant who led a squad for us in Vietnam

Hamm, Clark: B/2/12 veteran of Feb 25-26 battle

Harvey, Porter: Third Platoon and 196th Light Infantry veteran and Board Member

Heys, Jim: A/2/12 3rd Platoon infantryman

Janson, Rick: 3rd Platoon Infantryman

Jarvis, Ernie: 1st Platoon Machine Gunner and Squad Leader

Julian, Ralph LTC: Commander 2/22th at GOLD

Kawczak, Steve: Rifleman 3rd Platoon

Kelley, Walter: Company Commander's RTO on the battalion network

Kramarczyk, Casey: B/2/12 veteran of Feb 25-26 battle

Kirkup. Joe: 2/77th Field Artillery- Artilleryman RTO attached to A/2/12

LaRock, Paul: Third Platoon and 196th Light Infantry veteran

Livingston, Bob: Former Marine who joined A/2/12 in 1966 as squad leader

Lopez, Juan: Second Platoon Leader arriving in May, 1966

Mascaro, John: 2/77th Artillery Veteran of the Battle of Suoi Tre

May, Larry: A/2/12 replacement, later serving as a National Park Ranger

McLemore, Gene: 25th Inf. Division Artillery Liaison, Battle of Suoi Tre

More, Ron: HQ Company 4.2 Mortar Platoon Leader

Mosely, Charlie: NCO who trained us at Fort Lewis and deployed with us

Murphy, Bob: 1st Platoon Grenadier

Neyman, Charlie: Third Platoon Infantryman

Neal, Brian: A/2/12 basic graduate who was sent to HQ Recon Platoon after basic

Noel, Joseph: Weapons Platoon veteran who was first to lose his life in Vietnam

Olafson, Jim: Executive Officer, late 1966 to April, 1967

Osowiecki, Henry 'Ozzy': Squad Leader of A/2/12, Association Chief Executive Officer

Page, Charlie: Silver Star recipient for his valor at Loc Ninh

Palmer, Jon: Company Commander, A/2/12, December, 1965-April, 1967

Peckham, Al: Weapons Platoon Infantryman, Officer of Association

Peterson, Jerry: 2nd Wave replacement arriving so we could be sent home

Quick, Burnie: Future A/2/12 Cmdr. who served on staff of 3rd Bde, 4th ID

Shugart Walter Cpt: B/3/22 Commander at GOLD

Smith, Ed: Asst. Bde S-3 Dec 1966-Apr 67, Company Commander, A/2/12 April-July 1967

Springer, Sidney R.K.: First Sgt, A/2/12 December, 1965–April, 1967

Staib, Bob: First FAC to arrive at GOLD on March 21, 1967

Stenehjem, George J.: Commander, Centaurs, Air Cavalry troop of helicopters

Stone, John: 2nd Wave replacement who helped tell our story after we went home

John A. Vessey LTC: 2/77th Commander at GOLD

Walter, Larry: Former Marine who trained and led us in Vietnam

GLOSSARY OF TERMS, ABBREVIATIONS AND ACRONYMS

AAR – After-action report

AITA – dvanced Infantry Training

AO – Area of operation

AUT – Advanced Unit Training

APC – Armored Personnel Carrier

ARVN – Army of the Republic of Vietnam

BMC – Branch Medical Clinic

BT – Basic Training

BUT – Basic Unit Training

Bushmaster–Operation that checkerboard units to different locations

CAS–Close Air Support

Chieu Hoi–Government program to encourage defectors from the NLF

CIDG–Civilian Irregular Defense Group, local irregular military units

CMH–Center of Military History

CMH–Congressional Medal of Honor

COSVN–Central Office for South Vietnam (Central body of DRV, directing the war in South Vietnam

DEROS–Date Estimated Return from Overseas

DOD–Department of Defense

DRV–Democratic Republic of Vietnam, (North Vietnam)

FAC–Forward Air Controller for close air support to ensure that their attack hits the intended target and does not injure friendly troops

FNG–The insulting name given to New Guys coming into a unit

FOB–Forward Operating Base, operations centers sent forward to the target area

FSB–Forward Support Base, usually artillery bases to support infantry units

GVN–Government of South Vietnam

H&I–Harassment and interdiction fire

HQ–Headquarters

KIA–Killed in Action

KLICK–Shorthand for kilometer (1000 meters)

LAAGER–Night encampment location

LP–Listening Posts, positions in front of main units to alert main body

LRRP–Long Range Reconnaissance Patrol

MACV–Military Assistance Command, Vietnam

MEDCAP–Medical Civic Action Program

MIA–Missing in Action

MOH–Medal of Honor

MOS–Military Occupational Specialty

MASH – Mobile Army Surgical Hospital

NCO – Non-Commissioned Officer

NDP – Night Defensive Position

NLF – National Liberation Front

NLFA – National Liberation Front Army

NVA – North Vietnamese Army

OCS – Officer Candidate School

PAVN – People's Army of Vietnam

PF – Popular Forces

PLAF – Popular Liberation Armed Forces

POLE – Radio jargon indicating a casualty

PRC – Portable Radio Communications

QC – Vietnamese Military Police Corps

RAG – Radio jargon indicating a KIA

RP – Rendezvous point or Rally Point

R&R – Rest and Recreation

RPGA–shoulder-fired missile weapon that launches rockets equipped with an explosive warhead

RTO–Radio Telephone Operator

TOC–Tactical Operation Center

VTR–Vehicle, Tank Recovery

WIA–Wounded in Action

XO–Executive Officer

PHOTOGRAPHS

Sid Springer and General Hamilton Hawkins Howze, TMRB 1962

Jon and Sandy Palmer on their wedding day

These were the men drafted in New Bedford on December 13, 1965

Carol, my high school sweetheart, mom and my sister Annette

Photo taken of me during my Christmas pass home

HEADQUARTERS
4TH INFANTRY DIVISION AND FORT LEWIS
OFFICE OF THE COMMANDING GENERAL
FORT LEWIS, WASHINGTON 98432

Dear

Your son is now undergoing basic training here at Fort Lewis, Washington. He is taking his place alongside thousands of other young men helping to maintain our Nation's strength and readiness for any emergency.

At Fort Lewis, he will receive the best instruction possible. Although his life as an individual will be regulated to acquire the orderly habits needed in group living, you need have no anxiety for your son. The requirements are neither severe nor too demanding. Constant attention is given to the protection of health, and your son will receive adequate time for rest. The moral and spiritual needs of these young men are fully recognized, and ample recreational and religious facilities are available to them. They may be visited by parents, wives, or guardians. Two weeks leave will be given to each trainee at the end of the Basic Training Period.

You will find that your son's military service will benefit him greatly because it demands much of him, enhances his sense of responsibility, and develops his ability to both lead and get along with his fellow men. Your pride in him and his service, expressed in frequent letters to him, will be directly reflected in his accomplishments.

In the future, you will receive a letter from your son's battalion commander which will give you more specific information as to his assignment, when he may be visited, and how he may be contacted.

Sincerely,

A. S. COLLINS, JR.
Major General, United States Army
Commanding

Letter sent home from General Collins at beginning of basic training

THE 12TH INFANTRY
A BRIEF HISTORY

One of the first and foremost of America's regiments, the 12th Infantry, began its career on 3 July 1798 when the nation was threatened by European powers who were plundering our Merchant Marine. It was not until the war of 1812 that the men of the 12th saw action against the enemy. After campaigning in New York, the 12th was ordered to Fort McHenry, Maryland. There, after withstanding 25 hours of continuous bombardment, 500 men of the 12th repulsed the assualts of England's finest troops. It was this same bombardment that inspired Francis Scott Key to compose the lyrics of our National Anthem, "The Star Spangled Banner".

In 1846, when the nation was at war with Mexico, the 12th Infantry was again called to active duty. The Regiment landed at Vera Cruz and marched overland to join General Winfield Scott's Army at Puebla, about two-thirds of the way to Mexico City. After fighting its way across Mexico, the 12th was instrumental in the assualt to Mexico City, most notably by its charge at Contreras, a heavily fortified town guarding Mexico City.

When the War between the states began in 1861, the 12th Infantry was assigned to the Army of the Potomac and fought in ten of the eleven battles waged by that Army. The Army landed at Fort Monroe, Virginia, during March 1862 and advanced to the outskirts of Richmond. The Confederate Forces under General Robert E. Lee launched a ferocious attack on 21 July 1862. The brunt of the assualt was on Gaines Mill, the sector held by the 12th. The Regiment numbered 470 men and the concentrations massed against it were enormously greater. Lee's forces attacked repeatedly. The 12th, without breastworks, held every inch of ground for over 6 hours. This gallant action is commemorated on the regimental shield by two moline crosses which represent the iron fastening of the millstones at Gaines Mill. Other battles and campaigns in which they participated were the Second Battle of Manassas, Antietam Creek, Fredericksburg, Chancellersville, Gettysburg, the Virginia and Wilderness Campaigns, Spotsylvania, Cold Harbor and Petersburg.

The role of the 12th Infantry in opening the American West is represented on the regimental crest by a wigwam, the five poles of which represent five major Indian campaigns. During this period, in campaigns of fourty to fifty men, the regiment occupied outposts from the Mississippi to Pacific, from Fort Apache to Camp Hall, Idaho.

When the Spanish-American War broke out in 1898, the Regiment embarked for Cuba where the most memorable of their accomplishments was the storming and capture of El Caney, a fortress with thick walls 20 feet high. This action is represented on the regimental crest by the line of parapets extending across the shield. From Cuba the unit shipped to the Philippines where it engaged and held to quell the insurrection. The upper half of the shield honors these campaigns by an open field of solid gold on which rests a sea lion brandishing a sword, a device taken from the Philippine coat of arms.

12th Infantry Regiment and 4th ID unit histories handouts upon arrival in A/2/12

The 12th Infantry was next in action on Utah Beach, 6 June 1944. For six successive days, the order for attack was carried out. From the beachhead the Regiment fought on through Cherbourg, St. Lo, and Mortain where it inflicted a decisive defeat to a Nazi counterattack. After the bitter hedgerow combat at Mortain, the 12th continued to lead the way across Europe -- Paris, the Siegfried Line, Hurtgen Forest, Ardennes and finally the sweep through Bavaria. In the Ardennes, the 12th's epic defense of Luxemburg will above all be remembered. For its gallant action in containing the Nazi counterattack in this area of the Bulge, the Regiment received the Distinguished Unit Citation. The War Department General Order awarding this Citation included the following sentence: "The courage and fighting determination of each member of the Regiment in the stand along the Sauer River presented an inspiring example of the invincibility of free men at arms". To this might be added the invincibility of free men Ducti Amore Patriae. "Led By Love of Country".

The 12th Infantry was inactivated 27 February 1946 but was reactivated to join the 4th Infantry Division on 15 July 1947. With the 4th Division the 12th traveled to Germany to take its place among the NATO ground forces deterring the aggressive ambitions of the USSR. In September of 1956, the 12th Infantry returned to the United States and was assigned to Fort Lewis where in April 1957 it underwent a pentomic re-organization, becoming the 1st Battle Group, 12th Infantry, 4th Infantry Division. The 1st Battle Group, 12th Infantry continued to carry on traditions of the old 12th Infantry Regiment. It participated in almost every maneuver and exercise conducted by the 4th Division to include many firsts. In Operation Desert Rock the 12th Infantry became the first unit to train with the newly developed tactical atomic weapons. In Exercise Rocky Shoals a task force from the 12th Infantry was the first to have U.S. Army troops lifted by helicopter from the decks of a Navy Aircraft Carrier to an objective far behind the beachhead during the first major joint amphibious operation conducted since World War II. The 12th Infantry participated in Exercise Indian River and Exercise Dry Hills in Yakima, Washington, and in January 1960, after intensive training at Camp Denali, Alaska, successfully participated in Exercise "Little Bear", the unit's first Arctic maneuver.

The 12th Infantry continued in the forefront of the 4th Division during Exercise "Elkhorn" and in May 1962 in Exercise "Mesa Drive" at Yakima, Washington. The following winter the 12th Infantry returned to Central Alaska and distinguish itself against Alaska's 23rd Infantry, during Exercise "Timberline". Later that year the 12th Infantry participated in Exercise "Coulee Crest" in Yakima, Washington against the 5th Mechanized Division.

On 1 Oct 1963 the 1st Battle Group 12th Infantry, was reorganized under the Road Concept. From this Battle Group the First and Second Battalions of the 12th Infantry were formed.

Since this re-organiztion the 2d Battalion has participated in numerous exercises; "Polar Siege," a training field exercise conducted in Alaska, 30 Jan - 10 Feb June 1964; "March Hare", conducted on the Fort Lewis Reservation, 2-7 March 1964; "Coulee Cross" a training field exercise conducted at Yakima Firing Center (East of the Cascade Mountains), 28 September - 10 October 1965; "March Hare II", a training field exercise, conducted on the Fort Lewis Reservation 1-10 Apr 1965; and "Frontier Post 1" a training field exercise, conducted in the Olympic National Forest (In the Olympic Mountains) 17-26 May 1965.

12th Infantry Regiment and 4th ID unit histories

handouts upon arrival in A/2/12

HISTORY
FOURTH INFANTRY DIVISION

The "IVY" Division was first organized at Camp Greene, N. C., on December 3, 1917, under command of Major General George H. Cameron, and adopted its distinctive insignia, the ivy leaf, at that time. Composed of four green ivy leaves joined at the stem and opening at the four corners of a square on a brown background, the "I-Vy Leaf" suggests the Roman numeral "IV". Proud of it's motto "Steadfast and Loyal" the division became known by the Germans in World War I as "the men with the terrible green crosses.

The first casualties suffered by the Fourth occurred at sea on May 21, 1918, when the British Transport Maldovia, was torpedoed and sunk off the Isle of Wight by a German submarine, with the loss of 56 lives among division personnel.

The division landed in France on June 5, 1918, and participated in the Aisne-Marne, St. Mihiel and Meuse-Argonne Campaigns. Following the armistice, the division served on occupation duty for seven months, until its return to the United States for deactivation on August 1, 1919.

It was next reactivated at Fort Benning, Georgia, on June 3, 1940, and moved to Camp Gordon Johnston, Florida, for amphibious training in late 1943. It departed from the United States for World War II service January 18, 1944.

On D-Day, June 6, 1944, the 8th Infantry Regiment of the Division landed on Utah Beach and was one of the first Allied units to meet the Germans on the Normandy beaches, an achievement for which it received the Distinguished Unit Citation. After the landings, the 4th Division gained ground rapidly, and soon broke through the vital road center of St. Mere Englise to relieve the 82nd Airborne Division, which had then been isolated for thirty-six hours. By June 10, 1944, the division was near Le Thiel; and by June 25, 1944 had taken Touraville, a suburb of Cherbourg, except for a few isolated pockets of German resistance.

From July 1 to 20, 1944, the division held in the area near Periers, waiting for the tremendous First Army drive which was to be launched, and from July 24 to August 18, 1944, fought in the midst of the drive across France to aid in the liberation of Paris.

12th Infantry Regiment and 4th ID unit histories

handouts upon arrival in A/2/12

By the middle of September 1944, the 4th Division had fought its way into Belgium and the Siegfried Line; where a patrol from the 22nd Infantry Regiment filtered through German lines on September 11, to retrieve a bit of German soil for the President of the United States - -becoming the first American unit to cross the borders of the German Reich. The supposedly impregnable Seigfried Line was breached in four places by men of the division by September 14, 1944 as the "I-Vy Men" inched their way into Germany.

In December 1944, the division fought in the hell of the Heurtgen Forest, where one regiment claimed the annihilation of five German regiments during this constant 19-day attack.

It then moved to Luxembourg for a supposed rest, but was almost immediately called upon to hold firm at Dickweiler and Ostweiler against desperate German drives in the "Battle of the Bulge."

Exactly one year after leaving the United States, the division crossed the Sauer River on January 18, 1945, over-ran Fuhren and Viaden and quickly took all objectives by January 21, 1945 to move in pursuit of the Germans in their flight across their homeland. By the time of the cessation of hostilities in May 1945 the division had driven deeply into Germany to leave a trail of outstanding achievements such as its work in the Heurtgen Forest and the penetrations of the Seigfried Line. In addition to many individual awards won by division personnel, the 8th, 12th and 22nd Infantry Regiments had won Distinguished Unit Citations and the division had earned the Belgium Fourragere for action in the St. Hubert-St. Vith area for the battle of the Ardennes in the Echternach-Luxembourg area. This citation was made by Belgium Decree Number 1394 on November 20, 1945.

The division returned to the United States shortly after VE Day to begin retraining for duty in the Pacific Theater, but before it could be deployed to the Pacific, VJ Day was proclaimed and the unit was deactivated at Camp Butner, N. C. on March 5, 1946.

Reactivated as a training division on July 15, 1947, at Fort Ord, California, it was in October 1950 redesignated the Fourth Infantry Division at Fort Benning, Georgia, with the mission of preparing and training for overseas duty.

2

12th Infantry Regiment and 4th ID unit histories handouts upon arrival in A/2/12

It arrived in Europe in May-June 1951 to take its place in the NATO structure as a tactically ready organization; and participated in many field exercises which provided the first opportunity to work with its NATO allies. The division served in a dual role as a training division while occupying a vital position along the main avenues of approach of its zone of responsibility.

After five years in Germany, the 4th Infantry Division returned to the United States in September of 1956, with its colors assigned to Fort Lewis, Washington, where it provided basic and advanced training to several cycles of trainees. In April 1957, the division was reorganized to Pentomic structure adding the 2d Battle Groups of the 39th and 47th Infantries to its own 1st Battle Groups of the 8th, 12th and 22d Infantry Regiments; plus the 124th Signal Battalion, the 2d Reconnaissance Squadron, 8th Cavalry, the 14th Transportation Battalion and the 4th Aviation Company. Revision of the structure of Division Artillery to include the Honest John (7.62mm) Rocket has given the division both conventional and atomic artillery capability, with the rocket and the 8" howitzer furnishing the "Sunday Punch." In March 1959, Division Artillery was again reorganized to return to the increased fire-power concept that gives each battle group available direct and indirect artillery support; and to further this concept, a Support Company with armored and mortar fire capability was added to each Infantry battle group.

In early 1958, the 4th Infantry Division was honored by selection as a component of the Strategic Army Corps (STRAC), and assigned to 18th Airborne Corps and later III Corps, with the mission of establishing and maintaining itself in instant readiness to act as a "fire brigade" to quell aggression wherever and whenever needed.

It has, by continual training and regular maneuvers, tested the concepts of its Pentomic structure, with the 2d Battle Group, 39th Infantry, first testing these tactics during Exercise COLD BAY in Alaska. Exercise INDIAN RIVER (1958) and DRY HILLS(1959) tried the men of the division against Aggressor Forces at the Yakima Firing Center Maneuver Area in Central Washington; while Exercise ROCKY SHOALS (1958) found them in a beach-head attack against a theoretical aggressor who held the California coastline. Exercise DRAGON HEAD tried and proved the division on the east coast of North Carolina in late 1959.

Nineteen Hundred and Sixty proved a banner year in the Division's

3

12th Infantry Regiment and 4th ID unit histories

handouts upon arrival in A/2/12

role as the "Elite" force in our nations Strategic Army Corps. In January and February the 1st Battle Group, 12th Infantry, participated in Exercise COLD BEAR, conducted in the Big Delta country of Alaska. February and March saw the division's CPX Staff and several thousands of its troops engaged in Exercise BAY ISLAND, an Army-Air Force amphibious exercise conducted at Fort Lewis. Concurrent with Exercise BAY ISLAND, the 2d Battle Group, 39th Infantry was airlifted to Puerto Rico in Operation BIG SLAM/PUERTO PINE. May found the Ivy Division involved in a full scale maneuver, Exercise ELK HORN, on the rolling hills of Eastern Washington at the Yakima Firing Center. Troops of the 1st Battle Group, 12th Infantry participated in Exercise ICE CAP on the frozen tundra of Greenland during July and August. Exercise SOUTH WIND capped the Division's activities for 1960, maneuvering with the XVIII Airborne Corps at Eglin AFB, Florida.

In February 1961, the 1st Battle Group, 22d Infantry was airlifted to the torrid jungles of the far-off Philippine Islands for participation in Exercise LONG PASS. May found the division at the Yakima Firing Center for Exercise LAVA PLAINS, a joint Army/Air Force maneuver, and in September invading the picturesque San Juan Islands of Washington in the joint Army/Navy/Air Force amphibious Exercise SEA WALL. October and November 1961 was taken up with host activities for the Wisconsin National Guard's 32d Infantry Division and other Army Reserve and National Guard units called to duty during the defense forces build-up. In December much of the Division's equipment and many of its troops were prepositioned in Europe to bolster NATO defense forces.

1962 dawned with the Division involved in Exercise LONG THRUST, a tactical exercise which included the airlift of troops to Europe and maneuvers near the East-West German Borders. Four of the Division's five Battle Groups including the 8th, 22nd, 39th and 47th Infantries and their attached support elements participated. The first Battle Group, 8th Infantry and 2nd Battle Group, 47th Infantry, served duty in the divided city of Berlin.

The 1st Battle Group, 12th Infantry, was the leader in 1963 activities with its participation in the Alaska Army-Air Force Exercise TIMBERLINE conducted in the Big Delta area of our 50th State. Exercise COULEE CREST during the spring of 1963 saw the 4th Infantry Division pitted against the 5th Infantry Division (Mechanized) from Fort Carson, Colorado, in two weeks of mock war at Yakima Firing Center.

4

12th Infantry Regiment and 4th ID unit histories

handouts upon arrival in A/2/12

October 1963 saw another face lifting of the 4th Infantry Division - - this time to the ROAD concept (Reorganizational Objective Division). On October 1, the 4th shifted its organizational structure from that of a pentomic division to that of a ROAD division with three flexible brigades, adding increased firepower and mobility.

From January 16 to February 17, 1964, over 3,000 members of the 2d Brigade and attached units participated in Exercise POLAR SIEGE near Fairbanks, Alaska. Since the summer of 1964, men of the 4th Division have trained in the arctic wastelands of Alaska, over the open plains of Yakima Firing Center, in the rain drenched forests of the Olympic Mountains, off the shores of Solo Point, and on the vast Fort Lewis reservation. Troops of the 1st Brigade stormed ashore at Solo Point on Puget Sound conducting amphibious assault landings in September. In October, approximately 6,400 men of the 2nd Brigade and attached units crossed the mighty Columbia River during day and night warfare in Exercise COULEE CROSS at Yakima Firing Center in Eastern Washington. In February 1965, over 3,000 troops from the 1st Brigade and attached units journied to Alaska for the two week Exercise POLAR STRIKE, maneuvering in temperatures dipping as low as 50 below. In April 1965, the entire division moved to the post's training areas for a week-long field training Exercise MARCH HARE II. The Division's 2d Brigade saw training action in the Olympic National Forest in May as they participated in Exercise FRONTIER POST I, one of the largest counterinsurgency exercises in the United States. The 2d Battalion, 8th Infantry served as aggressors for the unconventional warfare training.

On November 1, 1965, the 4th Infantry Division reorganized again, adding two infantry battalions to its structure, at the same time realigning its armor and mechanized capability. Under this structure the division has eight infantry battalions, one mechanized battalion and one armor battalion.

Exercise BRAVE CAT involving a 900 man task force airlifted to Norway and back, was completed on November 10, 1965.

Major General Arthur S. Collins, Jr. commands the 4th Infantry Division and Fort Lewis.

(Up-dated as of 12 Nov 65)

5

12th Infantry Regiment and 4th ID unit histories handouts upon arrival in A/2/12

A/2/12 barracks in the rear, on the corner of Libby Ave and 3rd Division Drive

Lt. John Concannon and Johnny Martel

Sergeant First Class James Harris

Donald Evans

Bonnie and Donald's wedding photo

Jim Olafson as a sergeant, 1965

Jim Olafson arriving at Fort Lewis with his Volkswagen Beetle car

Most of Third Platoon, (on floor) Roger Corbin, (first row) Herman Shepherd, Sam Babcock, Larry Harris, (second row) Bill Kelly, John Watson, David Hartrum, John Ogan, John Faidley, Pete LaBrecque, Bill Comeau, (back row) Jim Heys, Jim Shulsky, Norm Soares, Vincent, Alan Stinson, Rick Janson, Larry Barton

HQ, 2/12th veterans, Recon or Mortar Platoon

Sgt. Raymond in his WW II photo and his C/1/12 commander, Ed Northrop, in 1967 when Raymond was killed in action

Henry Osowiecki, December, 1966

Henry's mother and father's wedding photo

Larry Walter

Bob Livingston

Peter Filous, Dau Tieng, December, 1966

Norm Smith, Rain Forest, Olympic National Park, May, 1966

Dave Cunningham, 1964 High School Photo and 1966 military photo

Third Platoon member, Gary Barney climbing mountain in Rain Forest and Sgt. Springer waiting for his beer to be poured

Author with his M14 rifle and PRC 10 radio, soon to be replaced with an M16 and PRC -25 radio

Assembling on the company street, Gary Barney and Henry Osowiecki facing the camera

Sixth Army Band playing for us on the sail date. Troop ship departs dock to an uncertain future

Members of A Company on deck of the Nelson M Walker troopship

Extraordinary panoramic photo taken by Jim Deluco, who would later go on to be an award-winning professional photographer. This photograph was taken of the Vietnam coast as we traveled south towards Vung Tau.

Troops of A/2/12 on LST preparing to land in Vietnam.

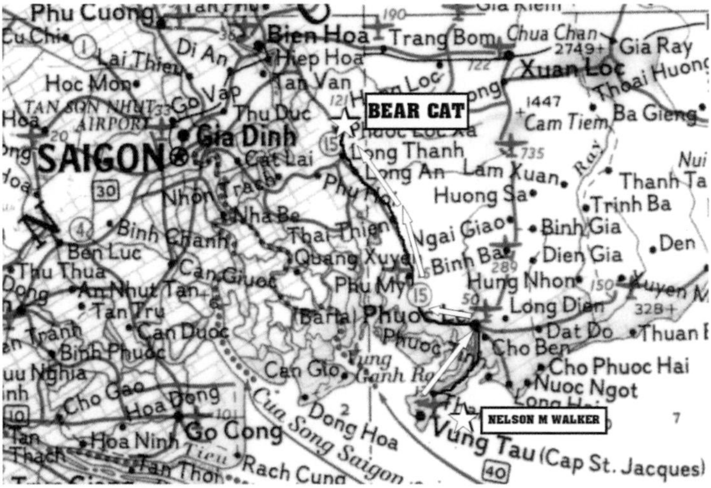

Map showing route to Bear Cat Base Camp

FSG Springer and the rest of A Company arriving back at the rendezvous point after the parade

Illustration showing the bombing of the presidential palace by two Douglas AD/A-1 "Skyraider" attack aircraft on Feb 27, 1962

Jim Olafson's photo of the Presidential Palace taken on the day before the parade

General Vo Nguyen Giap and General Nguyen Chi Thanh July 1967

Senior Colonel Hoang Cam, the 9th VC Division commander

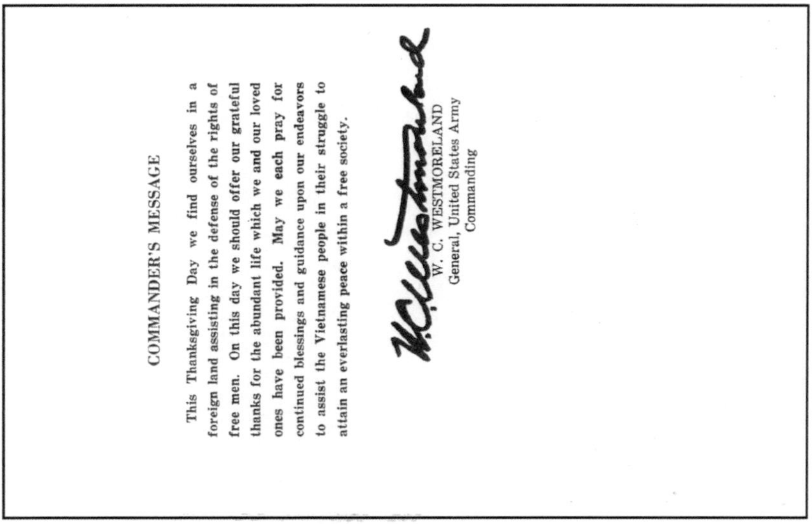

COMMANDER'S MESSAGE

This Thanksgiving Day we find ourselves in a foreign land assisting in the defense of the rights of free men. On this day we should offer our grateful thanks for the abundant life which we and our loved ones have been provided. May we each pray for continued blessings and guidance upon our endeavors to assist the Vietnamese people in their struggle to attain an everlasting peace within a free society.

W. C. WESTMORELAND
General, United States Army
Commanding

Thanksgiving Day Dinner

Shrimp Cocktail
With Cocktail Sauce and Crackers

Roast Tom Turkey — Giblet Gravy

Poultry Dressing

Snow Flaked Potatoes

Glazed Sweet Potatoes — Cranberry Sauce

Buttered
Peas and Corn

Crisp Relish Tray

Parkerhouse Rolls — Butter

Pumpkin Pie
With Whipped Cream

Mincemeat Pie — Old Fashion Fruit Cake

Fresh Chilled Fruit

Mixed Nuts

Assorted Candy

Tea — Coffee — Milk

PRAYER

Father of mercies and giver of all good, by whose power we were created, by whose bounty we are sustained, and by whose spirit we are transformed, accept, we beseech Thee, our prayer of thanks.

For those who love us and of whose love we would be more worthy; for those who believe in us and whose hopes we cannot disappoint; for every good gift of healing and happiness and renewal, we bless Thy name. We thank Thee for our homeland, for all that is just and true, noble and right, wise and courageous in our history. We praise Thee for our place in the community of nations and we invoke Thy blessings on men of goodwill wherever they may be and who labor for a world of justice, freedom, and fraternity.

Most of all, Eternal Father, we thank Thee for Thyself, the nearness of Thy presence and the warmth of Thy love, whereby our minds and hearts find joy and peace. Freely we have received, O God; freely let us give ourselves to Thy gracious purposes. For Thy love's sake. Amen.

PPC-Japan

Photos taken of Dau Tieng Village as we landed in 1966 and Dau Tieng today, the capital for Dau Tieng District 46

Marvin Fuller in foreground of the Battalion HQ where the Totem Pole was installed at Camp Rainier

Colonel Marshal B Garth and LTC Joe Elliott, December, 1966

Bill Comeau, Walter Kelley on the day we received our Combat Infantryman's Badgev

Captain Charles "Ed" Smith, Assistant S3 Officer 3rd Brigade, 4th ID

Joseph D Noel Square, Hope Rhode Island

Comparing the Michelin Plantation tree grouping to Noel's church

Porter Harvey, John Faidley, Larry Barton

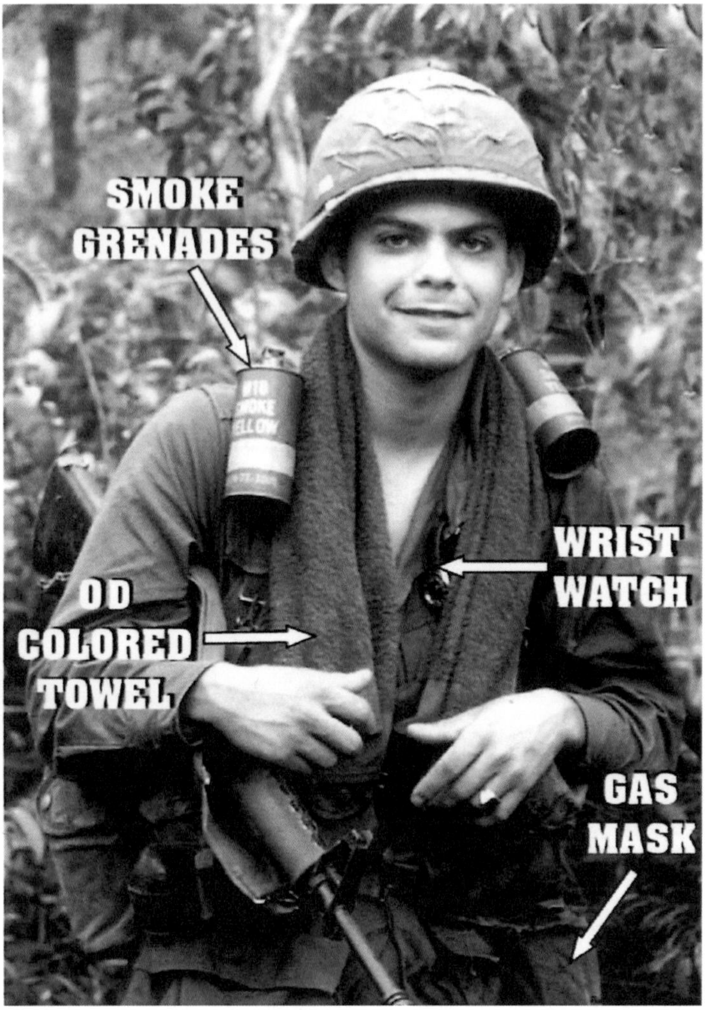

Jim Deluco displaying typical gear worn in January, 1967

Chief of Staff of the Army, Harold K. Johnson, Secretary of the Army, Stanley Resor, Bonnie Evans, Donald's wife, Elsie Evans, Donald's mother, Donald Evans Sr., Donald's dad and Richard and Harvey, Donald's brothers

Evans Auditorium, Fort Sam Houston

Evans Army Community Hospital, Fort Carson Colorado

Donald Evans painting at the hospital with Donald's two brothers at the dedication

2/12th troops boarding a C123 transport plane to take us to Trai Bi, Feb 2nd

A/2/12 set up a blocking position with fortified bunkers

Captain Ed Smith, displaying a rice polishing device captured during Operation Gadsden

Bob Gold's high school photo and Casey Kramarczyk in War Zone C

Joe Kirkup, 1967

SGT Robert Livingston examines type boxes used by the Viet Cong for sorting printing type, captured by members of Co "A", 2nd Ba, 12th, Inf, 4th Inf Div. 9 March 1967

Photos taken by Ken Eising on the security detail move north of the basecamp on March 15

Photos showing the bullet holes in the windshield and trash cans at the ambush site

Photo looking west from Suoi Da. It was taken as 2/12 troops were waiting for the order to line up on the airstrip. In the distance is the iconic Black Virgin Mt. which loomed 3,268 feet above Tay Ninh Province. The base of the mountain was located 2 ½ miles from where we stood at Suoi Da.

Some of the helicopters of the145th Combat Aviation Battalion preparing for the lift into LZ GOLD on March 19, 1967

Walt Shugart military photo and photo of author,

with Walt at the 2015 Suoi Tre Event

LTC Joe Elliott, at 2011 reunion and as 2/12th Commander, 1967

Photo taken from Huey helicopter as it traveled to Landing Zone GOLD

Photo taken showing some of the carnage left behind at LZ GOLD after the insertion took place

Photo of one of the destroyed choppers at LZ GOLD

Photo of James Dale Brewer, 196th LIB replacement sent to A/3/22 on March 19, 1967

High School photo of Captain Tonie Lee England Jr.

Colonel John Vessey being awarded the Silver Star for his leadership at FSB GOLD by General William Westmoreland

John Bender, West Point Photo

LTC Ralph W. Julian, Cmdr. 2/22nd, SGM Bill Austin, 3rd Brigade Operations Sergeant Major and Captain Walt Shugart, Cmdr. B/3/22

Colonel Marshal B. Garth, 3rd Brigade of the 4th Infantry Division Commander

Captain George Shoemaker, A/3/22nd Commander at FSB GOLD

Captain John Napper, C/2/12 Commander

Lou Urso, artilleryman from the 2/77th Fire Direction Control

Charles Ed Smith, Assistant S3, 3rd Bde, 4th ID

Bob Staib, original Forward Air Controller at the Battle of Suoi Tre

Captain Tonie England, left Capt. Walter Forbes, center, FACs for the 3rd Brigade of the 4th ID

Bob Deshaies, Artillery CDC

George J. Stenehjem, Centaurs, D Troop (Air), 3rd Squadron, 4th Cavalry, 25th Infantry Division

Colonel Garth with helmet off directing troops from the ground during the pursuit of the enemy northward

Dennis Baldauf, 1966 photo, and recreating the greeting at GOLD and photo of us at 2017 reunion

Official photo of the 3rd Brigade, 4th ID taken as the dead enemy bodies were collected for burial along the edge of LZ GOLD

General William Westmoreland speaking to the FSB GOLD defenders at noon of battle day

Photo taken during religious services soon after the battle (Mike Doolittle in the front row holding his helmet in his hand)

Representative of the 3rd Brigade 4th ID photos taken of killed enemy soldiers right after the battle

Additional photos of the brigade official photos taken on the day of the battle

President Lyndon Baines Johnson pinning a Silver Star for Gallantry medal at Cam Ranh Bay on Charlie Page in 1967

General Jim Bisson, who arrived as a PFC in Vietnam and retired as a Brigadier General

Captain Burnie Quick, A/2/12 Commander beginning on July 4, 1967

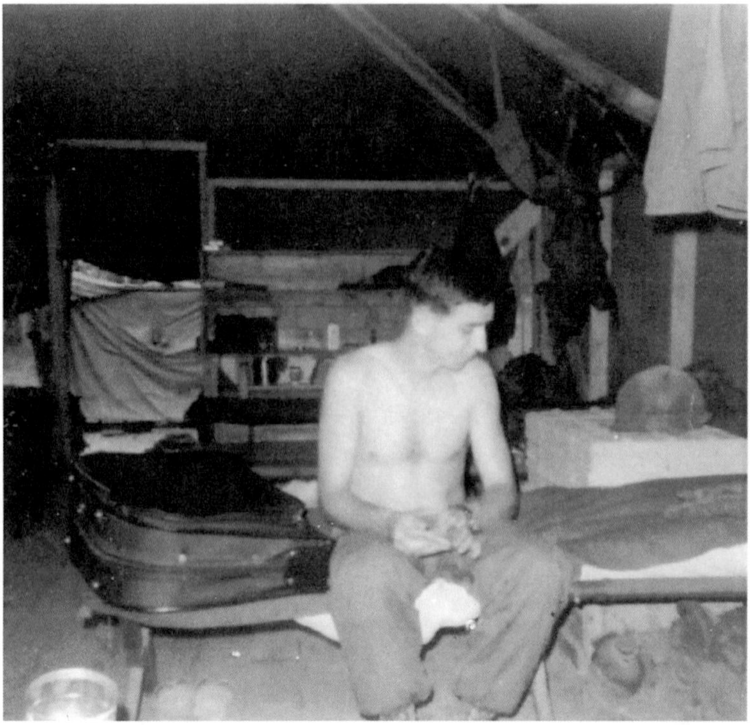

Author packing up to go home, Camp Rainier, Dau Tieng, Vietnam

Photo of Black Virgin Mt. taken by author as he proceeded to the chopper pad at Dau Tieng to catch his ride home

View from chopper flying to Saigon in a Huey helicopter

PHOTO AND CONTRIBUTOR CREDITS BY CHAPTER

Diamond In The Rough

Sid Springer and General Hamilton Hawki ns Howze, TMRB 1962

Source: Sid Springer, personal collection

The Prisoner Of Zenda

Jon and Sandy Palmer on their wedding day

Source: Jon and Sandy Palmer, personal collection

Greetings

These were the men drafted in New Bedford on December 13, 1965

Photo courtesy of Spinner Publications, New Bedford; Standard Times photographer, Milton Silvia collection

Carol, my high school sweetheart, mom and my sister Annette

Photo courtesy of Spinner Publications, New Bedford; Standard Times photographer, Milton Silvia collection

Christmas And Back

Photo taken of author during my Christmas pass home

Source: Bill Comeau, personal collection

Unit Esprit De Corp

Letter sent home from General Collins at beginning of basic training

Source: Bill Comeau, personal collection and: photo of MG Arthur S. Collins, courtesy of Fort Lewis Military Museum

12th Infantry Regiment and 4th ID unit histories handouts upon arrival in A/2/12

Source: Bill Comeau, personal collection. Handouts, Bill Comeau, personal collection,

Basic Training

A/2/12 barracks in the rear, on the corner of Libby Ave and 3rd Division Drive

Source: Map produced by Bill Comeau

Lt. John Concannon and Johnny Martel

Source: Linda Concannon, personal collection, Judy Martel, personal collection

Sergeant First Class James Harris

Source: James Harris, personal collection

Advanced Individual Training

Donald Evans College photo

Source: Richard Evans, Donald's brother, personal collection

Bonnie and Donald's wedding photo

Source: Bonnie Grauer, Donald's wife while he was in the service, personal collection

Jim Olafson, Leader

Jim Olafson as a sergeant, 1965

Source: Jim Olafson, personal collection

Jim Olafson arriving at Fort Lewis with his Volkswagen Beetle car

Source: Jim Olafson, personal collection

Advanced Infantry Unit Training

HQ, 2/12th veterans, Recon or Mortar Platoon Ken Eising, Brian Neal, Bill Bradbury, Ron More

Source: all photos provided from their personal collections

Most Of Third Platoon

Roger Corbin on floor
Herman Shepherd — Sam Babcock — Larry Harris Bill Kelly — John Watson — David Hartrum — John Ogan — John Faidley — Pete LaBrecque — Bill Comeau Jim Heys — Jim Shulsky — Norm Soares — Vincent — Alan Stinson — Rick Janson — Larry Barton

Source: Paul LaRock, personal collection

Sgt. Raymond in his WW II photo and his C/1/12 commander, Ed Northrop, in 1967 when Raymond was killed in action

Source: Sgt. Raymond photo source, Book, "Marching Together, The Story of Reed Barton veterans in WWII and Ed Northrop, personal collection

Henry Osowiecki

Henry Osowiecki, December, 1966

Source: Henry Osowiecki, personal collection

Henry's mother and father's wedding photo

Source: Henry Osowiecki, personal collection

Larry Walter

Larry Walter, 1966

Source: Larry Walter, personal collection

Bob Livingston

Bob Livingston, 1967

Source: Bob Livingston, personal collection

Peter Filous

Peter Filous, Dau Tieng, December, 1966

Source: Peter Filous, personal collection

Norm Smith

Norm Smith, Rain Forest, Olympic National Park, May, 1966

Source: Norm Smith, personal collection

David Cunningham

Dave Cunningham, 1964 High School Photo and 1966 military photo

Source: Dave Cunningham, personal collection

Making The Most Of Time

Third Platoon member, Gary Barney climbing mountain in Rain Forest and Sgt. Springer waiting for his beer to be poured

Source: Gary Barney and Rick Janson, personal collections

Author with his M14 rifle and PRC 10 radio, soon to be replaced with an M16 and PRC -25 radio

Source: Bill Comeau, personal collection

Saddle Up

Assembling on the company street, Gary Barney and Henry Osowiecki facing the camera

Source: Gary Barney, Henry Osowiecki, personal collections

Sixth Army Band playing for us on the sail date. Troop ship departs dock to an uncertain future

Source: Bill Comeau, Jim Deluco, personal collections

All-Inclusive Cruise Experience

L-R, Peter Filous, John Giovo, Gary Basquette, Roy Letta, rear, unknown, Henry Osowiecki and Jim Shulsky on the Nelson M Walker troopship

Source: Henry Osowiecki, personal collection

Panoramic photo was taken of the Vietnam coast as we traveled south towards Vung Tau

Source: Jim Deluco, personal collection

Welcome To The Party

Troops of A/2/12 on LST preparing to land in Vietnam

Source: Paul LaRock, personal collection

Map showing route to Bear Cat Base Camp

Source: author produced map

Touring Saigon clubs on day before parade. Left to right, unknown, Martel, Heys, Conefry, Comeau, and an Aussie who traded hats with me

Source: Bill Comeau, personal collection

Photos taken at rendezvous point after parade

Source: Bill Comeau, personal collection

History Behind The Parade

Illustration showing the bombing of the presidential palace by two Douglas AD/A-1 "Skyraider" attack aircraft on Feb 27, 1962

Illustration from website 'http://vnafmamn.com', no longer active on the internet

Jim Olafson's photo of the Presidential Palace taken on the day before the parade

Source: Jim Olafson, personal collection

Largest Operation To Date

General Vo Nguyen Giap and General Nguyen Chi Thanh July 1967

Source: webpage: https://www.danchimviet.info/con-hum-xam-nguyen-chi-thanh/01/2019/13542/

Senior Colonel Hoang Cam, the 9th VC Division Commander

Source: Gale Academic Onefile website

Phuoc Vinh

Map showing area north of Saigon

Source: 1967 National Geographic map of Vietnam modified by author

Major battle of Operation Attleboro

Map Source: author produced maps using topographic maps from the period

Thanksgiving menu 1966 (1)

Source: George Hanna, personal collection

Thanksgiving menu 1966 (2)

Source: George Hanna, personal collection

Dau Tieng

Photos taken of Dau Tieng Village as we landed in 1966 and Dau Tieng today, the capital for Dau Tieng District 46

Source: Left photo from Jim Deluco taken in 1967 and right photo taken by Jerry Peterson during a recent visit to Dau Tieng

Marvin Fuller in foreground of the Battalion HQ where the Totem Pole was installed at Camp Rainier

Source: Photo of battalion HQ building supplied by Ed Smith, A/2/12 CO Apr-Jul 1967, Fuller photo provided to Alpha Association from his personal collection

Colonel Marshal B Garth and LTC Joe Elliott, December, 1966

Photo source: 2/12th scrapbook of original battalion photos, acquired from Ray Cassidy, correspondent for

***Stars and Stripes* attached to 2/12th Inf. He discovered the collection at Camp Rainier in 1970 while locating and destroying battalion documents before transferring the base over to the ARVN**

Bill Comeau, Walter Kelley on the day we received our Combat Infantryman's Badge

Source: Bill Comeau, Walter Kelley, personal collections

First Is The Worst

Captain Charles "Ed" Smith, Assistant S3 Officer 3rd Brigade, 4th ID

Source: Ed Smith, personal collection

Joseph D Noel Square, Hope Rhode Island

Source: Bill Comeau, personal collection

Comparing the Michelin Plantation tree grouping to Noel's church

Source: Al Peckham and Darlene and Darryl Hogue, personal friends of the author

Return Of The Dragon

Jim Deluco displaying typical gear worn in January, 1967

Source: Jim Deluco, personal collection

Dual Camp Complex Located

Porter Harvey, John Faidley, Larry Barton

Source: Porter Harvey, Arlene Fritz, John's sister and Pat Hanna, Larry's sister, personal collections

The Aftermath

Chief of Staff of the Army, Harold K. Johnson, Secretary of the Army, Stanley Resor, Bonnie Evans, Donald's wife, Elsie Evans, Donald's mother, Donald Evans Sr., Donald's dad and Richard and Harvey, Donald's brothers

Source: Richard Evans, Donald's brother, personal collection

Evans Auditorium, Fort Sam Houston

Source: Bill Comeau, personal collection

Evans Army Community Hospital, Fort Carson Colorado

Source: www.cpr.org, online website

Donald Evans painting at the hospital with Donald's two brothers at the dedication

Source: Richard Evans, personal collection

Operation Gadsden

2/12th troops boarding a C123 transport plane to take us to Trai Bi, Feb 2nd

Source: Brian Neal, personal collection

A/2/12 set up a blocking position with fortified bunkers

Source: Dan Schlicter, personal collection

Captain Ed Smith, displaying a rice polishing device captured during Operation Gadsden

Source: Ed Smith, personal collection

Major Attack In War Zone C

Bob Gold's high school photo and Casey Kramarczyk in War Zone C

Source: Carleen Gold, Bob's widow and Casey Kramarczyk, personal collections

Border Defense

Joe Kirkup, 1967 photo

Source: Ed Smith, personal collection

SGT Robert Livingston examines type boxes used by the Viet Cong for sorting printing type, captured by members of Co "A", 2nd Ba, 12th, Inf, 4th Inf Div. 9 March 1967

Source: Official U.S. Army photograph archived at the National Archive Center, College Park Maryland

Last Visit Home

Photos taken by Ken Eising on the security detail move north of the basecamp on March 15

Source: Ken Eising, personal collection

Photos showing the bullet holes in the windshield and trash cans at the ambush site

Source: Jim Olafson, personal collection

Junction City II, D-Day Plus + 1

Photo looking west from Suoi Da. It was taken as 2/12 troops were waiting for the order to line up on the airstrip. In the distance is the iconic Black Virgin Mt. which loomed 3,268 feet above Tay Ninh Province. The base of the mountain was located 2 ½ miles from where we stood at Suoi Da.

Source: Ken Eising, personal collection

Some of the helicopters of the145th Combat Aviation Battalion preparing for the lift into LZ GOLD on March 19, 1967

Source: Ken Eising, personal collection

Walt Shugart military photo and photo of author, with Walt at the 2015 Suoi Tre Event

Source: Walt Shugart, and Bill Comeau, personal collections

LTC Joe Elliott, at 2011 reunion and as 2/12th Commander, 1967

Source: Bill Comeau, personal collection, and 2/12th photo book dated 1966-67

Photo taken from Huey helicopter as it traveled to Landing Zone GOLD

Source: Gene McLemore, personal collection

Photo taken showing some of the carnage left behind at LZ GOLD after the insertion took place

Source: Richard Little of the 118th AHC and posted on their 118AHC website

Photo of one of the destroyed choppers at LZ GOLD

Source: Gene McLemore, personal collection

Photo of James Dale Brewer, 196th LIB replacement sent to A/3/22 on March 19, 1967

Source: Charlie Brewer, James's brother, personal collection

High School photo of Captain Tonie Lee England Jr.

Source: Bob Staib, from his personal collection

The Base And Leadership

Colonel John Vessey being awarded the Silver Star for his leadership at FSB GOLD by General William Westmoreland

Source: Donna Vessey, John Vessey's daughter, personal collection

John Bender, West Point Photo

Source: Photo and story provided by the West Point Graduate Organization and edited by John Bender Jr.

LTC Ralph W. Julian, Cmdr. 2/22nd, SGM Bill Austin, 3rd Brigade Operations Sergeant Major and Captain Walt Shugart, Cmdr. B/3/22

Source: Ed Smith, personal collection

Colonel Marshal B. Garth, 3rd Brigade of the 4th Infantry Division Commander

Source: 2/12th scrapbook of original battalion photos, acquired from Ray Cassidy

Captain George Shoemaker, A/3/22nd Commander at FSB GOLD

Source: George Shoemaker, personal collection

Junction City II D-Day D+2

Captain John Napper, C/2/12 Commander

Source: John Napper, personal collection

Lou Urso, artilleryman from the 2/77th Fire Direction Control

Source: Lou Urso, personal collection

The Attack Commences

Charles Ed Smith, Assistant S3, 3rd Bde, 4th ID

Source: Ed Smith, personal collection

Bob Staib, original Forward Air Controller at the Battle of Suoi Tre

Source: Bob Staib, personal collection

Captain Tonie England, left Capt. Walter Forbes, center, FACs for the 3rd Brigade of the 4th ID

Source: Ed Smith, personal collection

The Combined Attack

Bob Deshaies, Artillery CDC

Source: Bob Deshaies, personal collection

George J. Stenehjem, Centaurs, D Troop (Air), 3rd Squadron, 4th Cavalry, 25th Infantry Division.

Source: Contributed by Bain Cowell, and authorized by Bruce Powell, Webmaster of centaursinvietnam.org

Colonel Garth with helmet off directing troops from the ground during the pursuit of the enemy northward

Source: Gene McLemore, personal collection

Dennis Baldauf, 1966 photo, and recreating the greeting at GOLD and photo of us at reunion 2017

Source: Dennis Baldauf, personal collection and author's personal collection

Official photo of the 3rd Brigade, 4th ID taken as the dead enemy bodies were collected for burial along the edge of LZ GOLD

Source: Official photo of the 3rd Brigade, 4th ID taken as the dead enemy bodies were collected for burial along the edge of LZ GOLD

The Aftermath

General William Westmoreland speaking to the FSB GOLD defenders at noon of battle day

Source, widely circulated photo distributed to American newspapers after the battle

Photo taken during religious services soon after the battle (Mike Doolittle in the front row holding his helmet in his hand)

Source: Mike Doolittle personal collection who was able to acquire the photo from the Associated Press release

Representative of the 3rd Brigade 4th ID photos taken of killed enemy soldiers right after the battle

Source, official photo collection shared with the author by Burnie Quick, Brigade HQ

Additional photos of the brigade official photos taken on the day of the battle

Source, official photo collection shared with the author by Burnie Quick, Brigade HQ

President Lyndon Baines Johnson pinning a Silver Star for Gallantry medal at Cam Ranh Bay on Charlie Page in 1967

Source: Charlie Page, personal collection

General Jim Bisson, who arrived as a PFC in Vietnam and retired as a Brigadier General

Source: Jim Bisson, personal collection

Back To The "World"

Captain Burnie Quick, A/2/12 Commander beginning on July 4, 1967

Source: John Stone, personal collection

Author packing up to go home, Camp Rainier, Dau Tieng, Vietnam

Source: Ron Del Orefice, personal collection

Photo of Black Virgin Mt. taken by author as he proceeded to the chopper pad at Dau Tieng to catch his ride home

Source: Bill Comeau, personal collection

View from chopper flying to Saigon in a Huey helicopter

Source: Bill Comeau, personal collection

ABOUT THE AUTHOR

Bill was raised in New Bedford, MA, an old mill town, during the 1950s and 60s. His dad was killed in a factory accident when he was 2 years old. He was drafted in the army on December 13th when he was 19 years old and sent to Fort Lewis Washington as part of the 4th Infantry buildup before the division was sent to Vietnam.

He has spent the last twenty-two years researching the history of his unit's time in Vietnam.

He helped form Alpha Association in the year 2000. This is a veterans organization manned by A/2/12 veterans who served in his company during his time in Vietnam.

This research was recognized by other notables and led up to him being named the Historian for the 12th Infantry Regiment Monument Project. For that initiative he was cited as a Distinguished Member of the 12th Infantry Regiment.

The US Infantry Association awarded the Order of St. Maurice Medal, Legionnaire level, to him for his contributions to the United States Infantry.

Locally, he is very active in Veteran organizations, and has several tributes awarded to him, including Franco-American Veteran of the Year for SE Massachusetts in 2017.

CPSIA information can be obtained
at www.ICGtesting.com
Printed in the USA
BVRC091034310822
645895BV00002B/2